ENCYCLOPAEDIA OF
ENGINEERING TECHNOLOGY

ENCYCLOPAEDIA OF
ENGINEERING TECHNOLOGY

Uday Tusti

ANMOL PUBLICATIONS PVT. LTD.
NEW DELHI - 110 002 (INDIA)

ANMOL PUBLICATIONS PVT. LTD.
H.O.: 4374/4B, Ansari Road, Daryaganj
New Delhi - 110 002
Ph.: 23278000, 23261597
B.O.: No. 1015, Ist Main Road, BSK III Stage
III Phase, III Block
Bangalore - 560 085 (India)
Visit us at: www.anmolpublications.com

Encyclopaedia of Engineering Technology

First Published, 2006

ISBN 81-261-2936-0

PRINTED IN INDIA

Printed at Mehra Offset Press, Delhi.

Preface

In recent years there have been dramatic changes in Engineering Technology on both national scale and international scale. New techniques and products have found global recognition. These changes have had a major impact on all our lives. Not only Technology becoming more complex, but even terminology is becoming complex day by day.

This volume, The Encyclopaedia of Engineering Technology covers all aspects of Engineering and its role in society. This Encyclopedia covers a range of various Engineering Technology disciplines such as Aerospace Engineering, Agriculture Engineering, Chemical Engineering, Biomedical Engineering, Civil Engineering, Computer Engineering, Control Systems Engineering, Electrical & Electronic Engineering, Environmental Engineering, Geotechnical Engineering, Industrial Engineering, Manufacturing Engineering, Mechanical Engineering, Mining Engineering, Nuclear Engineering, Petroleum Engineering, Sanitary Engineering. Each definition is given in detail and many are supported by formulas and diagrams.

This new encyclopedia integrates all of the diverse approaches that have gone into the making of the modern field of Engineering Technology. It is unique among available reference works in that it covers the most recent topics of research. Encyclopaedia of Engineering Technology is intended for all those interested in this field. It will be of immense value to students of all kinds, especially students of different fields of Engineering. It will also be a major source of reference for professionals and students, who require an authoritative and up-to-date guide to assist them in their work. In preparation of this book, the author has freely consulted large number of books and journals so no authenticity is claimed. Author is especially thankful to M/S Anmol Publications for shaping this book in its final form. Suggestions for further improvement of this book are not only welcome but also greatly appreciated.

Author

A display In radar, a display in which targets appear as vertical deflections from a line representing a time base. Also called A-scan or A-scope. Target distance is indicated by the horizontal position of the deflection from one end of the time base. The amplitude of the vertical deflection is a function of the signal intensity.

A station In loran, the designation applied to the transmitting station of a pair, the signal of which always occurs less than half a repetition period after the next preceding signal and more than half a repetition period before the next succeeding signal of the other station of the pair, designated a B-station.

Abducted A foot or any part of a foot is in an abducted position when its distal aspect is angled away from the midline of the body in a transverse plane and away from a sagittal plane passing through the proximal aspect of the foot or part, or away from some other specified reference point.

Abduction Motion occurring on the transverse plane during which the distal aspect of the foot moves away from the midline of body about a vertical axis of rotation at the aspect of the foot part. The axis of motion lies on the frontal and sagittal planes.

Abductus This is a fixation of a part in the position it would assume if abducted. It in a transverse plane fixation, in which the distal end is displaced away from the midline. Hallux Abductus is a fixed angulation of the hallux so that it is directed away from the midline of the body.

Aberration 1. In astronomy, the apparent angular displacement of the position of a celestial body in the direction of motion of the observer, caused by the combination of the velocity of the observer and the velocity of light.

2. In optics, a specific deviation from perfect imagery, as, for example: spherical aberration, coma, astigmatism, curvature of field, and distortion.

Abiotic environment The part of an ecosystem that includes the nonliving surroundings.

Ablate To carry away; specifically, to carry away heat generated by aerodynamic heating, from a vital part, by arranging for its absorption in a nonvital part, which may melt or vaporize, then fall away taking the heat with it.

Ablating material A material, especially a coating material, designed to provide thermal protection to a body in a fluid stream through loss of mass.

Ablating materials are used on the surfaces of some reentry vehicles to absorb heat by removal of mass, thus blocking the transfer of heat to the rest of the vehicle and maintaining temperatures within design limits. Ablating materials absorb heat by increasing in temperature and changing in chemical or physical state. The heat is carried away from the surface by a loss of mass (liquid or vapor). The departing mass also blocks part of the convective heat transfer to the remaining material in the same manner as transpiration cooling.

Ablating nose cone A nose cone designed to reduce heat transfer to the internal structure by the use of an ablating material.

Ablation The removal of surface material from a body by vaporization, melting, chipping, or other erosive process; specifically, the intentional removal of material from a nose cone or spacecraft during high-speed movement through a planetary atmosphere to provide thermal protection to the underlying structure.

Abnormal failure The failure of a component resulting from an induced condition that causes a malfunction of the device.

Abort 1. To cut short or break off an action, operation, or procedure with an aircraft, space vehicle, or the like, especially because of equipment failure, as to abort a mission, the launching was aborted.

2. An aircraft, space vehicle, or the like that aborts.

3. An act or instance of aborting.

Abrasion Wear or removal of the surface of a solid material as a result of relative movement of other solid bodies in contact with it

Abrasion resistance The ability of a construction element, such as a floor, pavement, roof or coating to resist degradation due to mechanical wear such as foot or wheel traffic and wind blown particles which tend to progressively remove materials from its surface.

Abrasive A hard and wear-resistant material (commonly a ceramic) that is used to wear, grind, or cut away other material.

Abrasive media Moderately to extremely hard materials, usually in a fine or very fine particulate form, used in abrasive blasting to remove surface contaminants or to grind and polish samples to the desired thickness or finish. Examples of moderately hard abrasive media are sand, iron shot, crushed iron slag, glass beads or ground nut shells. Examples of extremely hard abrasive media are industrial diamonds.

Absolute 1. Pertaining to a measurement relative to a universal constant or natural datum, as absolute coordinate system, absolute altitude, absolute temperature.

2. Complete, as in absolute vacuum.

Absolute accuracy The correctness of the indicated value in terms of its deviation from the true or absolute value.

Absolute coordinate system An inertial coordinate system which is fixed with respect to the stars.

In theory, no absolute coordinate system can be established because the reference stars are themselves in motion. In practice, such a system can be established to meet the demands of the problem concerned by the selection of appropriate reference stars.

Absolute delay 1. The time interval between the transmission of sequential signals. Also called delay.

2. Specifically, in loran, the time interval between transmission of a signal from the A-station and transmission of the next signal from the B-station.

Absolute humidity The amount of water vapor actually present in unit quantity of a gas, generally expressed as mass of water vapor per unit volume of gas + water vapor, e.g., as grains per cubic foot.

Absolute instrument An instrument whose calibration can be determined by means of physical measurements on the instrument.

Absolute manometer 1. A gas manometer whose calibration, which is the same for all ideal gases, can be calculated from the measurable physical constants of the instrument.

2. A manometer that measures absolute pressure.

Absolute pressure In engineering literature, a term used to indicate pressure above the absolute zero value of pressure that theoretically obtains in empty space or at the absolute zero of temperature as distinguished from gage pressure.

In high-vacuum technology, pressure is understood to correspond to absolute pressure, not gage pressure, and therefore the term absolute pressure is rarely used.

Absolute pressure transducer A transducer which measures pressure in relation to zero pressure (a vacuum on one side of the diaphragm).

Absolute system of units 1. A system of units in which a small number of units are chosen as fundamental, and all other units are derived from them.

2. Specifically, a system of electrical units put into effect by international agreement on 1 January 1948.

Prior to 1 January 1948 the international system was in effect; the two systems can be converted by the following relationships:

1 mean international ohm = 1.00049 absolute ohm

1 mean international volt = 1.00034 absolute volt.

"Electric units, called "international," for current and resistance had been introduced by the International Electrical Congress held in Chicago in 1893, and the definitions of the "international" ampere and the "international" ohm were confirmed by the International Conference of London in 1908.

Although it was already obvious on the occasion of the 8th CGPM (1933) that there was a unanimous desire to replace those "international" units by so-called "absolute" units, the official decision to abolish them was only taken by the 9th CGPM (1948), which adopted for the unit of electric current, the ampere".

Absolute temperature Temperature value relative to absolute zero.

Absolute temperature scale A temperature scale based upon the value zero as the lowest possible value. Thus, all obtainable temperatures are positive. The Kelvin and Rankine scales are absolute scales.

Absolute vorticity 1. The vorticity of a fluid particle expressed with respect to an absolute coordinate system.

2. The vertical component of the absolute vorticity (as defined above).

Absolute zero The theoretical temperature at which molecular motion vanishes and a body would have no heat energy; the zero point of the Kelvin and Rankine temperature scales.

Absolute zero may be interpreted as the temperature at which the volume of a perfect gas vanishes or, more generally, as the temperature of the cold source which would render a Carnot cycle 100 percent efficient. The value of absolute zero is now estimated to be - 273.15° Celsius, -459.67° Fahrenheit, 0° Kelvin, and 0° Rankine.

Absorbed water Mechanically held water in a porous material (as concrete, soil mass, aggregates) and having physical properties not substantially different from ordinary water at the same temperature and pressure.

Absorptance, absorbtance(symbol A, a, or) The ratio of the radiant flux absorbed by a body to that incident upon it. Also called absorption factor.

Total absorptance refers to absorptance measured over all wavelengths.

Spectral absorptance refers to absorptance measured at a specified wavelength.

Absorption 1. The process by which radiant energy is absorbed and converted into other forms of energy. Absorption takes place only after the radiant flux enters a medium and thus acts only on the entering flux not on the incident flux, some of which may be reflected at the surface of the medium. A substance which absorbs energy may also be a medium of refraction, diffraction, or scattering; these processes, however, involve no energy retention or transformation and are to be clearly differentiated from absorption.

2. In general, the taking up or assimilation of one substance by another.

3. In vacuum technology, gas entering into the interior of a solid.

Absorption band A range of wavelengths (or frequencies) in the electromagnetic spectrum within which radiant energy is absorbed by a substance.

When the absorbing substance is a polyatomic gas, an absorption band actually is composed of a group of discrete absorption lines which appear to overlap. Each line is associated with a particular mode of vibration or rotation induced in a gas molecule by the incident radiation.

The absorption bands of oxygen and ozone are often referred to in the literature of atmospheric physics.

The important bands for oxygen are: (a) the Hopfield bands, very strong, between about 670 and 1000 angstroms in the ultraviolet; (b) a diffuse system between 1019 and 1300 angstroms; (c) the Schumann-Runge continuum, very strong, between 1350 and 1760 angstroms; (d) the Schumann-Runge bands between 1760 and 1926 angstroms; (e) the Herzberg bands between 2400 and 2600 angstroms; (f) the atmospheric bands between 5380 and 7710 angstroms in the visible spectrum; and (g) a system in the infrared at about 1 micron. The important bands for ozone are: (a) the Hartley bands between 2000 and 3000 angstroms in the ultraviolet, with a very intense maximum absorption at 2550 angstroms; (b) the Huggins bands, weak absorption between 3200 and 3600 angstroms; (c) the Chappius bands, a weak diffuse system between 4500 and 6500 angstroms in the visible spectrum; and (d) the infrared bands centered at 4.7, 9.6 and 14.1 microns, the latter being the most intense.

Absorption coefficient (symbol) 1. A measure of the amount of normally incident radiant energy absorbed through a unit distance or by a unit mass of absorbing medium.

2. In acoustics, the ratio of the sound energy absorbed by a surface of a medium (or material) exposed to a sound field or sound radiation to the sound energy incident on the surface. The stated values of this ratio are to hold for an infinite area of the surface. The conditions under which measurements of absorption coefficients are made are to be stated explicitly.

Three types of absorption coefficients associated with three methods of measurement are: chamber absorption coefficient, obtained in a certain reverberation chamber; free-wave absorption coefficient, obtained when a plane, progressive, sound wave is incident on the surface of the medium; sabine absorption coefficient, obtained when the sound is incident from all directions on the sample.

Absorption cross section In radar, the ratio of the amount of power removed from a beam by absorption of radio energy by a target to the power in the beam incident upon the target.

Absorption emission pyrometer A thermometer for determining gas temperature from measurement of the radiation emitted by a calibrated reference source before and after this radiation has passed through and been partially absorbed by the gas. Both measurements are made over the same wavelength interval.

Absorption field Area through which septic tank effluent discharges into the surrounding ground. It comprises of a series of perforated pipes laid in shallow trenches backfilled with a predesigned arrangement of sand and gravel.

Absorption line A minute range of wavelength (or frequency) in the electromagnetic spectrum within which radiant energy is absorbed by the medium through which it is passing. Each line is associated with a particular mode of electronic excitation induced in the absorbing atoms by the incident radiation.

Absorption spectrum The array of absorption lines and absorption bands which results from the passage of radiant energy from a continuous source through a selectively absorbing medium cooler than the source.

The absorption spectrum is a characteristic of the absorbing medium, just as an emission spectrum is a characteristic of a radiator. An absorption spectrum formed by a monatomic gas exhibits discrete dark lines, whereas that formed by a polyatomic gas exhibits ordered arrays (bands) of dark lines, which appear to overlap. This type of absorption is often referred to as line absorption. The spectrum formed by a selectively absorbing liquid or solid is typically continuous in nature (continuous absorption).

Absorptive index The imaginary part of the complex index of refraction of a medium. It represents the energy loss by absorption and has a nonzero value for all media which are not dielectrics. Also called index of absorption .

Absorptive power The total flux of radiant energy absorbed in a unit area of absorbing substance; expressed, for example, in ergs per square centimeter per second or in watts per square centimeter.

Absorptivity The capacity of a material to absorb incident radiant energy, measured as the absorptance of a specimen of the material thick enough to be completely opaque, and having an optically smooth surface.

Absorptivity emissivity ratio In space applications, the ratio of absorptivity for solar radiation of a material to its infrared emissivity. Also called A/E ratio.

Abstract syntax notation (ASN.1) An OSI language used to define datatypes for networks. It is used within TCP/IP to provide conformance with the OSI model.

Abstraction reaction A reaction that removes an atom from a structure.

Abutment The outermost end supports on a bridge, which carry the load from the deck.

AC Inverter An electrical circuit which generates a sine-wave output (regulated and without breaks) using the DC current supplied by the rectifier-charger or the battery. The primary elements of the inverter are the DC/AC converter, a regulation system and an output filter.

AC Line filter A circuit filter placed in the ac line to condition or smooth out variations that are higher in frequency than the line frequency.

Accelerated corrosion test A test method designed to simulate, in a short time span, the destructive action of corrosion under normal long-term service conditions.

Accelerated weathering A test designed to simulate the deteriorating effect of natural outdoor weathering, with controlled, intensified and accelerated processes.

Acceleration The rate at which velocity changes. This change may be in either magnitude or direction. If the velocity does not change, the acceleration is zero; a decrease in velocity maybe known as negative acceleration, deceleration or retardation. If the velocity is measured in meters per second, the acceleration will be in meters per second per second, abbreviated to ms^{-2} or m/s^2. Acceleration due to gravity has a value of 9.81 m/s^2

Acceleration of gravity (symbol g) By the International Gravity Formula, g = 978.0495 [1 + 0.0052892 sin^2(p) - 0.0000073 sin^2 (2p)] centimeters per second squared at sea level at latitude p.

The standard value of gravity, or normal gravity, g, is defined as go=980.665 centimeters per second squared, or 32.1741 feet per second squared. This value corresponds closely to the International Gravity Formula value of g at 45° latitude at sea level.

Accelerator Substance used in small proportions to increase the speed of a chemical reaction. Accelerators are often used in the paint industry to speed up drying.

Accelerometer An instrument used to measure acceleration. Typically, they contain a small mass, connected to a stiff spring, with some electrical means of measuring the spring deflection when the mass is accelerated. The type of accelerometer used in gait analysis is usually very small, weighing only a few grams. It only measures acceleration in one direction, but they may be grouped together for two and three-dimensional measurements.

Acceptance An action by an authorized representative of the acquirer by which the acquirer assumes ownership of products as a partial or complete performance of contract.

Acceptance criteria The criteria a product must meet to successfully .complete a test phase or meet delivery requirements.

Acceptance test Formal testing conducted to determine whether or not a system satisfies its acceptance criteria and to enable the acquirer to determine whether or not to accept the system.

Acceptor In transistors, the P-type semiconductor, the electrode containing trivalent impurities (boron, gallium, or indium) to increase the number of holes which can accept electrons. Contrast with donor.

Access A users ability to communicate with a system.

Access hatch An airtight door system that preserves the pressure integrity of a reactor containment structure while allowing access to personnel and equipment.

Accidental error In experimental observations, an error which does not always recur when an observation is repeated under the same conditions. Contrast systematic error.

Acclimatization The adjustments of a human body or other organism to a new environment; the bodily changes which tend to increase efficiency and reduce energy loss.

Accommodation coefficient (symbol). The ratio of the average energy actually transferred between a surface and impinging gas molecules which are scattered by the surface to the average energy which would theoretically by transferred if the impinging molecules reached complete thermal equilibrium with the surface before leaving the surface or a = (Er-E i)/Es-Ei) where a is the accommodation coefficient, Er is the energy carried away from unit surface area per second by the scattered or re-evaporated molecules, Ei is the energy per unit surface area per second carried toward the surface by the impinging molecules, and Es is the energy per unit surface area per second which would be carried away by the molecules if the molecules reached complete thermal equilibrium with the surface before leaving.

Account A user ID and disk area restricted for the use of a particular person. Usually password protected.

Accumulator 1. A device or apparatus that accumulates or stores up, as: (a) a contrivance in a hydraulic system that stores fluid under pressure; (b) a device sometimes incorporated in the fuel system of a gas-turbine engine to store up and release fuel under pressure as an aid in starting; (c) an electrical storage battery.

2. In computer technology, a device which stores a number and upon receipt of another number adds it to the number already stored and stores the sum.

Accuracy The degree of correctness with which the measuring system yields the "true value" of a measured quantity, where the" true value" refers to an accepted standard, such as a standard meter or volt. Typically described in terms of a maximum percentage of deviation expected based on a full-scale reading.

Accustomization The process of learning the techniques of living with a minimum of discomfort in an extreme or new environment.

Acid In the Brønsted definition, an acid is a chemical species that can donate a proton to another species (a base). In the Lewis definition, an acid is a chemical species that can accept (and share) a pair of electrons from another species. Hydrochloric acid is a Brønsted acid; the proton it donates is a Lewis acid. A neutral Lewis acid and a neutral Lewis base can commonly form a dipolar bond.

Acid embrittlement Form of hydrogen embrittlement that may be induced in some metals by acid treatment or other acid exposure.

Acid-etch tube A glass tube charged with dilute hydrofluoric acid. Used to measure inclination: after being left in a borehole for 20 to 30 min, the inclination will be indicated by the angle of etch line on the tube. May be fitted in a clinometer. This method is called "acid-dip survey".

Acknowledgment (ACK) A positive response returned from a receiver to the sender indicating success. TCP uses acknowledgments to indicate the successful reception of a packet.

Aclinic line The line through those points on the earth's surface at which magnetic dip is zero. The aclinic line is a particular case of an isoclinic line. Also call dip equator, magnetic equator.

Acoustics The degree of sound. The nature, cause, and phenomena of the vibrations of elastic bodies; which vibrations create compressional waves or wave fronts which are transmitted through various media, such as air, water, wood, steel, etc.

Acoustic, acoustical Containing, producing, arising from, actuated by, related to, or associated with sound.

Acoustic is used to modify terms that designate an object, or physical characteristics, associated with sound waves; acoustical is used when the term being qualified does not designate explicitly something that has such properties, dimensions, or physical characteristics.

Acoustic delay line A device used in a communications link or a computer memory in which the signal is delayed by the propagation of a sound wave. Also call sonic delay line.

Acoustic mach meter A device which obtains data on sound propagation for the calculation of Mach number.

Some acoustics Mach meters measure transit time or velocity of a sound pulse; others measure an angle, as the angle of the Mach cone.

Acoustic radiation pressure A unidirectional, steady-state pressure exerted upon a surface exposed to a sound wave.

Acoustic refraction The process by which the direction of sound propagation is changed due to spatial variation in the speed of sound in the medium.

Acoustic streaming Unidirectional flow currents in a fluid that are due to the presence of sound waves.

Acoustic vibration With respect to operational environments, vibrations transmitted through a gas. These vibrations may be subsonic, sonic, and ultrasonic.

Acquirer An organization that procures products for itself or another organization.

Acquisition 1. The process of locating the orbit of a satellite or trajectory of a space probe so that tracking or telemetry data can be gathered.

2. The process of pointing an antenna or telescope so that it is properly oriented to allow gathering of tracking or telemetry data from a satellite or space probe.

Acquisition programme A directed, funded effort designed to provide a new, improved, or continuing system in response to a validated operational need.

Acrylic 1. Thermoplastic with reasonable optical clarity, plus good weather and shatter resistance.

2. One of the resins present in latex paints. It serves as bonding agent for the other ingredients.

Acrylic latex An aqueous dispersion of acrylic resins in latex paints.

Acrylic Resin (acrylic coating, acrylic plastic) Group of clear, thermoplastic resin polymerized from various acrylic monomers (acrylic acid ($C_3H_4O_2$) or methacrylic acid ($C_4H_6O_2$), or a derivative of either, especially an ester, e.g., methyl methacrylate), either alone or in combination. One such acrylic resin, polymethyl acrylate, which is a tough rubbery material, is used in various industrial finishes, lacquers, and pressure sensitive adhesives. Others, such as Lucite, are used in flat or corrugated panels.

Actinic Pertaining to electromagnetic radiation capable of initiating photochemical reactions, as in photography or the fading of pigments.

Because of the particularly strong action of ultra violet radiation on photochemical processes, the term has come to be almost synonymous with ultraviolet, as in actinic rays.

Actinogram The record of a recording actinometer.

Actinometer The general name for any instrument used to measure the intensity of radiant energy, particularly that of the sun. Actinometers may be classified, according to the quantities which they measure, in the following manner: (a) pyrheliometer, which measures the intensity of direct solar radiation; (b) pyranometer, which measure global radiation (the combined intensity of direct solar radiation and diffuse sky radiation); and (c) pyrgeometer, which measures the effective terrestrial radiation.

Activated sludge An active population of microorganisms used to treat wastewater, or the process in which the organisms are employed.

Activation energy Distinct states correspond to minima of a potential energy surface in a configuration space. In this classical picture, the activation energy for transforming state A into state B is the maximum increase in energy (relative to the ground state of A) encountered on a minimum-energy path from A to B. Energy here refers to potential energy; an analogous definition based on free energy can be constructed. When tunneling is considered, lower energy paths become possible, but an activation energy can be associated with the reaction (at a given temperature) via the relationship between temperature and reaction rate.

Activator 1. The activation agent of metal surface.

2. The curing agent of a two component product such as a coating system or an adhesive.

Active 1. Transmitting a signal, as active satellite . Antonym of passive.

2. = radioactive, as active sample.

3. = fissionable, as active material.

4. Receiving energy from some source other than a signal, as active element.

Active component 1. A component which adds energy to the signal it passes.

2. A device that requires an external source of power to operate upon its input signal(s).

3. Any device that switches or amplifies by the application of low-level signals. Examples of active devices which fit one or more of the above definitions: transistors, rectifiers, diodes, amplifiers, oscillators, mechanical relays and almost all IC's (Contrast with passive component)

Active homing The homing of an aerodynamic or space vehicle in which energy waves (as radar) are transmitted from the vehicle to the target and reflected back to the vehicle to direct the vehicle toward the target.

Active leg An electrical element within a transducer which changes its electrical characteristics as a function of the application of a stimulus.

Active listening. A technique used to help communication among team members and project personnel. Active listening involves paying careful attention to what is being said, then rephrasing that information and feeding it back to the originator to ensure that what you think you heard is what they meant.

Active marker systems A kinematic system used for gait analysis, where light emitting diodes (LED) are fixed over the joints and a special optoelectronic camera tracks them.

Active material The chemically reactive materials in an energy cell which react with each other converting from one chemical composition to another while generating electrical energy or accepting electric current from an external circuit.

Active open An operation performed by a client to establish a TCP connection with a server.

Active satellite A satellite which transmits a signal, in contrast to passive satellite .

Active tracking system A system which requires addition of a transponder, or transmitter on board the vehicle to repeat, transmit, or retransmit information to the tracking equipment, e.g. Dovap, Secor, Azusa, Miran, Minitrack.

Active transducer A transducer whose output is dependent upon sources of power, apart from that supplied by any of the actuating signals, which power is controlled by one or more of these signals.

Activity (ai) A thermodynamic term for the apparent or active concentration of a free ion in solution. It is related to concentration by the activity coefficient.

Activity based costing. Activity Based Costing is a costing and analysis method that associates resources and their costs to activities and then associates the costs of activities to cost objects (e.g., a product) based on a cost drivers which measure use of an activity by the cost object. These cost drivers, such as the number of persons performing work or the number of setups required per product reflect the consumption of activities by the products.

Activity based management A discipline that focuses on the management of activities as a route to improving the value received by the customer and the profit received by providing this value. This discipline includes cost driver analysis, activity analysis, and performance measurement.

Activity coefficient (fi) A ratio of the activity of species i(ai) to its molality (C). It is a correction factor which makes the thermodynamic calculations correct. This factor is dependent on ionic strength, temperature, and other parameters.

Individual ionic activity coefficients, f+ for cation and f- for an anion, cannot be derived thermodynamically. They can be calculated only by using the Debye-Huckel law for low concentration solutions in which the interionic forces depend primarily on charge, radius, and distribution of the ions and on the dielectric constant of the medium rather than on the chemical properties of the ions.

Mean ionic activity coefficient (f±) or the activity of a salt, on the other hand, can be measured by a variety of techniques such as freezing point depression and vapor pressure as well as paired sensing electrodes. It is the geometric mean of the individual ionic activity coefficients f± = (f+n+f-n-)1/n

Actual cash value The amount of money invested in the purchase and repairs of a used vehicle. Also known as ACV, this represents the amount of out of pocket expense a dealer or broker is "into" a car.

Actuating system A mechanical system that supplies and transmits energy for the operation of other mechanisms or systems.

Actuator A device used to transfer motion from one object to another. An actuator activates a movement or a process.

Adaptation luminance The average luminance (or brightness) of those objects and surfaces in the immediate vicinity of an observer. Also called adaptation brightness, adaptation level, adaptation illuminance.

The adaptation luminance has a marked influence on an observer's estimate of the visual range because, along with the visual angle of the object under observation, it determines the observer's threshold contrast. High adaptation luminance tends to produce a high threshold contrast, thus reducing the estimated visual range. This effect of the adaptation luminance is to be distinguished from the influence of background luminance.

Adaptation The adjustment, alteration, or modification of an organism to fit it more perfectly for existence in its environment.

Adaptation is applied particularly to evolutionary change.

Adapter 1. Any device or contrivance used or designed primarily to fit or adjust one thing to another, as: (a) a buckle or clip on a parachute harness, used in adjusting the harness to the wearer; (b) a joint attaching an afterburner to a turbine casing on a jet engine; (c) a fitting for connecting pipes, valves, etc., that have different types of threads.

2. Any device, appliance or the like used to alter something so as to make it suitable for a use for which it was not originally designed.

Adapter skirt A flange or extension of a space vehicle stage or section that provides a ready means for fitting some object to the stage or section.

Adaptive control system A control system which continuously monitors the dynamic response of the controlled system and automatically adjusts critical system parameters to satisfy preassigned response criteria, thus producing the same response over a wide range of environmental conditions.

ADC(Analog-to-Digital Converter) An electronic device which converts analog signals to an equivalent digital form, in either a binary code or a binary-coded-decimal code. When used for dynamic waveforms, the sampling rate must be high to prevent aliasing errors from occurring.

Adcock antenna A pair of vertical antennas separated by a distance of one-half wavelength or less, and connected in phase opposition to produce a radiation pattern having the shape of a figure eight.

Adder In a computer, a device which can form the sum of two or more numbers or quantities.

Additive Any material or substance added to something else. Specifically, a substance added to a propellant to achieve some purpose, such as a more even rate of combustion, or a substance added to fuels or lubricants to improve them or give them some desired quality, such as tetraethyl lead added to a fuel as an antidetonation agent, or graphite, talc, or other substances added to certain oils and greases to improve lubrication qualities.

Address A memory location in a particular machine's RAM. A numeric identifier or symbolic name that specifies the location of a particular machine or device on a network, and a means of identifying a complete network, subnetwork, or a node within a network.

Address mask (Subnet mask) A set of rules for omitting parts of a complete IP address in order to reach the target destination without using a broadcast message. The mask can, for example, indicate a subnetwork portion of a larger network. In TCP/IP, the address mask uses the 32 bit IP address.

Address resolution Mapping of an IP address to a machine's physical address. TCP/IP uses the Address Resolution Protocol (ARP) for this function.

Address space A range of memory addresses available to an application program.

Adducted A foot or part of a foot is in an adducted position when the distal aspect is angulated toward the midline of the body in a transverse plane and away from the proximal aspect of the foot or part, or away from some other specified reference point.

Adduction Motion occurring on the transverse plane during which the distal aspect of the foot moves toward the midline of the body about a vertical axis of rotation. The axis of motion lies on the frontal and sagittal planes.

Adductus This is a fixation of a part in the position it would assume if adducted. It is a transverse plane fixation, in which the distal end is displaced toward the midline. Metatarsal Adductus is a fixed position of the metatarsus, which is adducted, with reference to the lesser tarsus.

Adhesion The degree of attachment between two or more bodies during their service life, such as a: a paint film to the underlying material to which it is in contact (substrate) or b: the ability of a roofing membrane to remain attached to the substrate or to itself.

Adiabatic process A thermodynamic change of state of a system in which there is no transfer of heat or mass across the boundaries of the system. In an adiabatic process, compression always results in warming, expansion in cooling.

Adiabatic recovery temperature 1. The temperature reached by a moving fluid when brought to rest through an adiabatic process. Also called recovery temperature, stagnation temperature.

2. = adiabatic wall temperature.

3. The final and initial temperature in an adiabatic, Carnot cycle.

Adiabatic wall temperature The temperature assumed by a wall in a moving fluid stream when there is no heat transfer between the wall and the stream.

Admixture Liquid or finely ground solid materials combined into a particular concrete, mortar or grout mix design in predetermined, minute, and very controlled quantities. The intended result is a major change in the behavior of the resulting product.

Adsorbed water Water held on surfaces of a material: in a soil or rock mass, this is water held by physico-chemical forces, having physical properties substantially different from absorbed water or chemically combined water, at the same temperature and pressure.

Adsorption The adhesion of a thin film of liquid or gas to the surface of a solid substance. The solid does not combine chemically with the adsorbed substance.

Advanced wastewater treatment The removal of any dissolved or suspended contaminants beyond secondary treatment, often this is the removal of the nutrients nitrogen and/or phosphorus.

Advection The process of transport of an atmospheric property solely by the mass motion of the atmosphere; also, the rate of change of the value of the advected property at a given point.

Regarding the general distinction (in meteorology) between advection and convection, the former describes the predominantly horizontal, large-scale motions of the atmosphere whereas convection describes the predominantly vertical, locally induced motions.

Aeration Intimate contact of the atmosphere and water to add air (oxygen) to the water. The term is also applied to gas stripping where an undesirable gas is removed from the water.

Aeroballistics The study of the interaction of projectiles or high speed vehicles with the atmosphere. The problem of the effect of reentry on the trajectory of a vehicle is a problem in aeroballistics.

Aeroduct A ramjet type of engine designed to scoop up ions and electrons freely available in the outer reaches of the atmosphere or in the atmospheres of other spatial bodies, and by a metachemical process within the duct of this engine, expel particles derived from the ions and electrons as a propulsive jetstream.

Aerodynamic 1. The science that deals with the motion of air and other gaseous fluids, and of the forces acting on bodies when the bodies move through such fluids, or when such fluids move against or around the bodies, as, his research in aerodynamics.

2. (a) The actions and forces resulting from the movement or flow of gaseous fluids against or around bodies, as, the aerodynamics of a wing in supersonic flight. (b) The properties of a body or bodies with respect to these actions or forces, as, the aerodynamics of a turret or of a configuration.

3. The application of the principles of gaseous fluid flows and of their actions against and around bodies to the design and construction of bodies intended to move through such fluids, as a design used in aerodynamics.

Aerodynamic coefficient Any nondimensional coefficient relating to aerodynamic forces or moments, such as a coefficient of drag, a coefficient of lift, etc.

Aerodynamic heating The heating of a body produced by passage of air or other gases over the body; caused by friction and by compression processes and significant chiefly at high speeds.

Aerodynamic trail A condensation trail formed by adiabatic cooling to saturation (or slightly supersaturation) of air passing over the surfaces of high-speed aircraft.

Aerodynamic trails form off the tips of wings and propellers and other points of maximum pressure decrease. They are relatively rare and of short duration compared to exhaust trails.

Aerodynamic vehicle A device, such as an airplane, glider, etc., capable of flight only within a sensible atmosphere and relying on aerodynamic forces to maintain flight.

The term is used when the context calls for discrimination from space vehicle.

Aeroelasticity Any phenomenon which includes the mutual interaction between aerodynamic loads and structural deformation.

Aeroembolism 1. The formation or liberation of gases in the blood vessels of the body, as brought on by a too-rapid change from a high, or relatively high, atmospheric pressure to a lower one.

2. The disease or condition caused by the formation of gas bubbles (mostly nitrogen) in the body fluids. The disease is characterized principally by neuralgic pains, cramps, and swelling, and sometimes results in death. Also call decompression sickness.

Aeroemphysema A swelling condition caused by the formation of gas in the tissues of the body.

Aerogel A microporous, transparent silicate foam used as a glazing cavity fill material, offering possible U-values below 0.10 BTU/(h-sq ft-°F) or 0.56 W/(sq m-°C).

Aerolite A meteorite composed principally of stony material.

Aeronomy The study of the upper regions of the atmosphere where ionization, dissociation, and chemical reactions take place.

Aeropause A region of indeterminate limits in the upper atmosphere, considered as a boundary or transition region between the denser portion of the atmosphere and space.

From a functional point of view, it is considered to be that region in which the atmosphere is so tenuous as to have a negligible, or almost negligible, effect on men and aircraft, and in which the physiological requirements of man become increasingly important in the design of aircraft and auxiliary equipment.

Aerosinusitis An inflammatory reaction of one or more of the accessory nasal sinuses resulting from a difference in pressure between the gas in the sinus and the surrounding atmosphere. Also called sinus barotrauma .

Aerospace (From aeronautics and space). 1. Of or pertaining to both the earth's atmosphere and space, as in aerospace industries.

2. Earth's envelope of air and space above it; the two considered as a single realm for activity in the flight of air vehicles and in the launching, guidance, and control of ballistic missiles, earth satellites, dirigible space vehicles, and the like.

Aerospace engineering Aerospace engineering involves developing, designing, testing, and helping to manufacture commercial and military aircraft, missiles and spacecraft, and new technologies in commercial aviation, defense systems, and space exploration. Aerospace engineers have specialties within aerodynamics, propulsion, thermodynamics, structures, celestial mechanics, acoustics, and guidance and control systems.

Aerospace medicine That branch of medicine dealing with the effects of flight through the atmosphere or in space upon the human body and with the prevention or cure of physiological or psychological malfunctions arising from these effects.

Aerospace vehicle A vehicle capable of flight within and outside the sensible atmosphere.

Aerothermodynamics The study of aerodynamic phenomena at sufficiently high gas velocities that thermodynamic properties of the gas are important.

Aerothermodynamic border An altitude at about 100 miles, above which the atmosphere is so rarefied that the skin of an object moving through it at high speeds generates no significant heat.

Aerothermodynamic duct The full term for athodyd.

Aerothermoelasticity The study of the response of elastic structures to the combined effect of aerodynamic heating and loading.

AF Across the Flats. Measurement between 2 parallel faces on a nut. Indicates the size of spanner or socket required to tighten or loosen the nut.

Affinity A thermodynamic measurement of the strength of binding between molecules, say between an antibody and antigen. Each antibody/antigen pair has an association constant, Ka, expressed in L/mol.

Affinity constant The reciprocal of the dissociation constant; a measure of the binding energy of a ligand in a receptor.

Afterbody 1. A companion body that trails a satellite.

2. A section or piece of a rocket or spacecraft that enters the atmosphere unprotected behind the nose cone or other body that is protected for entry.

3. The afterpart of a vehicle.

Afterburning 1. Irregular burning of fuel left in the firing chamber or a rocket after fuel cutoff.

2. The function of an afterburner, a device for augmenting the thrust of a jet engine by burning additional fuel in the uncombined oxygen in the gases from the turbine.

Aftercooling 1. The cooling of a gas after compression.

2. The necessary cooling of a reactor core after its shutdown by pumping a liquid or gas through it to carry off the excess heat generated by continuing radioactive decay of fission products within the core.

Afterglow 1. A broad, high arch of radiance or glow seen occasionally in the western sky above the highest clouds in deepening twilight, caused by the scattering effect of very fine particles of dust suspended in the upper atmosphere.

2. The transient decay of a plasma after the power has been turned off.

The decay time involved is a direct consequence of the charged particle loss mechanisms, such as diffusion and recombination. The magnitude of these quantities is determined by measuring the decay time under controlled conditions.

Afterheat The heat generated in a reactor core after shutdown by continuing radioactive decay of fission products.

Age hardening Process of hardening by aging; in metals usually occurs after rapid cooling or cold working. It can be generally favorable (for concrete, mortar) or unfavorable (for metals, elastomers, etc.).

Aging 1. Change in the properties of certain metals and alloys that occurs at ambient (or somewhat elevated) temperatures after hot working or a heat treatment (quench aging in ferrous alloys, natural or artificial aging in ferrous and nonferrous alloys) or after a cold-working operation (strain aging). The change in properties does not involve a change in the chemical composition of the metal or alloy.

AGM (Absorbtive Glass Mat) battery A lead acid battery using a glass mat to promote recombination of the gases produced by the charging process.

Agonic line A line joining points at which the magnetic variation is zero. The agonic line is a particular case of an isogonic line.

Agravic Or pertaining to a condition of no gravitation.

Agravic illusion An apparent movement of a target in the visual field due to otolith response in zerogravity. Also called oculoagravic illusion .

Agreement state A state that has signed an agreement with the Nuclear Regulatory Commission under which the state regulates the use of byproduct, source, and small quantities of special nuclear material in that state.

Agricultural engineering Agricultural engineering involves every aspect of food production, processing, marketing, and distribution. Agricultural engineers design and develop agricultural and food processing equipment, irrigation systems, grain storage facilities, feed mills, and farm structures.

Air 1. The mixture of gases comprising the earth's atmosphere.

2. The realm or medium in which aircraft operate.

Air bag The air bag, also known as a Supplemental Inflatable Restraint System, is a passive safety device, supplemental to safety belts, that inflates to provide a cushion to absorb impact forces during moderate to severe frontal collisions. This system can help to lessen the chance of contact with the steering wheel, instrument panel and windshield. The air bag is actuated automatically by sensors located in the front of the vehicle. To maximize effectiveness, seat and shoulder belts must always be used in conjunction with this system.

Air barrier Intentionally designed and constructed barrier meant to prevent both the exfiltration of indoor air to the outside and the infiltration of outdoor air into the building environment. This applies whether the air is humid or relatively dry. Air leakage will cause the deposition of moisture in both cavity and solid walls, as well as energy losses and rain infiltration.

Air breakup The breakup of a test reentry body after reentry into the atmosphere. Air breakup is sometimes provided for, as by the firing of a cartridge inside the reentry body, so as to retard the fall of certain pieces and increase the chances of their recovery.

Air drying Curing method of a film coating in which drying occurs by simple air exposure and without heat or the presence of a catalyst; the process of curing happens either by oxidation or bysolvent evaporation.

Air Impermeability 1. Airtightness

2. A major requirement of an air barrier is that it offer a high resistance to air flow. While absolute air impermeability may not be required, materials such as glass, sheet metal, gypsum board, cast-in-place concrete and a properly supported polyethylene sheet offer a much higher resistance to air flow than do more porous materials such as concrete blocks, fibre board sheathing, and expanded polystyrene insulation. A second major consideration is that individual panels be joined into an airtight assembly. The joints between gypsum boards can be taped quickly and effectively, sheet metal panels can be lapped with tape, precast panels can be sealed with rope and sealants, etc.

Air infiltration The amount of air leaking in and out of a building through cracks in walls, windows and doors.

Air injection A system that injects air into the exhaust ports of the engine for combustion of unburned hydrocarbons in the exhaust gases, thus producing "cleaner" exhaust emissions.

Air leakage Air flow in and out of a building, which occurs under two conditions: there must be an opening in the building envelope and there must be an air pressure difference across the wall at that location. The opening need not be straight or direct through the wall: it can follow a convoluted path inside the wall. As for the pressure difference, it can result from one or combination of three possible causes: wind, stack effect, and fan pressurization.

Air leakage can cause a number of problems: spalling masonry, ice build-up under soffits, frozen pipes, condensation in cavities, rain leaks, high energy costs, and indoor humidity problems. The seeds for many of these pitfalls are sown during the design phase, largely because of major confusion concerning the function of the air and vapour barriers.

Air leakage rating A measure of the rate of infiltration around a window or skylight, or through a wall in the presence of a specific pressure difference. It is expressed in units of cubic feet per minute per square foot of window area (cfm/sq ft) or cubic feet per minute per foot of window perimeter length (cfm/ft). The lower an envelope component's air leakage rating, the better its airtightness.

Air light Light from sun and sky which is scattered into the eyes of an observer by atmospheric suspensoids (and, to slight extent, by air molecules) lying in the observer's cone of vision. That is, air light reaches the eye in the same manner that diffuse sky radiation reaches the earth's surface.

Air light is not be confused with airglow .

Air lock 1. A stoppage or diminution of flow in a fuel system, hydraulic system, or the like, caused by a pocket of air or vapor.

2. A chamber capable of being hermetically sealed that provides for passage between two places of different pressure, as between an altitude chamber and the outside atmosphere.

Air position indicator (API) An airborne computing system which presents a continuous indication of the aircraft position on the basis of aircraft heading, airspeed, and elapsed time.

Air sampling The collection of samples of air to measure the radioactivity or to detect the presence of radioactive material, particulate matter, or chemical pollutants in the air.

Air shower A grouping of cosmic-ray particles observed in the atmosphere; a cascade shower in the atmosphere. Also called shower.

Primary cosmic rays slowed down in the atmosphere emit bremsstrahlung photons of high energy. Each of these photons produces secondary electrons which generate more photons and the process continues until the available energy is absorbed.

Air sounding The act of measuring atmospheric phenomena or determining atmospheric conditions at altitude, especially by means of apparatus carried by balloons or rockets.

Air space Of or pertaining to both the atmosphere and space.

Because this adjective is pronounced as the noun airspace is, it is subject to misunderstanding. Aerospace is commonly used instead.

Air vane A vane that acts in the air, as contrasted to a jet vane which acts within a jetstream.

Airborne radioactivity area A room, enclosure, or area in which airborne radioactive materials, composed wholly or partly of licensed material, exist in concentrations that 1. exceed the derived air concentration limits or

2. would result in an individual present in the area without respiratory protection exceeding, during those hours, 0.6 percent of the annual limit on intake or 12 derived air concentration-hours.

Aircraft Any structure, machine, or contrivance, especially a vehicle, designed to be supported by the air, being borne up either by the dynamic action of the air upon the surfaces of the structure or object, or by its own buoyancy; such structures, machines, or vehicles collectively, as, fifty aircraft.

Aircraft, in its broadest meaning, includes fixed-wing airplanes, helicopters, gliders, airships, free and captive balloons, ornithopters, flying model aircraft, kites, etc., but since the term carries a strong vehicular suggestion, it is more often applied, or recognized to apply, only to such of these craft as are designed to support or convey a burden in or through the air.

Aircraft loading apron A paved or unpaved surface for loading cargo aircraft; loading personnel for medical evacuation and transient aircraft operations; or providing an apron area for fueling aircraft, arming and disarming aircraft weapons, loading and unloading ammunition, special handling or decontamination of chemical, biological, radiological (CBR) warfare items, and for special security operations.

Aircraft maintenance parking apron A paved or unpaved apron for parking fixed or rotary wing aircraft awaiting maintenance.

Aircraft runway holding apron A paved or unpaved surface providing an aircraft holding area accessible from a taxiway. It is located near the intersection of taxiways and ends of runways and is provided for pre-takeoff engine and instrument checks. For inventory purposes, only the prepared surface is included.

Airfield pavements: aprons Prepared surfaces, other than runways and taxiways, where aircraft are parked or moved about the airfield. They are designed to support specific types of aircraft and to meet operational requirements such as maintenance and loading activities.

Airflow A flow or stream of air. An airflow may take place in a wind tunnel, in the induction system of an engine, etc., or a relative airflow can occur, as past the wing or other parts of a moving craft; a rate of flow, measured by mass or volume per unit of time.

Airfoil An aerodynamic device designed to improve traction by increasing the downforce on the car. The use of airfoils (also called wings) increases the cornering capability and improves stability at speed, but often at the expense of additional aerodynamic drag.

Airframe The assembled structural and aerodynamic components of an aircraft or rocket vehicle that support the different systems and subsystems integral to the video.

The word airframe, a carryover from aviation usage, remains appropriate for rocket vehicles since a major function of the airframe is performed during flight within the atmosphere. There is disagreement as to whether the nose cone and combustion chambers are included in the term airframe while they are attached to the vehicle.

Airglow The quasi-steady radiant emission from the upper atmosphere as distinguished from the sporadic emission of the auroras.

Airglow is a chemiluminescence due primarily to the emission of the molecules O_2 and N_2, the radical OH, and the atoms O and Na. Emissions observed in airglow could arise from three-body atom collisions forming molecules, from two-body reactions between atoms and molecules, or from recombination of ions.

Historically, airglow has referred to visual radiation. Some recent studies use airglow to refer to radiation outside the visual range.

Airless spray A type of spraying system in which paint is atomized by using high hydraulic pressure, rather than compressed air.

Airscoop A hood or open end of an air duct or a similar structure, projecting into the airstream about a vehicle in such a way as to utilize the motion of the vehicle in capturing air to be conducted to an engine, a ventilator, etc.

Airsickness Motion sickness occurring in flight.

Airspace Specifically, the atmosphere above a particular portion of the earth, usually defined by the boundaries of an area on the surface projected upward.

Airspace is sometimes particularized by altitude, as the airspace above 20,000 feet.

Aitken dust counter An instrument developed by John Aitken for determining the dust content of the atmosphere. In operation, a sample of air is mixed, in an expandable chamber, with a larger volume of dust-free air containing water vapor. Upon a sudden expansion, the chamber cools adiabatically below its dewpoint, and the droplets form with the dust particles as nuclei (Aitken nuclei). A portion of these droplets settle on a ruled plate in the instrument and are counted with the aid of a microscope. Also called Aitken nucleus counter.

Aitken nuclei The microscopic particles in the atmosphere which serve as condensation nuclei for droplet growth during the rapid adiabatic expansion produced by an Aitken dust counter. These nuclei are both solid and liquid particles whose diameters are of the order of tenths of microns or even smaller.

The Aitken nuclei play an important role in atmospheric electrical processes, for they are the particles which capture (by adsorption or other surface electrical processes) small ions and thereby form large ions. In air containing large numbers of Aitken nuclei, the small ion population is small, the large ion population is large, and the air conductivity is low.

ALARA Acronym for "as low as (is) reasonably achievable." Means making every reasonable effort to maintain exposures to ionizing radiation as far below the dose limits as practical, consistent with the purpose for which the licensed activity is undertaken, taking into account the state of technology, the economics of improvements in relation to state of technology, the economics of improvements in relation to benefits to the public health and safety, and other societal and socioeconomic considerations, and in relation to utilization of nuclear energy and licensed materials in the public interest.

Albedo The ratio of the amount of electromagnetic radiation reflected by a body to the amount incident upon it, often expressed as a percentage, as, the albedo of the earth is 34% .

The concept defined above is identical with reflectance. However, albedo is more commonly used in astronomy and meteorology and reflectance in physics.

Albedo is sometimes used to mean the flux of the reflected radiation as, the earth albedo is 0.64 calorie per square centimeter. This usage should be discouraged.

The albedo is to be distinguished from the spectral reflectance which refers to one specific wavelength (monochromatic radiation).

Usage varies somewhat with regard to the exact wavelength interval implied in albedo figures; sometimes just the visible portion of the spectrum is considered, sometimes the totality of wavelengths in the solar spectrum.

Albedometer An instrument used for the measurement of the reflecting power, the albedo, of a surface. A pyranometer adapted for the measurement of radiation reflected from the earth's surface is sometimes employed as an albedometer.

Alclad Composite material having an aluminum alloy core clad on one or both surfaces with a metallurgically bonded aluminum or aluminum alloy coating that is anodic to the core and thus electrochemically protects the core against corrosion.

Alcohol Group of solvents including ethanol, methanol, and isopropyl; they tend to have relatively high evaporation rates and fairly low solvent strength.

Aldehyde An organic compound with a carbonyl at one end of a hydrocarbon group.

Alford loop A multielement antenna, having approximately equal and in-phase currents uniformly distributed along each of its peripheral elements, producing a substantially circular radiation pattern in the plane of polarization.

Alfvén mach number The ratio of the local flow velocity to the local Alfvén speed.

Alfvén speed The speed at which Alfvén waves are propagated along the magnetic field.

For a perfectly conducting fluid with a mass density of 1 kilogram per cubic meter in a magnetic field of 10,000 gauss, the Alfvén speed is about 1,000 meters per second while the speed of sound in air is about 300 meters per second.

Alfvén wave A transverse wave in a magnetohydrodynamic field in which the driving force is the tension introduced by the magnetic field along the lines of force. Also called magnetohydrodynamic wave.

Alga (plural, algae) Any plants of a group of unicellular and multicellular primitive organisms that include the Chlorella, Scenedesmus, and other genera.

The green algae and blue-green algae, for example, provide a possible means of photosynthesis in a closed ecological system, also a source of food.

Algorism The art or system of calculating with any species of notation, as in arithmetic with nine figures and a zero. Also called algorithm.

Different algorisms have been used in the design of computing machines.

Algorithm A set of well-defined mathematical rules or operations for solving a problem in a finite number of steps.

Aliasing If the sample rate of a function (fs) is less than two times the highest frequency value of the function, the frequency is ambiguously presented. The frequencies above (fs/2) will be folded back into the lower frequencies producing erroneous data.

Alidade That part of an optical measuring instrument comprising the optical system, indicator, vernier, etc.

In modern practice the term is used principally to refer to a telescope mounted over a compass or compass repeater to facilitate observation of bearings, and to a surveying instrument consisting of a telescope mounted over a compass rose, for measuring directions.

Alignment Generally refers to wheel alignment, which is the proper adjustment of the car's front and rear suspension for camber, toe, caster and ride height.

Aliphatic hydrocarbons A class of organic solvents which are composed of open chains of carbon atoms. Aliphatics are relatively weak solvents. Mineral spirits and VM & P Naphtha are aliphatic solvents.

Alive (Live) Electrically connected to a source of voltage difference or electrically charged so as to have a voltage different from that of earth; the term may be used in place of 'current-carrying' where the intent is clear, to avoid repetition of the longer term.

Alkali A compound which dissolves in water producing negatively charged hydroxide ions. Alkaline solutions are strongly basic and neutralise acids forming a salt and water.

Alkali metal Any of the highly reactive metals (such as lithium, sodium, potassium, rubidium, cesium, and francium) found in the first column of the periodic table. They act as bases (form strongly alkaline hydroxides), hence the name.

Alkaline 1. Having a pH value of between 7 and 14.

2. Having properties of an alkali. (2) Having a pH greater than 7.

Alkaline battery A battery which uses an aqueous alkaline solution for its electrolyte.

Alkaline earth metal A metal in group IIA of the periodic table (beryllium, magnesium, calcium, strontium, barium, radium).

Alkalinity The capacity of a water to neutralize acids.

Alkali-silica reaction (ASR) Reaction of alkalis with aggregate with various forms of poorly crystalline reactive silica: opal, chert, flint and chalcedony and also tridymite, crystoblite and volcanic glasses. Aggregate containing such materials (e.g., some cherty gravels) may cause deterioration of concrete when present in amounts of 1% to 5%. Concrete made of these aggregates is characterized by the early onset of a relatively rapid expansion. Cracking of structures is often observed within 10 years of construction.

All wheel drive Often confused with Four-Wheel Drive (4WD), this drive system features four, full-time active drive wheels to reduce wheel slippage and provide greater driver control over the vehicle. All-Wheel Drive automatically splits engine torque between the front and rear wheels as needed, improving on-road traction in unfavorable road conditions. Unlike Four-Wheel Drive, All-Wheel Drive is an on-road system and is not designed for off-road use. AWD does not require the driver to actively engage the system. It is operational at all times, and requires no switches, lights or visor instructions for system operation.

Alliance contract A contract that generally relates to a specific and discrete set of services such as design or maintenance.

Not as extensive as an Integrated Service Contract.

Alligatoring Roofing or pavement defect characterized by randomized cracking of the surfacing bitumen, producing a pattern of cracks similar to an alligator's hide; these cracks may or may not extend through the whole thickness of the surfacing bitumen.

Allotriomorphic crystal A crystal having a normal lattice structure, but with an imperfect outward shape due to the influence of its surroudings.

Allotropy The possibility of existence of two or more different crystal structures for a substance (generally an elemental solid).

Alloy Substance composed of two or more elements, of which at least one is a metal. These elements are intimately united usually by being dissolved in each other when molten.

Alloy 11 A compensating alloy used in conjunction with pure copper as the negative leg to form extension wire for platinum-platinum rhodium thermocouples Types R and S.

Alloy 200/226 The combination of compensating alloys used with tungsten vs. tungsten 26% rhenium thermocouples as extension cable for applications under 200°C.

Alloy 203/225 The combination of compensating alloys used with tungsten 3% rhenium vs. tungsten 150 rhenium thermocouples as extension cable for applications under 200°C.

Alloy 405/426 The combination of compensating alloys used with tungsten 5% rhenium vs. tungsten 26% rhenium thermocouples as extension cable for applications under 870°C.

Alloy wheels A generic term used to describe any non-steel road wheel. The most common alloy wheels are cast aluminum. Technically, an alloy is a mixture of two or more metals. These wheels are known for their light weight and strength.

Alpha decay The radioactive transformation of a nuclide by alpha-particle emission. Also called alpha disintegration.

The decay product is the nuclide having a mass number four units smaller and an atomic number two units smaller than the original nuclide.

Alpha particle A positively charged particle ejected spontaneously from the nuclei of some radioactive elements. It is identical to a helium nucleus that has a mass number of 4 and an electrostatic charge of +2. It has low penetrating power and a short range (a few centimeters in air). The most energetic alpha particle will generally fail to penetrate the dead layers of cells covering the skin and can be easily stopped by a sheet of paper. Alpha particles are hazardous when an alpha-emitting isotope is inside the body.

Also An electrical circuit which provides a continuous quantitative output (as opposed to a digital output which may be a series of pulses or numbers) in response to its input.

Anechoic chamber A room whose walls do not reflect either electromagnetic or acoustic waves.

Alternating current A periodic current the average value of which over a period is zero. Unless distinctly specified otherwise, the term refers to a current that reverses at regularly recurring intervals of time and which has alternately positive and negative values.

Altimeter An instrument for measuring height above a reference datum; specifically, an instrument similar to an aneroid barometer that utilizes the change of atmospheric pressure with altitude to indicate the approximate elevation above a given point or plane used as a reference.

Altitude (symbol h) 1. In astronomy, angular displacement above the horizon; the arc of a vertical circle between the horizon and a point on the celestial sphere, measured upward from the horizon.

Angular displacement below the horizon is called negative altitude or dip.

2. Height, especially radial distance as measured above a given datum, as average sea level.

In space navigation altitude designates distance from the mean surface of the reference body as contrasted to distance, which designates distance from the center of the reference body.

Altitude acclimatization A physiological adaptation to reduced atmospheric and oxygen pressure.

Altitude chamber A chamber within which the air pressure, temperature, etc., can be adjusted to simulate conditions at different altitudes; used for experimentation and testing.

Altitude difference In navigation, the difference between computed and observed altitudes, or between precomputed and sextant altitudes. It is labeled T (toward) or A (away) as the observed (or sextant) altitude is greater or smaller than the computed (or precomputed) altitude. Also called altitude intercept, intercept.

Altitude sickness In general, any sickness brought on by exposure to reduced oxygen tension and barometric pressure.

Many writers have proposed specific definitions for this term but the definitions are highly variable.

Altitude wind tunnel A wind tunnel in which the air pressure, temperature, and humidity can be varied to simulate conditions at different altitudes.

In an altitude wind tunnel for testing engines, provision is made for exchanging fresh air for exhaust-laden air during operation.

ALU (Arithmetic Logic Unit) The part of a CPU where binary data is acted upon with mathematical operations.

Alumel An aluminum nickel alloy used in the negative leg of a Type K thermocouple (Trade name of Hoskins Manufacturing Company).

Alumina A ceramic used for insulators in electron tubes or substrates in thin-film circuits. It can withstand continuously high temperatures and has a low dielectric loss over a wide frequency range. Aluminum oxide (Al_2O_3)

Aluminum roof coating Fibered or unfibered cutback asphalt coating pigmented with microscopic aluminum flakes.

Alveolar air The respiratory air in the alveoli (air sacs) deep within the lungs.

Alveolar oxygen pressure The oxygen pressure in the alveoli. The value is about 105 millimeters of mercury.

Alveoli The terminal air sacs deep within the lungs. The inhaled oxygen diffuses across the thin alveolar membranes (the walls of the air sacs) into the blood stream, and at the same time carbon dioxide diffuses from the blood into the alveoli and is exhaled through the lungs.

AM 1 The air mass 1 spectrum of a light source is equivalent to that of sunlight at the earth's surface when the sun is at zenith.

Ambient (symbol , used as a subscript) Surrounding; especially, of or pertaining to the environment about a flying aircraft or other body but undisturbed or unaffected by it, as in ambient air, or ambient temperature.

Ambient compensation The design of an instrument such that changes in ambient temperature do not affect the readings of the instrument.

Ambient conditions The conditions around the transducer (pressure, temperature, etc.).

Ambient noise The pervasive noise associated with a given environment, being usually a composite of sounds from sources both near and distant.

Ambient temperature The average temperature of the environment immediately surrounding the power supply. For forced air-cooled units, the ambient temperature is measured at the air intake.

Ambiguity In navigation, the condition arising when a given set of observations defines more than one point, direction, line of position, or surface of position.

Amide A molecule containing an amine bonded to a carboxyl group; the resulting bond has substantial double-bond character. Also termed a peptide; amide bonds link amino acids in proteins.

Ampere (A) The unit of electric current; the constant current which, if maintained in two straight, parallel conductors of infinite length, of negligible circular cross sections, and placed 1 meter apart in a vacuum will produce between these conductors a force equal to 2×10^{-7} newtons per meter of length.

Ampere hours (Ah) The unit of measure used for comparing the capacity or energy content of a batteries with the same output voltage. For automotive (Lead Acid) batteries the SAE defines the Amphour capacity as the current delivered for a period of 20 hours until the cell voltage drops to 1.75 V.

Ampere's law (AM Ampere) The line integral of the magnetic flux around a closed curve is proportional to the algebraic sum of electric currents flowing through that closed curve; or, in differential form.

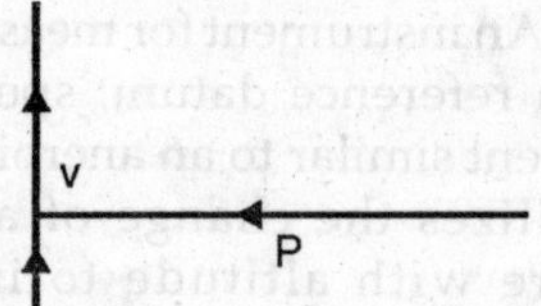

Fig. Ampere's law

Ampere hour capacity The quantity of electricity measured in ampere-hours (Ah) which may be delivered by a cell or battery under specified conditions.

Ampere hour efficiency The ratio of the number of ampere-hours delivered during the discharge of a secondary cell or battery to the number of ampere-hours necessary to restore the initial state of charge under specified conditions.

Amperometric sensor Amperometric sensors involve a heterogeneous electron transfer as a result of an oxidation/reduction of an electro-active species at a sensing electrode surface. A current is measured at a certain imposed voltage of the sensing electrode with respect to the reference electrode. Analytical information is obtained from the current-concentration relationship at that given applied potential.

Amplidyne A special type of direct current generator used as a power amplifier in which the output voltage responds to changes in field excitation; used extensively in servo systems.

Amplifier A device which enables an input signal to control a source of power, and thus is capable of delivering at its output an enlarged reproduction of the essential characteristics of the signal.

Typical amplifying elements are electron tubes, transistors, and magnetic circuits.

Amplifier, differential An amplifier which has available both an inverting and a non-inverting input, and which amplifies the difference between the two inputs.

Amplifier, inverting An amplifier whose output is 180 degrees out of phase with its input. Such an amplifier can be used with degenerative feedback for stabilization purposes.

Amplifier, Non inverting An amplifier whose output is in phase with its input.

Amplifier, operational A dc amplifier whose gain is sufficiently large that its characteristics and behavior are substantially determined by its input and feedback elements. Operational amplifiers are widely used for signal processing and computational work.

Amplitude 1. The maximum value of the displacement of a wave or other periodic phenomenon from a reference position.

2. Angular distance north or south of the prime vertical; the arc of the horizon, or the angle at the zenith between the prime vertical and a vertical circle, measured north or south from the prime vertical to the vertical circle.

The term is customarily used only with reference to bodies whose centers are on the celestial horizon, and is prefixed E or W, as the body is rising or setting, respectively; and suffixed N or S to agree with the declination. The prefix indicates the origin, and the suffix indicates the direction of measurement. Amplitude is designated as true, magnetic, compass, or grid as the reference direction is true, magnetic, compass, or grid east or west, respectively.

Amplitude modulation 1. In general, modulation in which the amplitude of a wave is the characteristic subject to variation.

2. Specifically, in telemetry those systems of modulation in which each component frequency, f, of the transmitted intelligence produces a pair of sideband frequencies at carrier frequency plus f and carrier minus f.

In special cases: (a) the carrier may be suppressed, (b) either the lower or upper sets of sideband frequencies may be suppressed; (c) the lower set of sideband frequencies may be produced by one or more channels of information and the upper set of sideband frequencies may be produced by one or more other channels of information; (d) the carrier may be transmitted without intelligence carrying sideband frequencies.

Amplitude modulated indicator One of two general classes of radar indicators, in which the sweep of the electron beam is deflected vertically or horizontally from a base line to indicate the existence of an echo from a target. The amount of deflection is usually a function of the echo signal strength. Also called deflection-modulated indicator.

Amplitude span The Y-axis range of a graphic display of data in either the time or frequency domain. Usually a log display (dB) but can also be linear.

Ampoule battery A battery in which the electrolyte is stored in a separate chamber from the cell electrodes until the battery is needed.

Anabolism Biosynthesis, the production of new cellular materials from other organic or inorganic chemicals.

Anacoustic zone The region above an altitude of about 100 miles where the distance between the air molecules is greater than the wavelength of sound, and sound waves can no longer be propagated.

Anaerobes A group of organisms that do not require molecular oxygen. These organisms, as well as all known life forms, require oxygen. These organisms obtain their oxygen from inorganic ions such as nitrate or sulfate or from protein.

Analog In computers, pertaining to the use of physical variables such as voltage, distance, rotation, etc. To represent numerical variable as in analog computer, analog output.

Analogue (Analog) circuit An electronic circuit in which an electrical value (usually voltage or current, but sometimes frequency, phase) represents something in the physical world.The magnitude of the electrical value varies with with the intensity of an external physical quantity.

Analog computer A computing machine working on the principle of measuring, as distinguished from counting, in which the input data is analogous to a measurement continuum, such as linear lengths, voltages, resistances, etc., which can be manipulated by the computer.

Analog computers range in complexity from a slide rule to electrical computers used for solving mathematical problems.0

Analog output Transducer output in which the amplitude is continuously proportional to a function of the stimulus. Distinguished from digital output.

Analog to digital converter (ADC) A device which will convert an analog voltage sample to an equivalent digital code of some finite resolution. Also called digitizer, encoder.

Analyte A chemical species targeted for qualitative or quantitative analysis.

Analytical photography Photography, either motion picture or still, accomplished to determine (by qualitative, quantitative, or any other means) whether a particular phenomenon does or does not occur.

Differs from metric photography in that measurements are not a prime requisite.

Anatomical position This describes a person standing erect, with the feet together and the arms by the sides of the body, with the palms forward. Anatomists and clinicians use the anatomical position to build hypothetical biomechanical models of normalcy in which to describe movement of the centre of gravity.

Anchor pattern Substrate preparation: the surface profile obtained by abrasive treatments such as sand blasting or power tool cleaning; may also refer to the distance between peaks and valleys of the blast profile.

Anchorage A secure fixing, usually made of reinforced concrete to which the cables are fastened.

Andon board A visual control device in a production area, such as a lighted overhead display. It communicates the current status of the production system and alerts team members to emerging problems.

Anemometer An instrument for measuring and/or indicating the velocity of air flow.

Aneroid A thin, disk-shaped box or capsule, usually metallic, partially evacuated of air and sealed, which expands and contracts with changes in atmospheric or gaseous pressure.

The aneroid is the sensing and actuating element in various meters or gages, such as barometers, altimeters, manifold-pressure gages, etc; it is also the triggering or operating element in various automatic mechanisms. A device similar to an aneroid, but open to outside pressures, such as the capsule in an airspeed indicator, is not commonly called an aneroid .

Angel A radar echo caused by a physical phenomenon not discernible to the eye.

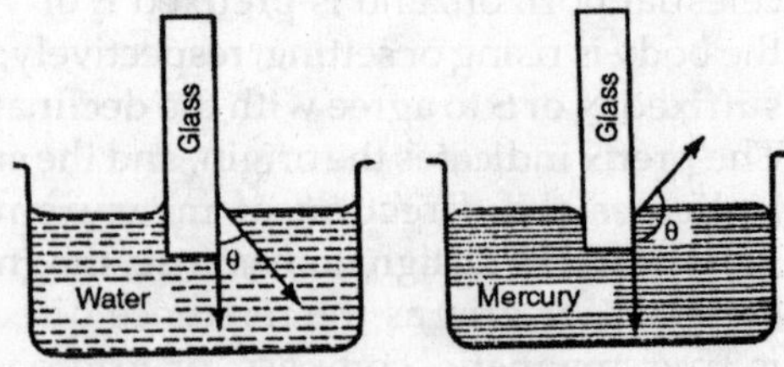

Fig. Angle of contact

Angels are usually coherent echoes and sometimes of great signal strength (up to 40 decibels above the noise level). They have been ascribed to insects flying through the radar beam, but have also been observed under atmospheric conditions which indicate there must be other causes. Studies indicate that a fair portion of them are caused by strong temperature or moisture gradients, or both, such as might be found near the boundaries of bubbles of especially warm or moist air. They frequently occur in shallow layers at or near temperature inversions within the lowest few thousand feet of the atmosphere.

Angle modulation Modulation in which the angle of a sine-wave carrier is the characteristic varied from its normal value.

Phase and frequency modulation are particular forms of angle modulation.

Angle of arrival A measure of the direction of propagation of electromagnetic radiation upon arrival at a receiver (most commonly used in radio). It is the angle between the plane of the phase front and some plane of reference, usually the horizontal, at the receiving antenna. This angle is a function of the index of refraction gradient of the medium through which the energy is traveling, and the relative positions of the transmitter and receiver.

Angles of arrival can be measured for both the direct and reflected components of a wave using a multiple-antenna receiving system called an interferometer.

Angle of attack The angle between a reference line fixed with respect to an airframe and a line in the direction of movement of the body.

Angle of climb The angle between the flight path of a climbing vehicle and the local horizontal.

Angle of depression The angle in a vertical plan between the local horizontal and a descending line. Also called depression angle .

Angle of deviation The angular change in direction of a ray of light, or other electromagnetic radiation, upon entering another medium.

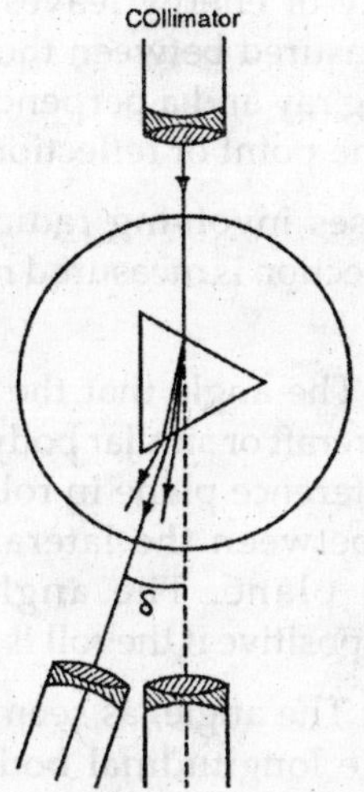

Fig. Angle of deviation

Angle of elevation The angle in a vertical plane between the local horizontal and an ascending line, as from an observer to an object. Also called elevation angle.

A negative angle of elevation is usually called an angle of depression.

Angle of incidence 1. The angle at which a ray of energy impinges upon a surface, usually measured between the direction of propagation of the energy and a perpendicular to the surface at the point of impingement, or incidence.

In some cases involving radio waves, the angle of incidence is measured relative to the surface.

2. = angle of attack.

Angle of pitch 1. The angle, as seen from the side, between the longitudinal body axis of an aircraft or similar body and a chosen reference line or plane, usually the horizontal plane. This angle is positive when the forward part of the longitudinal axis is directed above the reference line.

2. Same as blade angle.

Angle of reflection The angle at which a reflected ray of energy leaves a reflecting surface, measured between the direction of the outgoing ray and a perpendicular to the surface at the point of reflection.

In some cases involving radio waves, the angle of reflection is measured relative to the surface.

Angle of roll The angle that the lateral body axis of an aircraft or similar body makes with a chosen reference plane in rolling; usually the angle between the lateral axis and a horizontal plane. The angle of roll is considered positive if the roll is to starboard.

Angle of yaw The angle, as seen from above, between the longitudinal body axis of an aircraft, rocket, or the like and a chosen reference direction. This angle is positive when the forward part of the longitudinal axis is directed to starboard. Also called yaw angle .

Angstrom Ten to the minus tenth meters (10^{-10}) or one millimicron, a unit used to define the wave length of light. Designated by the symbol ‰.

Ångström compensation pyrheliometer An instrument developed by K. Ångström for the measurement of direct solar radiation. The radiation receiver station consists of two identical manganin strips whose temperatures are measured by attached thermocouples. One of the strips is shaded, whereas the other is exposed to sunlight. An electrical heating current is passed through the shaded strip so as to raise its temperature to that of the exposed strip. The electric power required to accomplish this is a measure of the solar radiation.

Ångström pyrgeometer An instrument developed by K. Ångström for measuring the effective terrestrial radiation. It consists of four manganin strips, of which two are blackened and two are polished. The blackened strips are allowed to radiate to the atmosphere while the polished strips are shielded. The electrical power required to equalize the temperature of the four strips is taken as a measure of the outgoing radiation.

Angular acceleration The rate of change of angular velocity.

Angular distance 1. The angular difference between two directions, numerically equal to the angle between two lines extending in the given directions.

2. The arc of the great circle joining two points, expressed in angular units.

3. Distance between two points, expressed in wave lengths at a specified frequency.

It is equal to the number of waves between the points multiplied by if expressed in radians, or multiplied by 360° if measured in degrees.

Angular frequency The frequency of a periodic quantity expressed in radians per second. It is equal to the frequency in cycles per second multiplied by . Also called circular frequency.

Angular momentum Quantity of rotational motion.

Linear momentum is the quantity obtained by multiplying the mass of a body by its linear speed. Angular momentum is the quantity obtained by multiplying the moment of inertia of a body by its angular speed. The momentum of a system of particles is given by the sum of the momentums of the individual particles which make up the system or by the product of the total mass of the system and the velocity of the center of gravity of the system. The momentum of a continuous medium is given by the integral of the velocity over the mass of the medium or by the product of the total mass of the medium and the velocity of the center of gravity of the medium.

Angular resolution Specifically, the ability of a radar to distinguish between two targets solely by the measurement of angles.

It is generally expressed in terms of the minimum angle by which targets must be spaced to be separately distinguishable.

Angular speed 1. Change of direction per unit time, as of a target on a radar screen. Also called angular rate.

2. = angular velocity.

Angular velocity (symbol) The change of angle per unit time; specifically, in celestial mechanics, the change in angle of the radius vector per unit time.

Anion Particles in the electrolyte of a galvanic cell carrying a negative charge and moving toward the anode during operation of the cell.

Annealing Generic term used to denote a controlled heat treatment process, followed by cooling performed on material in the solid state. It is used to remove stresses, to induce softness and to alter the ductility and toughness of metals, glass, or other materials by allowing them to recrystallize. The temperature of the operation and the rate of cooling will depend upon the metal and the purpose of the annealing.

Annealing twin A twin formed in a crystal during recrystallization.

Annual limit on intake (ALI) The derived limit for the amount of radioactive material taken into the body of an adult worker by inhalation or ingestion in a year. ALI is the smaller value of intake of a given radionuclide in a year by the reference man that would result in a committed effective dose equivalent of 5 rems (0.05 sievert) or a committed dose equivalent of 50 rems (0.5 sievert) to any individual organ or tissue.

Anode 1. The positive element such as the plate of a vacuum tube; the element to which the principal stream of electrons flows.

2. In a cathode-ray tube, the electrodes connected to a source of positive potential. These anodes are used to concentrate and accelerate the electron beam for focusing. (Graf)

Anodic protection A technique to reduce the corrosion rate of a metal by polarizing it into its passive region, where dissolution rates are low.

Anodic reaction Electrode reaction equivalent to a transfer of positive charge from the electronic to the ionic conductor; an anodic reaction is an oxidation process.

Anodizing Forming a conversion coating on a metal surface by anodic oxidation; most often applied to aluminum.

Anomalistic month The average period of revolution of the moon from perigee to perigee, a period of 27 days 13 hours 18 minutes 33.2 seconds.

Anomalistic period The interval between two successive perigee passages of a satellite in orbit about a primary. Also called perigee-to-perigee period .

Anomalistic year The period of one revolution of the earth about the sun from perihelion to perihelion; 365 days 6 hours 13 minutes 53.0 seconds in 1900 and increasing at the rate of 0.26 second per century.

Anomalous dispersion Dispersion of electromagnetic radiation characterized by a decrease in refractive index with increase in frequency.

Anomalous propagation The propagation of energy when it arrives at a destination via a path significantly different from the normally expected path.

The term is usually applied to the transmission of various forms of energy through the atmosphere when, in addition to the line-of-sight path, the energy is refracted by density discontinuities at one or more levels in atmosphere. Therefore, it propagates to a point that could not be reached via a line-of-sight path. In radio and radar studies, it refers to the abnormal refraction of a beam of radio energy, usually applied to superstandard propagation rather than to substandard propagation. In either case, anomalous propagation results from an unusual vertical distribution of temperature and moisture in the atmosphere. The anomalous propagation of sound refers to the downward refraction of an oblique sound wave from an explosion, the refraction occurring in the region of increasing temperature with height in the lower mesosphere. The anomalous propagation of sound has been used as a method for determining upper air temperatures and winds.

Anoxia A complete lack of oxygen available for physiological use within the body.

Anoxia is popularly used as a synonym for hypoxia . This usage should be avoided.

Anoxic process A process which occurs only at very low levels of molecular oxygen or in the absence of molecular oxygen.

Antalgic gait A gait pattern specifically modified to reduce the amount of pain a person is experiencing. The term is usually applied to a rhythmic disturbance in which as short a time as possible is spent on the painful limb and a correspondingly longer time is spent on the healthy side.

Antenna A metallic aerial for sending and receiving electromagnetic waves. A transmitting antenna converts electrical signals from a transmitter into an electromagnetic wave which it then emanates. A receiving antenna intercepts an electromagnetic wave and converts it back into electrical signals that can be decoded by a receiver. Central television, pole and wire, and switching station antennas, as well as satellite dishes, are also included within this category code.

Antenna array A system of antennas coupled together to obtain directional effects, or to increase sensitivity.

Antenna effect A weakening of the effectiveness of the directional properties of a loop antenna by the capacitance of the loop to the ground. Also called height effect. In usual direction-finding practice on ground waves, antenna effect would be manifested: (a) if in phase, by an angular displacement of the nulls from 180° displacement and (b), if in quandrature, by a residual signal obscuring the nulls. The in-phase effect is often used to eliminate the 180° ambiguity (i.e., to permit sense finding).

Antenna field A group of antennas placed in a geometric configuration.

Antenna pair Two antennas located on a base line of accurately surveyed length.

Antenna temperature In radio astronomy, a measure of the power absorbed by the antenna. In an ideal, loss-free radio telescope, the antenna temperature is equal to the brightness temperature if the intensity of the received radiation is constant within the main lobe. If the angular dimension of the source is small compared to the main lobe, the antenna temperature is equal to the brightness temperature multiplied by the ratio of the solid angle subtended by the source to the effective solid angle of the antenna.

Anterior An anatomical term meaning situated before or to the front; designated the forward part of the foot.

Anthropogenic Of, made, or caused by human activity or actions.

Anti lock brake system (ABS) On a vehicle equipped with Anti-Lock Brakes, the wheels are equipped with speed sensors. When a sensor determines that a wheel is decelerating so rapidly that lockup may occur, the electro-Hydraulic Control Unit (EHCU) is activated. The EHCU then modulates the brake pressure in the appropriate brake lines by means of the solenoid-operated valves. This is intended to prevent wheel lockup and help the vehicle maintain directional stability during potentially hazardous braking situations.

Anti matter Matter consisting of anti-particles.

Anti particle Any particle with a charge of opposite sign to the same particle in normal matters.

Thus, the proton has a positive charge; the antiproton, a negative charge. When a particle and its anti-particle collide, both may disappear with the creation of lighter particles; this process is called annihilation.

Anticipated transient without scram (ATWS) ATWS is one of the "worst case" accidents, consideration of which frequently motivates the NRC to take regulatory action. The accident could happen if the system that provides a highly reliable means of shutting down the reactor (scram system) fails to work during a reactor event (anticipated transient). The types of events considered are those used for designing the plant.

Anticyclonic Having a sense of rotation about the local vertical opposite to the rotation of the earth; that is, clockwise in the northern hemisphere, counterclockwise in the southern hemisphere, undefined at the equator; the opposite of cyclonic.

Antiferromagnetism A phenomenon where complete magnetic moment cancellation occurs as a result of antiparallel coupling of adjacent atoms or ions; the macroscopic solid possesses no net magnetic moment.

Antigravity A hypothetical effect that would arise from cancellation by some energy field of the effect of the central force field of the earth or other body.

Antinode 1. Either of the two points on an orbit where a line in the orbit plane, perpendicular to the line of nodes, and passing through the focus, intersects the orbit.

2. A point, line, or surface in a standing wave where some characteristic of the wave field has maximum amplitude. Also called loop.

Anti-reset windup This is a feature in a three-mode PID controller which prevents the integral (auto reset) circuit from functioning when the temperature is outside the proportional band.

Antiresonance For a system in forced oscillation, the condition existing at a point when any change, however small, in the frequency of excitation causes an increase in the response at this point.

Apastron That point of the orbit of one member of a binary star system at which the stars are farthest apart. That point at which they are closes together is called periastron.

Aperture 1. An indexed shape with a specified x and y dimension, or line-type with a specified width, used as a basic element or object by a photoplotter in plotting geometric patterns on film. The index of the aperture is its Position (a number used in an aperture list to identify an aperture) or D code

2. A small, thin, trapezoidal piece of plastic used to limit and shape a light source for plotting light patterns on film, and mounted in a mechanical disk called an " aperture wheel " which in turn is mounted on the lamp head of a vector photoplotter. An aperture is mostly opaque, but with a transparent portion that controls the size and shape of the light pattern. A vector photoplotter plots images from a CAD database on photographic film in a darkroom by drawing each line with a continuous lamp shined through an annular-ring aperture, and creating each shape (or pad) by flashing the lamp through a specially sized and shaped aperture.

3. A line of textual data in an aperture list describing the index names (D code and position), the shape, the usage (flash or draw) and the X and Y dimensions of an aperture. Some aperture lists leave out certain of those types of data. For example, laser photoplotters don't need to know whether an aperture is a flash or draw, so a

modern-day aperture list might leave that datum out.

Aperture list 1. An ASCII text data file which describes the size and shape of the apertures used by a photoplotter for any one photoplot.

2. A print-out of this file.

3. A binary version of this file. (Also called "aperture table.")

Aperture ratio The ratio of the useful diameter of a lens to its focal length. It is the reciprocal of the f-number.

In application to an optical instrument, rather than to a lens, numerical aperture is more commonly used. The aperture ratio is then twice the tangent of the angle whose sine is the numerical aperture.

Aperture wheel A component of a vector photoplotter , it is a metal disk having cut-outs with brackets and screw holes arranged near its rim for attaching apertures. Its center hole is attached to a motorized spindle on the lamp head of the photoplotter. When a D code denoting a particular position on the wheel is retreived from a Gerber file by the photoplotter, the wheel is caused to rotate so that the aperture in that position is placed between the lamp and the film.

In preparation for a photoplotting, the aperture wheel is set up by a technician who reads a printed aperture list, selects the correct aperture from a set of them stored in a box with compartments and, using a small screw driver, installs the aperture onto the position on the wheel which is called for on the list. This process is subject to human error and is one of the disadvantages of vector photoplotters as compared with laser photoplotters.

Aphelion That point in a solar orbit which is most distant from the sun. The point nearest the sun is called perihelion.

A-Pillar In the side view, the foremost roof support of a vehicle, located in most instances between the outer edge of the windshield and the leading edge of the front door upper. Also known as an A-Post.

Apogee 1. That point in a geocentric orbit which is most distant from the earth. That orbital point nearest the earth is called perigee.

By extension, apogee and perigee are also used in reference to orbits about other planets and natural satellites.

2. Of a satellite or rocket: To reach its apogee, as in the Vanguard apogees at 2,560 miles.

Apostilb A unit of luminance equal to international candles per square centimeter.

Apparent additional mass A fictitious mass of fluid added to the mass of the body to represent the force required to accelerate the body through the fluid.

The apparent additional mass has inertia and momentum equal to the apparent increase of the inertia and momentum of the body.

Apparent In astronomy, observed.

True values are reduced from apparent (observed) values by eliminating those factors such as refraction, light time, etc., which affected the observation.

Apparent force A force introduced in a relative coordinate system in order that Newton laws of motion be satisfied in this system. This force must be equal and opposite to an acceleration in an inertial coordinate system, in which Newton laws are (by definition) satisfied. Examples are the coriolis force, and the centrifugal force incorporated in gravity.

Apparent motion Motion relative to a specified or implied reference point which may itself be in motion. Also called relative motion.

In astronomy apparent motion usually refers to movement of celestial bodies as observed from the earth.

Apparent position The position on the celestial sphere at which a heavenly body (or a space vehicle) would be seen from the center of the earth at a particular time. Compare astrometric position.

The apparent position of a body is displaced from the true position at the time of observation by the motion of the body during the time it takes light to travel from the body to the earth and by aberration.

Most ephemerides tabulate apparent position of the sun, moon, and planets.

Apparent solar day The duration of one rotation of the earth on its axis, with respect to the apparent sun. It is measured by successive transits of the apparent sun over the lower branch of a meridian. The length of the apparent solar day is 24 hours of apparent time and averages the length of the mean solar day, but varies somewhat from day to day.

Apparent time Time based upon the rotation of the earth relative to the apparent or true sun. This is the time shown by a sundial. Apparent time may be designated as either local or Greenwich, as the local or Greenwich meridian is used as the reference.

Apparent wander Apparent change in the direction of the axis of rotation of a spinning body, as a gyro, due to rotation of the earth. Often shortened to wander.

The horizontal component of apparent wander is called drift, and the vertical component is called topple.

Application layer The highest layer in the OSF model. It establishes communications rights and can initiate a connection between two applications.

Application program A computer program that accomplishes specific tasks, such as word processing.

Application programming interface (API) A set of routines available to developers and applications to provide specific services used by the system, usually specific to the application's purpose. They act as access methods into the application.

Application rate The quantity (mass, volume or thickness) of coating material applied per unit area.

Approval Written notification by an authorized representative of the acquirer that the developers plans, design, or other aspects of the project appear to be sound and can be used as the basis for further work. Such approval does not shift responsibility from the developer to meet contractual requirements.

Approved data That issue or version of data which has been identified by the developer/ supplier as being the master issue or version of data subject to configuration management which has been submitted for acquirer approval, and which has been approved by the acquirer.

Approximate absolute temperature scale (AA) A temperature scale with the ice point at 273° and boiling point of water at 373°. It is intended to approximate the Kelvin temperature scale with sufficient accuracy for many sciences, notably meteorology, and is widely used in the meteorological literature. Also called tercentesimal thermometric scale .

Appulse 1. The near approach of one celestial body to another on the celestial sphere, as in occultation, conjunction, etc.

2. A penumbral eclipse of the moon.

Apron Specifically, a protective device specially designed to cover an area surrounding the fuel inlet on a rocket or spacecraft.

Aps, apus International Astronomical Union abbreviations for Apus .

Apsis (plural apsides) In celestial mechanics, either of the two orbital points nearest or farthest from the center of attraction. Also called apse. The apsides are the perihelion and aphelion in the case of an orbit about the sun, and the perigee and apogee in the case of an orbit about the earth. The line connecting these two points is called line of apsides. The nearest point is the lower apsis while the farthest point if the higher apsis.

Arago point One of the three commonly detectable points along the vertical circle through the sun at which the degree of polarization of diffuse sky radiation goes to zero; a neutral point.

The Arago point, so named for its discoverer, is customarily located at about 20° above the antisolar point; but it lies at higher altitudes in turbid air. The latter property makes the Arago distance a useful measure of atmospheric turbidity.

Arc 1. A part of a curved line, as a circle.

2. A luminous glow which appears when an electric current passes through ionized air or gas.

3. An auroral arc.

Arc discharge A luminous, gaseous, electrical discharge in which the charge transfer occurs continuously along a narrow channel of high ion density. An arc discharge requires a continuous source of electric potential difference across the terminals of the arc.

Arc discharge is to be distinguished from corona discharge, point discharge, and spark discharge.

Arc spectrum The spectrum of a neutral atom, designated by the Roman numeral I following the symbol for the element, and He I.

Arch bridge A curved structure that converts the downward force of its own weight, and of any weight pressing down on top of it, into an outward force along its sides and base.

Arch dam A dam with an arched shape that resists the force of water pressure; requires less material than a gravity dam for the same distance.

Architect A person who designs all kinds of structures; must also have the ability to conceptualize and communicate ideas effectively — both in words and on paper — to clients, engineers, government officials, and construction crews.

Archive A repository of files available for access at an Internet site. Also, a collection of files, often a backup of a disk or files saved to tape to allow them to be transferred.

Area divider A raised, double wood member attached to a properly flashed wood base plate that is anchored to the roof deck. It is used to relieve the stresses of thermal expansion and contraction in a roof mebrane where no expansion joints have been provided.

Area rule A prescribed method of design for obtaining minimum zero-lift drag for a given aerodynamic configuration, such as a wing-body configuration, at a given speed. For a transonic body, the area rule is applied by subtracting from, or adding to, its cross-sectional area distribution normal to the air stream at various stations so as to make its cross-sectional area distribution approach that of an ideal body of minimum drag; for a supersonic body, the sectional areas are frontal projections of areas intercepted by planes inclined at the Mach angle.

Areal velocity In celestial mechanics, the area swept out by the radius vector per unit time.

The areal velocity is constant for a central force.

Argument of perigee In celestial mechanics, the angle or arc, as seen from a focus of an elliptical orbit, from the ascending node to the closest approach of the orbiting body to the focus. The angle is measured in the orbital plane in the direction of motion of the orbiting body.

Ari, Arie International Astronomical Union abbreviations for Aries.

Arithmetic unit That part of a computer which performs arithmetic operations. Also called arithmetic element.

Aromatic A form of bonding in which ring compounds share electrons over more than two atoms. The electrons are delocalized. This leads to unusual ring stability.

Aromatic hydrocarbons A class of relatively strong organic solvents which contain an unsaturated ring of carbon atoms. Examples are benzene, toluene and xylene.

ARPA (Advanced Research Projects Agency) A government agency that originally funded the research on the ARPANET (became DARPA in the mid-1970s).

Arpanet An experimental communications network funded by the government that eventually developed into the Internet.

Array context An array value is required (either a normal array or an associative array, both of which are lists).

Arrhenius equation The equation representing the rate constant as k = AeEa/RT where A represents the product of the collision frequency and a steric factor, and eEa/RT is fraction of collisions with sufficient energy to produce a reaction.

Arrhythmia Absence of rhythm, as, for example, in heart beat.

Arrow wing An aircraft wing of V-shaped planform, either tapering or of constant chord, suggesting a stylized arrowhead.

Article Message submitted to a UseNet newsgroup. Unlike an e-mail message that goes to a specific person or group of persons, a newsgroup message goes to directories (on many machines) that can be read by any number of people.

Artificial antenna A device which has the equivalent impedance characteristics of an antenna and the necessary power-handling capabilities, but which does not radiate nor intercept radiofrequency energy. Also called dummy antenna .

Artificial feel A control feel simulated by mechanisms incorporated in the control system of an aircraft or spacecraft where the forces acting on the control surfaces are not transmitted to the cockpit controls, as in the case of an irreversible control system or a power boosted system.

Artificial gravity A simulated gravity established within a space vehicle by rotation or acceleration.

Artificial horizon 1. A gyro-operated flight instrument that shows the pitching and banking attitudes of an aircraft or spacecraft with respect to a reference line horizon, within limited degrees of movement, by means of the relative position of lines or marks on the face of the instrument representing the aircraft and the horizon.

2. A device, such as a spirit level, pendulum, etc., that establishes a horizontal reference in a navigation instrument.

Artificial satellite A manmade satellite.

Artwork Artwork for printed circuit design is photoplotted film (or merely the Gerber files used to drive the photoplotter), NC Drill file and documentation which are all used by a board house to manufacture a bare printed circuit board.

Asbestos Any of several naturally occuring, fibrous, impure silicate materials; have been implicated as causes of certain cancers, and that have been used in the past as fireproofing and insulating materials

Ascendent The negative of the gradient. The ascendent of a function is a vector with magnitude equal to the maximum spatial rate of change of that function at a given point at a given time. It is directed toward increasing values of the function along the line of maximum change, and is represented by, where F is the function and the del-operator.

Ascending node That point at which a planet, planetoid, or comet crosses to the north side of the ecliptic; that point at which a satellite crosses to the north side of the equatorial plane of its primary. Also called northbound node. The opposite is descending node or southbound node.

ASIC Application-specific integrated circuit; an IC designed for a custom requirement.

Aspects The apparent positions of celestrial bodies relative to one another; particularly, the apparent positions of the moon or a planet relative to the sun.

Aspect ratio The ratio of the square of the span of an airfoil to the total airfoil area, or the ratio of its span to its mean chord.

An airfoil of high aspect ratio is of relatively long span and short chord; one of low aspect ratio is of relatively short span and long chord.

Asphalt A dark brown to black cementitious material in which the predominating constituents are bitumens, which may occur naturally or are obtained in petroleum refining. It is used as a waterproofing agent, sealant or pavement.

Asphalt felt A mat of organic or inorganic fibers, impregnated with asphalt or coal tar pitch, or impregnated and coated with asphalt.

Asphalt mastic Mixture of asphaltic material and graded mineral aggregate that can be poured when heated, but requires mechanical manipulation to apply when cool.

Aspiration condenser An ion counter collecting element consisting of a cylindrical condenser which when charged produces a radial field which collects ions from the aspirated air.

Assemble In computer terminology, to organize the subroutines into a complete program.

Assembled battery A battery composed of two or more cells.

Assembler In recent popular usage, any nanomachine, usually assumed to offer magical, universal capabilities in an atom-sized package. In the author's usage, any programmable nanomechanical system able to perform a wide range of mechanosynthetic operations.

Assembly A number of parts or subassemblies or any combination thereof joined together to perform a specific function and capable of disassembly.

Assembly drawing A drawing depicting the locations of components, with their reference designators , on a printed circuit. Also called "component locator drawing."

Assembly house A manufacturing facility for attaching and soldering components to a printed circuit.

Assembly language A machine oriented language in which mnemonics are used to represent each machine language instruction. Each CPU has its own specific assembly language.

Assignable cause Assignable Cause is a source of variation which is not due to chance and, therefore, can be identified and eliminated. An assignable cause is often signaled by an excessive number of data points outside a control limit and/or a non-random pattern within the control limits. Also called "special cause".

Associated corpuscular emission The full complement of secondary charged particles (usually limited to electrons) associated with an X-ray or gamma ray beam in its passage through matter.

The full complement of electrons is obtained after the radiation has traversed sufficient matter to bring about equilibrium between the primary photons and secondary electrons. Electronic equilibrium with the secondary photons is intentionally excluded.

Associativity A link between two different functions in a CAD system that assures that a change made in one area is reflected in all other areas. For example, a change to a solid model will be reflected in its drawing and related CAM program. Bi-directional associativity indicates that updates happen in both directions between functions. For example, a change to a drawing will be reflected in its solids model.

Assumptions (for IPEs, IPEEs, and PRAs) In the context of PRAs, assumptions are those parts of the mathematical models that the analyst expects will hold true for the range of solutions used for making decisions. Without assumptions, even the most powerful computers may not be able to provide useful solutions for the models.

Asteroid One of the many small celestial bodies revolving around the sun, most of the orbits being between those of Mars and Jupiter. Also called planetoid, minor planet.

The term minor planet is preferred by many astronomers but asteroid continues to be used in astronomical literature, especially attributively, as in asteroid belt. All asteroids with determined orbits (except for a few discovered during World War II) are numbered for identification in the order of their discovery. The Ephemerides of the Minor Planets published by the U.S.S.R. Academy of Sciences lists all numbered asteroids, data concerning them, and their predicted positions. The daily positions of the first four minor planets are tabulated in the American Ephemeris and Nautical Almanac. Orbits have been determined for approximately 1700 asteroids. The names are usually feminine but masculine names have been used for asteroids closer to or farther away from the Sun than the majority. The first asteroid to be given a masculine name, Eros (number 443) was the first to be discovered inside the orbit of Mars. The Trojan asteroids, named for heroes of the Trojan war, are in the orbit of Jupiter.

Astre fictif A point on the celestial sphere used as a reference in measuring time intervals.

Astro A prefix meaning star or stars and, by extension, sometimes used as the equivalent of celestial, as in astro nautics.

Astroballistics The study of the phenomena arising out of the motion of a solid through a gas at speeds high enough to cause ablation; for example, the interaction of a meteoroid with the atmosphere.

Astroballistics uses the data and methods of astronomy, aerodynamics, ballistics, and physical chemistry.

Astrobiology The study of living organisms on celestial bodies other than the earth.

Astrocompass An instrument used to determine direction by sighting heavenly bodies of known position.

Astrodome A transparent dome in the fuselage or body of an aircraft or spacecraft intended primarily to permit taking celestial observations in navigating. Also called a navigation dome, astral dome.

Astrodynamics The practical application of celestial mechanics, astroballistics, propulsion theory, and allied fields to the problem of planning and directing the trajectories of space vehicles.

Astrodynamics is sometimes used as a synonym for celestial mechanics. This usage should be discouraged.

Astrolabe 1. In general, any instrument designed to measure the altitudes of celestial bodies.

2. Specifically, an instrument designed for very accurate celestial altitude measurements, as in survey work.

Astrometric position The position of a heavenly body (or space vehicle) on the celestial sphere corrected for aberration but not for planetary aberration. Compare apparent position.

Astrometric positions are used in photographic observations where the position of the observed body can be measured in reference to the positions of comparison stars in the field of the photograph.

Astrometry The branch of astronomy dealing with geometrical relations of the celestial bodies and their real and apparent motions. The techniques of astrometry, especially the determination of accurate position by photographic means, are used in tracking satellite and space probes.

Astron machine An experimental thermonuclear device where a magnetic filed is generated by a relativistic ring of electrons and shaped into a magnetic mirror configuration. The hot electrons serve as a heat source to heat the ions.

Astronaut 1. A person who rides in a space vehicle.

2. Specifically, one of the test pilots selected to participate in Project Mercury, Project Gemini, Project Apollo, or any other United States program for manned space flight.

Astronavigation The plotting and directing of the movement of a spacecraft from within the craft by means of observations on celestial bodies. Sometimes contracted, to astrogation or called celestial navigation.

Astronomical Of or pertaining to astronomy or to observations of the celestial bodies. Also called astronomic. Astronomers have long preferred astronomical. Geodesists usually use astronomic as an intended parallel to geodetic. The Coast and Geodetic Survey uses astronomic in their publications insofar as is compatible with established practice.

Astronomical constants 1. The elements of the orbits of the bodies of the solar system, their masses relative to the sun, their size, shape, orientation, rotation, and inner constitution, and the velocity of light.

2. = system of astronomical constants.

The astronomical constants used in the calculations of The American Ephemeris and Nautical Almanac, as well as other national ephemerides, were adopted at various times between 1896 and 1930. Although the system was known to contain many inconsistencies, the International Astronomical Union recommended their continued use in 1952. Space-related research has provided data for the computation of a more accurate system, and in January 1964 The Working Group on the System of Astronomical Constants recommended a new system of constants to be introduced into the national and international ephemerides at the earliest practicable date.

Astronomical coordinates Coordinates defining a point on the surface of the earth, or of the geoid, in which the local direction of gravity is used as a reference. Sometimes called geographic coordinates.

Astronomical day A mean solar day beginning at mean noon, 12 hours later than the beginning of the civil day of the same date. Astronomers now generally use the civil day.

Astronomical equator A line on the surface of the earth connecting points having 0° astronomical latitude. Sometimes called terrestrial equator. When the astronomical equator is corrected for station error, it becomes the geodetic equator.

Astronomical latitude Angular distance between the direction of gravity and the plane of the celestial equator. Sometimes called geographic latitude.

Astronomical latitude corrected for the meridional component of station error becomes geodetic latitude .

Astronomical longitude The angle between the plane of the reference meridian and the plane of the celestial meridian. Sometimes called geographic longitude.

Astronomical longitude corrected for the prime-vertical component of station error divided by the cosine of the latitude becomes geodetic longitude.

Astronomical meridian A line connecting points having the same astronomical longitude. Also called terrestrial meridian . Because the deflection of the vertical varies from point to point, the astronomical meridian is an irregular line. When the astronomical meridian is corrected for station error, it becomes the geodetic meridian.

Astronomical parallel A line connecting points having the same astronomical latitude. Because the deflection of the vertical caries from point to point, the astronomical parallel is an irregular line. When the astronomical parallel is corrected for station error, it becomes the geodetic parallel.

Astronomical position 1. A point on the earth whose coordinates have been determined as a result of observation of celestial bodies.

The expression is usually used in connection with position on land determined with great accuracy for survey purposes.

2. A point on the earth, defined in terms of astronomical latitude and longitude.

Astronomical refraction 1. The angular difference between the apparent zenith distance of a celestial body and its true zenith distance, produced by refraction effects as the light from the body penetrates the atmosphere. Also called atmospheric refraction, astronomical refraction error.

For bodies near zenith the astronomical refraction is only about 0.1 minute, but for bodies near the horizon it becomes about 30 minutes or more and contributes measurably to the length of the apparent day.

2. Any refraction phenomenon observed in the light originating from a source outside of the earth's atmosphere; as contrasted with terrestrial refraction. This is applied only to refraction caused by inhomogeneities of the atmosphere itself, and not to that caused by ice crystals suspended in the atmosphere.

Astronomical scintillation Any scintillation phenomena, such as irregular oscillatory motion, variation of intensity, and color fluctuation observed in the light emanating from am extraterrestrial source; to be distinguished from terrestrial scintillation primarily in that the light source for the latter lies somewhere within the earth's atmosphere. Also called stellar scintillation.

Astronomical scintillation is typically strongest for celestial objects lying at large zenith distances and is not easily observed by eye for objects whose zenith distances are under 30°. Nonperiodic vibratory motions of stellar images with frequencies of the order of 1 to 10 cycles per second create a troublesome problem of seeing in astronomical work. The size of the schlieren producing vibratory scintillations has been estimated to be of the order of centimeters, and chromatic scintillations of celestial objects appear to be produced by parcels whose dimensions are of the order of decimeters or, perhaps, meters. Hence, astronomical scintillation is primarily a consequence of the high-frequency, short-wavelength type of atmospheric turbulence.

Astronomical time Mean time reckoned from the upper branch of the meridian.

Astronomical triangle The navigational triangle, either terrestrial or celestial, used in the solution of celestial observations.

Astronomical unit (AU) A unit of length, usually defined as the distance from the earth to the sun, 149,599,000 kilometers.

This value for the AU was derived from radar observations of the distance of Venus. The value given in astronomical ephemerides, 149,500,000 kilometers, was derived from observations of the minor planet Eros.

Astronomy The science concerning the location, magnitudes, motions, and constitution of celestial bodies and structures.

Astrophysics A branch of astronomy concerning the physical properties of celestial bodies, such as luminosity, size, mass, density, temperature, and chemical composition.

Asymmetry potential The potential developed across the glass membrane with identical solutions on both sides. Also a term used when comparing glass electrode potential in pH 7 buffer.

Asynchronous A communication method where data is sent when it is ready without being referenced to a timing clock, rather than waiting until the receiver signals that it is ready to receive.

Asynchronous computer An automatic computer in which succeeding operations are started by signals indicating the completion of the previous operation, rather than by signals from a master synchronizer. Contrast to synchronous computer.

Atactic A type of polymer chain configuration where side groups are randomly positioned on one side of the chain or the other.

Ataxia A lack of co-ordination of muscular action due to disease of the central nervous system. Ataxic Gait Patterns are unsteady.

Athetosis A disorder of the central nervous system, resulting in continual uncoordinated movements of the limbs.

Athodyd A type of jet engine consisting essentially of a duct or tube of varying diameter and open at both ends, which admits air at one end, compresses it by the forward motion of the engine, adds heat to it by the combustion of fuel, and discharges the resulting gases at the other end to produce thrust.

The ramjet is an athodyd; the pulsejet, especially the earlier type, is usually not considered an athodyd.

Atmosphere 1. The envelope of air surrounding the earth; also the body of gases surrounding or comprising any planet or other celestial body. Compare biosphere, geosphere, hydrosphere, lithosphere.

2. = standard atmosphere.

3. (atm) A unit of pressure equal to 14.7 pounds per square inch.

Atmospherics The radiofrequency electromagnetic radiation originating, principally, in the irregular surges of charge in thunderstorm lightning discharges. Atmospherics are heard as a quasi-steady background or crackling noise (static) in ordinary amplitude-modulated radio receivers. Also called atmospheric interference, strays, sferics.

Since any acceleration of electric charge leads to emission of electromagnetic radiation, and since the several processes involved in propagation of lightning lead to very large charge accelerations, the lightning channel acts like a huge transmitter, sending out broad band radiation; the 10-kilocycle range propagates best and is used in detecting atmospherics. Atmospherics may occasionally be detected at distances in excess of 2000 miles from their source. Advantage has been taken of this in using radio direction-finding equipment to locate active thunderstorm areas in remote regions and in between weather reporting stations.

Atmospheric braking The action of slowing down an object entering the atmosphere of the earth or other planet from space, by using the drag exerted by air or other gas particles in the atmosphere; the action of the drag so exerted.

Atmospheric duct An almost horizontal layer in the troposphere, extending from the level of a local minimum of the modified refractive index as a function of height, down to the level where the minimum value is again encountered, or down to earth's surface if the minimum value is not encountered again.

Atmospheric ducts may act as waveguides for radio and radar waves.

Atmospheric electric field 1. The electric field strength of the atmosphere at any specified point in space and time.

2. The distribution of electrical potential in the atmosphere regarded merely from a geometric point of view as a typical scalar field (rarely used).

Atmospheric electricity 1. Electrical phenomena, regarded collectively, which occur in the earth's atmosphere.

2. The study of electrical processes occurring within the atmosphere.

Atmospheric entry The penetration of any planetary atmosphere by any object from outer space; specifically, the penetration of the earth's atmosphere by a manned or unmanned capsule or spacecraft.

Atmospheric optics The study of the optical characteristics of the atmosphere and of the optical phenomena produced by the atmosphere's suspensoids and hydrometeors. It embraces the study of refraction, reflection, diffraction, scattering, and polarization of light, but is not commonly regarded as including the study of any other kinds of radiation. Also called meteorological optics.

Atmospheric pressure The pressure at any point in an atmosphere due solely to the weight of the atmospheric gases above the point concerned.

Atmospheric radiation Infrared radiation emitted by or being propagated through the atmosphere.

Atmospheric radiation, lying almost entirely within the wavelength interval of from 3 to 80 microns, provides one of the most important mechanisms by which the heat balance of the earth-atmosphere system is maintained. Infrared radiation emitted by the earth's surface (terrestrial radiation) is partially absorbed by the water vapor of the atmosphere which in turn remits it, partly upward, partly downward. This secondarily emitted radiation is then, in general, repeatedly absorbed and reemitted, as the radiant energy progresses through the atmosphere. The downward flux, or counterradiation, is of basic importance in the greenhouse effect; the upward flux is essential to the radiative balance of the planet.

Atmospheric refraction Refraction resulting when a ray of radiant energy passes obliquely through an atmosphere.

It may be called "astronomical refraction" if the ray enters the atmosphere from outer space, or "terrestrial refraction" if it emanates from a point on or near the surface of the earth.

Atmospheric shell Any one of a number of strata or layers of the earth's atmosphere. Also called atmospheric layer, atmospheric region.

Temperature distribution is the most common criterion used for denoting the various shells. The troposphere (the region of change) is the lowest 10 or 20 kilometers of the atmosphere, characterized by decreasing temperature with height. The top of the troposphere is called the tropopause. Above the tropopause, the stratosphere, a region in which the temperature generally increases with altitude, extends to the stratopause, the top of the inversion layer, at about 50 to 55 kilometers. Above the stratosphere, the mesosphere, a region of generally decreasing temperatures with height extends to the mesopause, the base of an inversion layer at about 80 to 85 kilometers. The region above the mesopause, in which temperature generally increases with height, is the thermosphere.

The distribution of various physicochemical processes is another criterion. The ozonosphere, lying roughly between 10 and 50 kilometers, is the general region of the upper atmosphere in which there is an appreciable ozone concentration and in which ozone plays an important part in the radiative balance of the atmosphere; the ionosphere, starting at about 70 or 80 kilometers, is the region in which ionization of one or more of the atmospheric constituents is significant; the neutrosphere is the shell below this which is, by contrast, relatively unionized; and the chemosphere, with no very definite height limits, is the region in which photochemical reactions take place.

Dynamic and kinetic processes are a third criterion. The exosphere is the region at the top of the atmosphere, above the critical level of escape, in which atmospheric particles can move in free orbits, subject only to the earth's gravitation.

Composition is a fourth criterion. The homosphere is the shell in which there is so little photodissociation or gravitational separation that the mean molecular weight of the atmosphere is sensibly constant; the heterosphere is the region above this, where the atmospheric composition and mean molecular weight are not constant. The boundary between the two is probably at the level at which molecular oxygen begins to be dissociated, and this occurs in the vicinity of 80 or 90 kilometers.

The term mesosphere has been given another definition which does not fit into any logical set of criteria, i.e., the shell between the exosphere and the ionosphere. This use of mesosphere has not been widely accepted.

Atmospheric tide Defined in analogy to the oceanic tide as an atmospheric motion on a worldwide scale, in which vertical accelerations are neglected (but compressibility is taken into account). Also called atmospheric oscillation. Both the sun and moon produce atmospheric tides; and there exist both gravitational tides and thermal tides. The harmonic component of greatest amplitude, the 12-hour or semidiurnal solar atmospheric tide, is both gravitational and thermal in origin, the fact that it is greater than the corresponding lunar atmospheric tide being ascribed usually to a resonance in the atmosphere with a free period very close to the tidal period. Other tides of 6, 8, and 24 hours have been observed.

Atom percent (at%) Concentration specification on the basis of the number of moles (or atoms) of a particular element relative to the total number of moles (or atoms) of all elements within an alloy.

Atom The smallest particle of an element that cannot be divided or broken up by chemical means. It consists of a central core of protons and neutrons, called the nucleus. Electrons revolve in orbits in the region surrounding the nucleus.

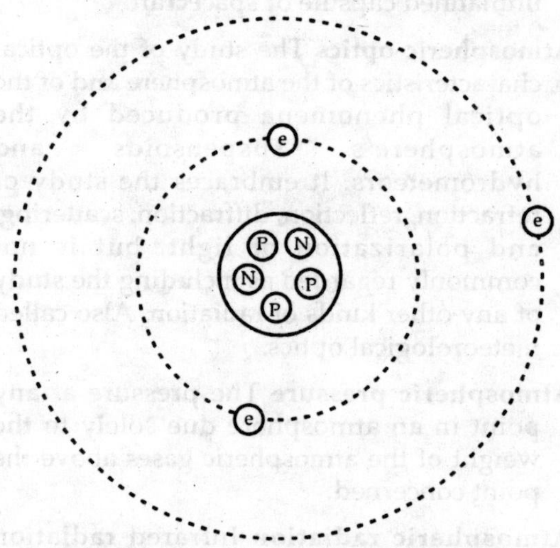

Fig. Basic model of the atom

Atomic energy Energy released in nuclear reactions. Of particular interest is the energy released when a neutron initiates the breaking up or fissioning of an atom's nucleus into smaller pieces (fission) or when two nuclei are joined together under millions of degrees of heat (fusion). It is more correctly called nuclear energy.

Atomic force microscope A device in which the deflection of a sharp stylus mounted on a soft spring is monitored as the stylus is moved across a surface. If the deflection is kept constant by moving the surface up and down by measured increments, the result (under favorable conditions) is an atomic-resolution topographic map of the surface. Also termed a scanning force microscope.

Atomic mass The mass of a neutral atom of a nuclide usually expressed in atomic mass units.

The atomic mass unit, amu, is exactly one-twelfth of the mass of a neutral atom of the most abundant isotope of carbon, C12=12.0000.

Atomic number (symbol Z) An integer that expresses the positive charge of the nucleus in multiples of the electronic charge e. It is the number of electrons outside the nucleus of a neutral (un-ionized) atom and, according to widely accepted theory, the number of protons in the nucleus.

An element of atomic number Z occupies the Zth place in the periodic table of the elements. Its atom has a nucleus with a charge +Ze, which is normally surrounded by Z electrons, each of charge -e.

For example, the carbon isotope $6C^{14}$ has an atomic number of 6 and an atomic mass of 14.

Atomic packing factor (APF) The fraction of the volume of a unit cell that is occupied by "hard sphere" atoms or ions.

Atomic particle One of the particles of which an atom is constituted, as an electron, neutron, or a positively charged nuclear particle.

Atomic rocket A projected rocket engine in which the energy for the jetstream is to be generated by atomic fission or fusion.

Atomic weight (at. Wt.) The weight of an atom according to a scale of atomic weight units, awu, valued as one-twelfth the mass of the carbon atom (C12 = 12.00000).

Thus expressed, the atomic weight to the nearest integer is identical with the mass number.

Atomiser A device that is used for reducing liquids to a fine spray.

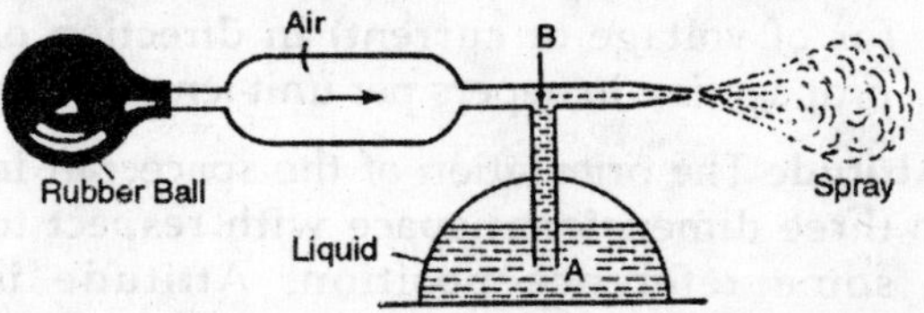

Fig. Atomiser

Attached growth reactor A reactor in which the microorganisms are attached to engineered surfaces within the reactor. Examples of attached growth reactors are the trickling filter and the rotating biological contactor.

Attached shock wave An oblique or conical shock wave that appears to be in contact with the leading edge of an airfoil or the nose of a body in a supersonic flow field. Also called attached shock.

Attachment The process in which two particles collide and stick together forming a single complex particle. The most common attachment process is the formation of a negative ion from electron attachment to an atom or molecule. Some negative ions are unstable, however, and cannot survive. The usual measure for this process is the attachment coefficient, which on the average is the fraction of a large 1 in 10,000 to 1 in 1,000.

Attenuation The process by which the number of particles or photons entering a body of matter is reduced by absorption and scattered radiation.

Attenuation constant 1. A measure of the rate of attenuation per unit length; the rate of flux-density (or power) reduction as energy (visual, electromagnetic, acoustic) propagates from its source. Also called attenuation factor, decay constant. Compare attenuation coefficient.

For free-space transmission of radar frequency energy, the attenuation constant is usually expressed in decibels per mile or kilometer (db/mi or db/km).

2. Specifically, of a traveling plane wave at a given frequency, the relative rate of decrease of amplitude of a field component (or of voltage or current) in direction of propagation in nepers per unit length.

Attitude The orientation of the spacecraft in three dimensional space with respect to some reference position. Attitude is commonly expressed in terms of yaw, pitch, and roll, and is used frequently when describing the orientation of airplanes and spacecraft.

Attitude control 1. The regulation of the attitude of an aircraft, spacecraft, etc.

2. A device or system that automatically regulates and corrects attitude, especially of a pilotless vehicle.

Attitude gyro 1. A gyro-operated flight instrument that indicates the attitude of an aircraft or spacecraft with respect to a reference coordinate system throughout 360° of rotation about each axis of the craft.

This instrument is similar to the artificial horizon, but has greater angular indication.

2. Broadly, any gyro-operated instrument that indicates attitude.

Attitude jet A jetstream used to correct or alter the attitude of a flying body either in the atmosphere or in space; the nozzle that directs this jetstream.

The jet may be continuous or intermittent. A vernier engine is sometimes used to produce it.

Attribute A form of a command-line switch as applied to tags in the HTML language. HTML commands or tags can be more specific when attributes are used. Not all HTML tags utilize attributes.

Attributes testing A reliability test procedure where the items under test are classified according to qualitative rather than quantitative characteristics.

Audio frequency range The range of frequencies to which the human ear is sensitive, approximately 15 cycles per second to 20,000 cycles per second. Also called audiorange.

Audio Pertaining to the audiofrequency range.

The word audio may be used as a modifier to indicate a device or system intended to operate at audiofrequencies, e.g., audioamplifier.

Audit An independent examination of a work product/process or set of work products/processes to assess compliance with specifications, standards, contractual agreements, or other criteria.

Auditory sensation area In acoustics, the frequency region enclosed by the curves defining the threshold of pain and the threshold of audibility.

Augmentation The apparent increase in the semidiameter of a celestial body, as observed from the earth, as its altitude increases, due to reduced distance from the observer.

The term is used principally in reference to the moon.

Augmentation correction A correction due to augmentation, particularly that sextant altitude correction due to the apparent increase in the semidiameter of a celestial body as its altitude increases.

Augmenter tube A tube or pipe, usually one of several, through which the exhaust gases from an aircraft reciprocating engine are directed especially to provide additional thrust.

Aurora The sporadic radiant emission from the upper atmosphere over middle and high latitudes. It is believed to be due primarily to the emission from nitrogen - atomic N I and N II, molecular, N2, and ionic N^{2+}; atomic oxygen (O I and O II); atomic sodium (Na I); the hydroxyl radical (OH); and hydrogen. Compare airglow.

According to various theories, auroras seem definitely to be related to magnetic storms and the influx of charged particles from the sun. The exact details of the nature of the mechanisms involved are still being investigated, but release of trapped particles from Van Allen belt apparently plays an important part. The aurora is most intense at times of magnetic storms (when it is also observed farthest equatorward), and shows a periodicity which is related to the sun's 27-day rotation period and the 11-year sunspot cycle. The distribution with height shows a pronounced maximum near 100 kilometers. The lower limit is probably near 80 kilometers.

The aurora can often be clearly seen, and it assumes a variety of shapes and colors which are characteristic patterns of auroral emission.

The following is the general classification and abbreviations of the forms of the auroras adopted by the International Union of Geodesy and Geophysics in 1930 for reporting of visual observations. The classification was modified slightly and expanded in 1963.

I. Forms without ray structure:

HA (abbr for homogeneous quiet area). These can appear near the horizon, and between the arc and the horizon a dark segment is often seen. These arcs can be narrow or broad, and are very often diffuse along the upper border but sharp along the lower one.

HB (abbr for homogeneous bands). These forms do not have the regular shape of the arcs; they are more rapidly moving phenomena. The lower border is often irregular and sharp. The breadth can vary from a very narrow band to a band which is so large that it resembles a curtain hanging down. These bands very often turn into bands with ray structure.

PA (abbr for pulsating arcs). Parts of an arc flash up and disappear regularly within a period of about 20 seconds. This form quite often stands isolated in the sky without other auroras.

DS (abbr for diffuse luminous surfaces). These either appear like a diffuse veil or glow over great parts of the heavens without distinct boundaries, often appearing after intense displays of rays and curtains, or as more isolated feeble luminous streaks which sometimes bear a striking resemblance to clouds. Sometimes large areas of the heavens can be discolored by a green, violet, or red diffuse light.

PS (abbr for pulsating surfaces). Diffuse patches appear and disappear rhythmically at the same place, retaining the same irregular shape, When the patches are lying near the magnetic zenith the contours can be more sharp, and form a sort of corona. These forms appear often in connection with flaming auroras.

G (abbr for feeble glow near the horizon resembling the dawn). Of white or redlike color, this form is often the upper part of an arc whose lower border is below the horizon.

II. Forms with ray structure: These forms consist of short or long rays which can be arranged in different ways.

RA (abbr for arcs with ray structure). A homogeneous arc which has remained quiet and unaltered for a rather long time may become sharp and luminous along the lower border and they very rapidly change into an arc of rays. The rays can be short or long.

RBI (abbr for bands with ray structure). These resemble the bands mentioned under HB but are constituted of a series of rays which are arranged close to each other along the band, or they can appear more scattered. Often a series of parallel bands appear. When a band is near the magnetic zenith is may have the form of a corona.

D (abbr for draperies). If the ray become very long the band appears like a curtain or drapery whose lower border is often more luminous. Several parallel curtains frequently appear at the same time. Near the zenith the curtain may have a fanlike form on account of the perspective.

R (abbr for rays). The rays can be isolated, narrow or broad, short or long. They may appear in great segments or like masses or rays, very often resembling curtains.

C (abbr for corona). When the rays approach the magnetic zenith they seem, on account of the perspective, to converge to this point and form a corona. This may be formed by long rays or by short ones, it may be completed or incomplete. A corona can also be formed by bands, draperies, or more diffuse forms near the magnetic zenith.

III. Flaming auroras (abbr F). A characteristic, rapidly moving form, consisting of strong waves of light which move upwards, one after the other, in the direction of the magnetic zenith. The waves have the form of detached arcs which move upwards normally to the direction of the arc; they can be compared to invisible waves illuminating broad rays and patches which appear and disappear rhythmically when the waves pass them. The flaming aurora frequently appears after strong displays of rays and curtains and is often followed by the formation of a corona.

Auroral zone A roughly circular band around either geomagnetic pole above which there is a maximum of auroral activity. It lies about 10 to 15° of geomagnetic latitude from the geomagentic poles.

The auroral zone broadens and extends equatorward during intense auroral displays. The northern auroral zone is centered along a line passing near Point Barrow, Alaska, through the lower half of Hudson Bay, slightly off the southern tip of Greenland, through Iceland, northern Norway and northern Siberia. Along this line auroras are seen on an average of 240 nights a year. The frequency of auroras falls off both to the north and to the south of this line but more rapidly to the south. The most severe blackouts occur in the auroral zone.

Austenitizing Forming austenite by heating a ferrous alloy above its upper critical temperature - to within the austenite phase region from the phase diagram.

Authentication The procedure (essentially approval) used by the approval authority in verifying that specification content is acceptable. Authentication does not imply acceptance or responsibility for the specified item to perform successfully.

Authorized carrier frequency A specific carrier frequency authorized for use, from which the actual carrier frequency is permitted to deviate, solely because of frequency instability, by an amount not to exceed the frequency tolerance.

Authorized person A qualified person who, by nature of his duties or occupation, is obliged to approach or handle electrical equipment or, a person who, having been warned of the hazards involved, has been instructed or authorized to do so by someone in authority.

Auto ranging input An input voltage sensing circuit in the power supply that automatically switches to the appropriate input voltage range (90-132 VAC or 180-264 VAC). Sometimes known as Auto-select input.

Auto router Automatic router, a computer program that routes a PC board design (or a silicon chip design) automatically.

Autoconvective lapse rate The environmental lapse rate of temperature in an atmosphere in which the density is constant with height (homogenous atmosphere), equal to g/R, where g is the acceleration of gravity and R the gas constant. For dry air the autoconvective lapse rate is approximately 3.4×10^{-4} °C per centimeter. Also called autoconvection gradient.

Autocorrelation In statistics the simple linear internal correlation of members of a time series (ordered in time or other domains).

Automatic data processing system An electronic system that includes an electronic data processing system plus auxiliary and connecting communications equipment.

Automatic direction finder (ADF) A radio direction finder which automatically and continuously provides a measure of the direction of arrival of the received signal. Data are usually displayed visually.

Automatic frequency control (AFC) An arrangement whereby the frequency of an oscillator is automatically maintained within specified limits.

Automatic gain control (AGC) A process by which gain is automatically adjusted as a function of input or other specified parameter.

Automatic locking front hubs Found in some four-wheel drive vehicles, this allows the driver to engage, or "lock," the front axle hubs without leaving the vehicle.

Automatic pilot Equipment which automatically stabilizes the attitude of a vehicle about its pitch, roll, and yaw axes. Also called autopilot .

Automatic reset 1. A feature on a limit controller that automatically resets the controller when the controlled temperature returns to within the limit bandwidth set.

2. The integral function on a PID controller which adjusts the proportional bandwidth with respect to the set point to compensate for droop in the circuit, i.e., adjusts the controlled temperature to a set point after the system stabilizes.

Automatic stability Stability achieved with the controls operated by automatic devices, as by an automatic pilot.

Automatic tracking Tracking in which a servomechanism automatically follows some characteristic of the signal; specifically, a process by which tracking or data acquisition systems are enabled to keep their antennas continually directed at a moving target without manual operation.

Autosyn A remote-indicating instrument or system based upon the synchronous-motor principle, in which the angular position of the rotor of one motor at the measuring source is duplicated by the rotor of the indicator motor, used, e.g., in fuel-quantity or fuel-flow measuring systems, position-indicating systems, etc.

Autotrophic Organisms which utilize inorganic carbon for synthesis of protoplasm. Ecologists narrow the definition further by requiring that autotrophs obtain their energy from the sun. In microbiologist parlance, this would be a photoautotroph.

Autumnal equinox 1. That point of intersection on the celestial sphere of the ecliptic and the celestial equator occupied by the sun as it changes from north to south declination, on or about September 23. Also called September equinox, first point of Libra.

2. That instant the sun reaches the point of zero declination when crossing the celestial equator from north to south.

Auxiliary building Building at a nuclear power plant, frequently located adjacent to the reactor containment structure, that houses most of the reactor auxiliary and safety systems, such as radioactive waste systems, chemical and volume control systems, and emergency cooling water systems.

Auxiliary feedwater Backup water supply used during nuclear plant startup and shutdown to supply water to the steam generators during accident conditions for removing decay heat from the reactor.

Auxiliary fluid ignition A method of ignition of a liquid-propellant rocket engine in which a liquid which is hypergolic with either the fuel or the oxidizer is injected into the combustion chamber to initiate combustion. Aniline is used as an auxiliary fluid with nitric acid and some organic fuels to initiate combustion.

Auxiliary landing gear That part or parts of a landing gear, as an outboard wheel, which is intended to stabilize the craft on the surface but which bears no significant part of the weight.

Auxiliary power unit (APU) A power unit carried on an aircraft or spacecraft which can be used in addition to the main sources of power of the craft.

Auxiliary supply A power source supplying power other than load power as required for the proper functioning of a device.

Availability The product metric that defines the percentage of time that a product is available and operational for customer use.

Available capacity The total capacity, Ah or Wh, that will be obtained from a cell or battery at defined discharge rates and other specified discharge or operating conditions.

Avalanche The cumulative process in which charged particles accelerated by an electric field produce additional charged particles through collision with natural gas molecules or atoms.

Average deviation In statistics, the average or arithmetic mean of the deviations, taken without regard to sign, from some fixed value, usually the arithmetic mean of the data. Also called mean deviation .

Average planar linear heat generation rate (APLGHR) The average value of the linear heat generation rate of all the control rods at any given horizontal plane along a fuel bundle.

Avogadro's number The number of atoms in exactly 12 grams of pure 12C, equal to 6.022 × 10^{23}.

Axial flow compressor A rotary compressor having interdigitated rows or stages of rotary and of stationary blades through which the flow of fluid is substantially parallel to the rotor's axis of rotation. Compare centrifugal compressor.

Axiomatic design. Axiomatic Design recognizes four domains. The needs of the customer are identified in customer domain and are stated in the form of required functionality of a product in functional domain. Design parameters that satisfy the functional requirements are defined in physical domain, and, in process domain, manufacturing variables define how the product will be produced. Solution alternatives are created by mapping the requirements specified in one domain to a set of characteristic parameters in an adjacent domain. The mapping between the customer and functional domains is defined as concept design; the mapping between functional and physical domains is product design; the mapping between the physical and process domains corresponds to process design. The output of each domain evolves from abstract concepts to detailed information in a top-down or hierarchical manner. Two design axioms provide a rational basis for evaluation of proposed solution alternatives and the subsequent selection of the best alternative. The first axiom is the independent axiom, and it states that a good design maintains the independence of the functional requirements. The second axiom is the information axiom and it establishes information content as a relative measure for evaluating and comparing alternative solutions that satisfy the independence axiom.

Axis (plural axes) 1. A straight line about which a body rotates, or along which its center of gravity moves (axis of translation).

2. A straight line around which a plane figure may rotate to produce a solid; a line of symmetry.

3. One of a set of reference lines for a coordinate system.

Axis of motion The axis of motion is perpendicular to the plane in which the joint motion occurs. The closer the axis of the motion is to the body plane, the less movement there is in that body plane. An axis of rotation is a kinematic parameter that is used to describe the characteristics of the motion of one rigid body (body segment) relative to a reference rigid body (another body segment) - e.g movement of the calcaneus relative to the talus. The axis of rotation is a straight line whose particles all have zero velocity during the angular displacement of one rigid body relative to the reference rigid body and can lie within or outside the physical boundaries of either rigid body. Axes of rotation are part of the methodology we use to model human joint motion.

Axle ratio The ratio between the rotational speed (RPM) of the drive shaft and that of the driven wheel. Gear reduction in final drive is determined by dividing the number of teeth on the ring gear by the number of teeth on the pinion gear.

Azimuth 1. Horizontal direction or bearing. Compare azimuth angle.

2. In navigation, the horizontal direction of a celestial point from a terrestrial point, expressed as the angular distance from a reference direction, usually measured from 0° at the reference direction clockwise through 360°.

An azimuth is often designated as true, magnetic, compass, grid, or relative as the reference direction is true, magnetic, compass, grid north, or heading, respectively. Unless otherwise specified, the term is generally understood to apply to true azimuth, which may be further defined as the arc of the horizon, or the angle at the zenith, between the north part of the celestial meridian or principal vertical circle and a vertical circle, measured from 0° at the north part of the principal vertical circle clockwise through 360°.

3. In astronomy, the direction of a celestial point from a terrestrial point measured clockwise from the north or the south point of the meridian plane.

4. In surveying, the horizontal direction of an object measured clockwise from the south point of the meridian plane.In surveying, an azimuth of a celestial body is called an astronomic azimuth.

Azimuth angle 1. Azimuth measured from 0° at the north or south reference direction clockwise or counterclockwise through 90° or 180°.

Azimuth angle is labeled with the reference direction as a prefix and the direction of measurement from the reference direction as a suffix. Thus, azimuth angle S 144° W is 144° west of south, or azimuth 324°. When azimuth angle is measured through 180°, it is labeled N or S to agree with the latitude and E or W to agree with the meridian angle.

2. In surveying, an angle in triangulation or in traverse through which the computation of azimuth is carried.

Azimuth error An error in the indicated azimuth of a target detected by radar, resulting from horizontal refraction. Compare range error.

Inasmush as significant gradients of index of refraction are very uncommon in the atmosphere, these errors almost invariably are negligible. Seacoast areas may give rise on occasion to appreciable horizontal bending of radio waves because of the contrast of refractive index values between the air over land and the air over water.

Azimuth marker 1. A scale encircling the plan position indicator (PPI) scope of a radar on which the azimuth of a target from the radar may be measured.

2. Reference limits inserted electronically at 10° or 15° intervals which extend radially from the relative position of the radar on an off center PPI scope. These are employed for target azimuth determination when the radar position is not at the center of the PPI scope and hence the fixed azimuth scale on the edge of the scope cannot be employed.

On such markers north is usually 0°, east 90°, etc. Occasionally, on ship or airborne radars, 0° is used to indicate the direction in which the craft is heading, in which cases the relative bearing, not azimuth, of the target is indicated.

Azran Azimuth and range.

This term was coined in the field of radar, and has since been extended in application to the locating of any object (or target) by means of polar coordinates.

Azusa A short-baseline, continuous wave, phase comparison, single-station, tracking system operating at C-band and giving two direction cosines and slant range which can be used to determine space position and velocity.

B

B display In radar, a rectangular display in which targets appear as blips with bearing indicated by the horizontal coordinate and distance by the vertical coordinate. Also called B-scan or B-scope .

BA A device which ensures that the wearer has a continuously available supply of uncontaminated air through a face mask; helmet or mouthpiece. Often referred to as a BA set.

Babble The aggregate crosstalk from a large number of communications channels.

Babinet point One of the three commonly detectable points of zero polarization of diffuse sky radiation, neutral points, lying along the vertical circle through the sun; the other two are the Arago point and Brewster point.

The Babinet point typically lies only 15° to 20° above the sun, and hence is difficult to observe because of solar glare. The existence of this neutral point was discovered by Babinet in 1840.

Back pressure Pressure exerted backward; in a field of fluid flow, a pressure exerted contrary to the pressure producing the main flow.

Backbone A set of nodes and links connected together comprising a network, or the upper layer protocols used in a network. Sometimes the term is used to refer to a network's physical media.

Background Any effect in a sensor or other apparatus or system above which the phenomenon of interest must manifest itself before it can be observed.

Background counts In radiation counters, responses of the counting system caused by radiation coming from sources other than that to be measured.

Background luminance In visual-range theory, the luminance (brightness) of the background against which a target is viewed. Compare adaptation luminance.

Background noise 1. In recording and reproducing, the total system noise independent of whether or not a signal is present. The signal is not to be included as part of the noise.

2. In receivers, the noise in the absence of signal modulation on the carrier.

Ambient noise detected, measured, or recorded with the signal part of the background noise. Included in this definition is the interference resulting from primary power supplies, that separately is commonly described as hum.

Background radiation Radiation from cosmic sources; naturally occurring radioactive materials, including radon (except as a decay product of source or special nuclear material) and global fallout as it exists in the environment from the testing of nuclear explosive devices. It does not include radiation from source, byproduct, or special nuclear materials regulated by the Nuclear Regulatory Commission. The typically quoted average individual exposure from background radiation is 360 millirems per year.

Backlash Dead space or unwanted movement in a control system.

Backout An undoing of things already done during a countdown, usually in reverse order.

Backup 1. An item kept available to replace an item which fails to perform satisfactorily.

2. A redundant component in a system which is not the normally active (or prime, primary) component. Also known as second-string.

Backup power supply A power supply used to provide alternate system power in the event the primary power source fails or is unable to continue providing adequate system power.

Backward scatter The scattering of radiant energy into the hemisphere of space bounded by a plane normal to the direction of the incident radiation and lying on the same side as the incident ray; the opposite of forward scatter. Also called back scattering .

Atmospheric backward scatter depletes 6 to 9 percent of the incident solar beam before it reaches the earth's surface.

In radar usage, backward scatter refers only to that radiation scattered at 180° to the direction of the incident wave.

Backward wave In traveling-wave tubes, a wave whose group velocity is opposite to the direction of electron-stream motion.

Baffle A plate, grating, or the like used especially to block, hinder, or divert a flow or to hinder the passage of something, as: (a) A plate used to conduct, or help to conduct, a flow of cooling air around an engine cylinder. (b) A plate, wall, or the like in a fuel tank or other liquid container, used especially to prevent sloshing of the contents. (c) A ridge or wall on the top of a piston in a two-stroke-cycle engine, used to deflect the incoming mixture upward and divert it from the exhaust port. (d) A plate in the forward section of a pitot tube, used to reduce turbulence in the tube and to prevent dirt, moisture, etc., from entering the system.

Bailout bottle A personal supply of oxygen usually contained in a cylinder under pressure and utilized when the individual has left the central oxygen system as in a parachute jump.

Bainite An austenitic transformation product found in some steels and cast irons; it forms at temperatures between those at which pearlite and martensite transformations occur; the microstructure consists of alpha-ferrite and a fine dispersion of cementite.

Balance 1. The equilibrium attained by an aircraft, rocket, or the like when forces and moments are acting upon it so as to produce steady flight, especially without rotation about its axes; also used with reference to equilibrium about any specified axis, as, an airplane in balance about its longitudinal axis.

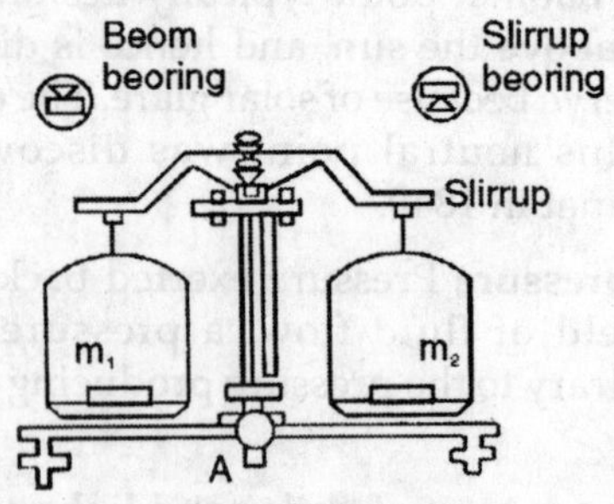

Fig. Balance

2. A weight that counterbalances something, especially on an aircraft control surface, a weight installed forward of the hinge axis to counterbalance the surface aft of the hinge axis.

Balance shaft A shaft designed so that, as it turns, it counter rotates the rotational direction of the engine crankshaft in a manner that reduces or cancels out some of the vibration produced by the engine.

Balance sheet A formal statement of the financial position of a company on a partic ular day, normally presented to shareholders once a year. Everything owned by the company (i.e. its assets) must be equal to the sum of the company's debts (liabilities) and the value of its shares and retained earnings (net worth)

Balanced amplifier An amplifier circuit in which there are two identical signal branches connected so as to operate in phase opposition and with input and output connections each balanced to ground. Also called push-pull amplifier.

Balanced circuit A circuit, the two sides of which are electrically alike and symmetrical with respect to a common reference point, usually ground.

Balanced detector A demodulator for frequency-modulation systems. In one form the output consists of the rectified difference of the two voltages produced across two resonant circuits, one circuit being tuned slightly above the carrier frequency and the other slightly below.

Balanced modulator A device in which the carrier and modulating signal are so introduced that, after modulation takes place, the output contains the two sidebands without the carrier.

Balanced scorecard A comprehensive performance measurement technique that considers four areas of performance in a balanced way: 1. Customer perspective - how customers see us,

2. Internal perspective - what we must excel at,

3. Innovation & learning - how we continue to improve and create value,

4. Financial perspective - how we meet shareholder needs.

Balancing the line The process of evenly distributing both the quantity and variety of work across available work time, avoiding overburden and underuse of resources. This eliminates bottlenecks and downtime, which translates into shorter flow time.

Ball grid array (BGA). A flip-chip type of package in which the internal die terminals form a grid-style array, and are in contact with solder balls (solder bumps), which carry the electrical connection to the outside of the package. The PCB footprint will have round landing pads to which the solder balls will be soldered when the package and PCB are heated in a reflow oven. Advantages of the ball grid array package are 1. that its size is compact and

2. Its leads do not get damaged in handling (unlike the formed "gull-wing" leads of a QFP') and thus has a long shelf life. Disadvantages of the BGA are 1. they, or their solder joints, are subject to stress-related failure. For example, the intense vibration of rocket-powered space vehicles can pop them right off the PCB,

2. they can not be hand-soldered (they require a reflow oven), making first-article prototypes a bit more expensive to stuff,

3. except for the outer rows, the solder joints can not be visually inspected and

4. they are difficult to rework.

Ball joint A flexible joint consisting of a ball within a socket. Ball joints act as pivots which allow turning of the front wheels and compensate for changes in the wheel and steering geometries that occur while driving.

Ball lightning A relatively rare form of lightning, consisting of a reddish, luminous ball, of the order of 1 foot in diameter, which may move rapidly along solid objects or remain floating in midair. Hissing noises emanate from such balls, and they sometimes explode nosily but may also disappear noiselessly. Also called globe lightning.

It has been suggested that ball lightning is a temporarily stable plasma.

Ballistics The science that deals with the motion, behavior, and effects of projectiles, especially bullets, aerial bombs, rockets, or the like; the science or art of designing and hurling projectiles so as to achieve a desired performance.

Ballistic body A body free to move, behave, and be modified in appearance, contour, or texture by ambient conditions, substances, or forces, as by the pressure of gases in a gun, by rifling in a barrel, by gravity, by temperature, or by air particles.

A rocket with a self-contained propulsion unit is not considered a ballistic body during the period of its guidance or propulsion.

Ballistic camera A ground-based camera using multiple exposures on the same plate to record the trajectory of a rocket.

Ballistic condition A condition affecting the behavior of a vehicle in flight.

Ballistic conditions include the velocity, weight, shape, and size of the vehicle; likewise the density and temperature of the ambient element, the magnetic field, etc.

Ballistic density A representation of the atmospheric density actually encountered by a projectile in flight, expressed as a percentage of the density according to the standard artillery atmosphere.

Thus, if the actual density distribution produced the same effect upon a projectile as the standard density distribution, the ballistic density would be 100 percent .

Ballistic missile A missile designed to operate primarily in accordance with the laws of ballistics.

A ballistic missile is guided during a portion of its flight, usually the upward portion, and is under no thrust from its propelling system during the latter portion of its flight; it describes a trajectory similar to that of an artillery shell.

Ballistic temperature That temperature (in° F) which, when regarded as a surface temperature and used in conjunction with the lapse rate of the standard artillery atmosphere, would produce the same effect on a projectile as the actual temperature distribution encountered by the projectile in flight.

Ballistic trajectory The trajectory followed by a body being acted upon only be gravitational forces and the resistance of the medium through which it passes.

A rocket without lifting surfaces is in a ballistic trajectory after its engines cease operating.

Ballistic vehicle A nonlifting vehicle; a vehicle that follows a ballistic trajectory.

Ballistic wind That constant wind which would produce the same effect upon the trajectory of a projectile as the actual wind encountered in flight. Ballistic winds can be regarded as made up of range wind and crosswind components.

Balloon type rocket A liquid-fuel rocket, such as Atlas, that requires the pressure of its propellants (or other gases) within it to give it structural integrity.

Band elimination filter A wave filter that attenuates one frequency band, neither the critical nor cutoff frequencies being zero or infinite.

Band gap energy (Eg) For semiconductors and insulators, the energies that lie between the valence and conduction bands; for intrinsic materials, electrons are forbidden to have energies within this range.

Band of position An area extending to either side of a line of position of imperfect accuracy, within which a craft is considered to be located.

Band pass filter A wave filter that has a single transmission band extending from a lower cutoff frequency greater than zero to a finite upper cutoff frequency.

Bandwidth 1. In an antenna, the range of frequencies within which its performance, in respect to some characteristic, conforms to a specified standard.

2. In a wave, the least frequency interval outside of which the power spectrum of a time-varying quantity is everywhere less than some specified fraction of its value at a reference frequency.

3. The number of cycles per second between the limits of a frequency band.

4. In information theory, the information-carrying capacity of a communications channel.

Bang address A type of e-mail address that separates host names in the address with exclamation points. Used for mail sent to the UUCP network, where specifying the exact path of the mail (including all hosts that pass on the message) is necessary. The address is in the form of machine!machine!userID, where the number of machines listed depends on the the account userID

Bang bang control Flicker control, especially as applied to rockets.

Bang-bang in this term is imitative, arising from the noise made by control mechanisms slamming first to one side, then to the other, in this sort of control.

Bar A unit of pressure equal to 106 dyne per square centimeter (106 barye), 1000 millibars, 29.53 inches of mercury.

Some writers have used bar as equivalent to barye (1 dyne per square centimeter).

Baralyme A commercial trade name for a type of carbon dioxide absorber, a mixture of calcium hydroxide and barium hydroxide.

Bárány chair (After Robert Bárány, 1876-1936, Swedish physician). A kind of chair in which a person is revolved to test his susceptibility to vertigo.

Baroclinity The state of stratification in a fluid in which surfaces of constant pressure (isobaric) intersect surfaces of constant density (isosteric). The number, per unit area, of isobaric-isosteric solenoids intersecting a given surface is a measure of the baroclinity. Also called baroclinicity, barocliny.

Barotropy is the state of zero baroclinity. Since the presence of solenoids (baroclinity) complicates the dynamics of the fluid, there has been much investigation of the extent to which an atmosphere, though obviously a baroclinic fluid, can be dynamically treated as barotropic.

Barometer, aneroid A thin disk of metal covering the aperture of a box from which the air has been exhausted. Variation in atmospheric pressure causes a bulging of the disk, which shifts a pointer over a scale and so indicates the pressure.

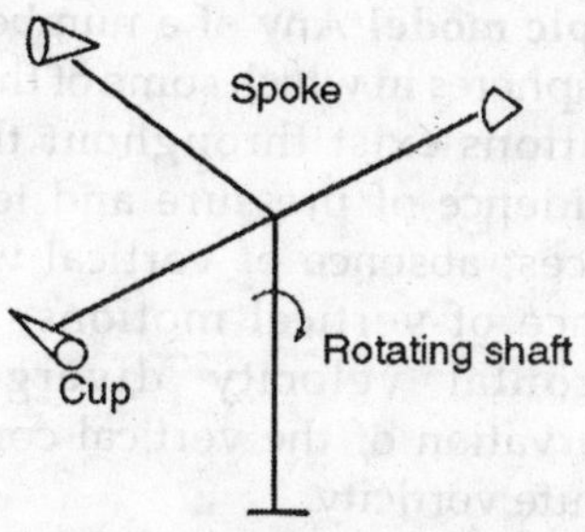

Fig. Typical Anemometer

Barometric wave Any wave in the atmospheric pressure field. The term is usually reserved for short-period variations not associated wit cyclonic-scale motions or with atmospheric tides.

Barosphere The atmosphere below the critical level of escape.

Baroswitch (From barometric switch). 1. Specifically, a pressure-operated switching device used in a radiosonde. In operation, the expansion of an aneroid capsule causes an electrical contact to scan a radiosonde commutator composed of conductors separated by insulators. Each switching operation corresponds to a particular pressure level. The contact of an insulator or a conductor determines whether temperature, humidity, or reference signals will be transmitted.

2. Any switch operated by a change in atmospheric pressure.

Barotropic disturbance 1. An atmospheric wave in a two-dimensional nondivergent flow, the driving mechanism for which lies in the variation of vorticity of the basic current and/or in the variation of the vorticity of the earth about the local vertical. When the basic current is uniform, the wave is a Rossby wave. Also called barotropic wave.

2. An atmospheric wave of cyclonic scale in which troughs and ridges are approximately vertical.

Barotropic model Any of a number of model atmospheres in which some of the following conditions exist throughout the motion: coincidence of pressure and temperature surfaces; absence of vertical wind shear; absence of vertical motions; absence of horizontal velocity divergence; and conservation of the vertical component of absolute vorticity.

Barrier coat A coating used to isolate a paint system either from the surface to which it is applied or a previous coating for the purpose of increasing adhesion or insuring compatibility

Barye The pressure unit of the centimeter-gram-second system of physical units; equal to one dyne per square centimeter (0.001 millibar). Sometimes called bar or microbar.

Base A proton acceptor. A compound containing hydrogen which dissociates in aqueous solution producing negatively charged hydroxide (OH-) or other ions. Alkalis are bases and a basic solution has a pH greater than 7.0.

Base Coat/clear coat A paint system that adds a final clear-coat paint layer over primer and color coats to provide a deep, "wet-look" shine that resists fading.

Base line 1. Any line which serves as the basis for measurement of other lines, as in a surveying triangulation, measurement of auroral heights, etc.

2. The geodesic line between two stations operating in conjunction for the determination of a line of position, as the two stations constituting a loran rate.

3. In radar, the line traced on amplitude-modulated indicators which corresponds to the power level of the weakest echo detected by the radar. It is retraced with every pulse transmitted by the radar, but appears as a nearly continuos display on the scope. Target signals show up as perpendicular deviations from the base line; range is measured along the base line; signal strength is indicated by the magnitude of the deviations; and the type of target usually can be determined by the appearance of the deviations.

Base point In computer terminology, the character, or the location of an implied symbol, which separates the integral part of an expression in positional notation from the fractional part; the point which marks the place between the zero and negative powers of the base. Also called radix point.

Base pressure In aerodynamics, the pressure exerted on the base, or extreme aft end, of a body, as of a cylindrical or boattailed body or of a blunt-trailing-edge wing, in a fluid flow.

Base timing sequencing (BT sequencing) The control of the time sharing of a single transponder between several ground transmitters through the use of suitable coded timing signals.

Baseband A type of channel where data transmission is carried across only one communications channel, supporting only one signal transmission at a time. Ethernet is a baseband system.

Baseplate Mounting platform for power supply components.

Baseplate temperature The temperature at the hottest spot on the mounting platform of the supply.

Basic A high-level programming language designed at Dartmouth College as a learning tool. Acronym for Beginner's All-purpose Symbolic Instruction Code.

Basic research Fundamental scientific research concerned solely with scientific principles as opposed to applied scientific research which is concerned with the commercial applica tion of those principles.

Bath Tub Curve Bath-Tub Curve represents the failure rate of components over the life of the product. Its upward slope at the beginning and end suggests that most components fail either right away (at the beginning of the product life) or towards the end of the expected product life.

Battery limits The perimeter surrounding the processing area and included process equipment of a process plant.

Battery management system (BMS) Electronic circuits designed to monitor the battery and keep it within its specified operating conditions and to protect it from abuse during both charging and discharging.

Battery monitoring Sometimes confused with BMS of which it is an essential part, these circuits monitor the key operating parameters (current, voltage, temperature, SOC, etc.) of the battery and provide information to the user.

Battery Two or more electrochemical cells electrically interconnected in an appropriate series/parallel arrangement to provide the required operating voltage and current levels. Under common usage, the term 'battery' is often also applied to a single cell.

Battery voltage The total voltage between the positive and negative terminals of the battery.

Baud A unit of data transmission speed equal to the number of bits (or signal events) per second; 300 baud = 300 bits per second.

Baud rate Serial communications data transmission rate expressed in bits per second.

Baumé scale (Be) Either of two scales sometimes used to graduate hydrometers; one scale is for liquids heavier than water, the other for liquids lighter than water.

Bauxite An ore of aluminum consisting of moderately pure hydrated alumina - $Al_2O_3 \times 2H_2O$.

Bay window An arrangement of three or more individual window units, attached so as to project from the building at various angles. In a three-unit bay, the center section is normally fixed, with the end panels operable as single-hung or casement windows.

Bayard alpert ionization gage A type of ionization vacuum gage using a tube with an electrode structure designed to minimize X-ray induced electron emission from the ion collector.

Bayesian estimation A mathematical formulation, using Bayes' theorem, by which the likelihood of an event can be estimated taking explicit consideration of certain contextual features (such as amount of data, nature of decision, etc.).

Bayesian prior A way to express the context of a Bayesian estimation in which initial data are updated as new data become available.

BCD, parallel A digital data output format where every decimal digit is represented by binary signals on four lines and all digits are presented in parallel. The total number of lines is 4 times the number of decimal digits.

BCD, Three state An implementation of parallel BCD, which has 0, 1 and high-impedance output states. The high-impedance state is used when the BCD output is not addressed in parallel connect applications.

Beacon 1. A light, group of lights, electronic apparatus, or other device that guides, orients, or warns aircraft, spacecraft, etc. In flight.

2. A structure, building, or station where such a device is mounted or located.

Beacon delay The amount of inherent delay within a beacon, i.e., the time between the arrival of a signal and the response of the beacon.

Beacon skipping A condition where transponder return pulses from a beacon are missing at the interrogating radar.

Beacon skipping can be caused by interference, overinterrogation of beacon, antenna nulls, or pattern minimums.

Beacon stealing Loss of beacon tracking by one radar due to (interfering) interrogation signals from another radar.

Beacon tracking The tracking of a moving object by means of signals emitted from a transmitter or transponder within or attached to the object.

Bead A wood strip against which a swinging sash closes, as in a casement window. Also, a finishing trim at the sides and top of the frame to hold the sash, as in a fixed sash or a double-hung window. Also referred to as bead stop.

Beam 1. A ray or collection of focused rays of radiated energy.

2. A beam of radio waves used as a navigation aid.

3. =electron beam.

4. A body, one of whose dimensions is large compared with the others, whose function is to carry lateral loads (perpendicular to the long dimension) and bending movements.

Beam lead A metal beam (flat metallic lead which extends from the edge of a chip much as wooden beams extend from a roof overhang) deposited directly onto the surface of the die as part of the wafer processing cycle in the fabrication of an integrated circuit. Upon separation of the individual die (normally by chemical etching instead of the conventional scribe-and-break technique), the cantilevered beam is left protruding from the edge of the chip and can be bonded directly to interconnecting pads on the circuit substrate without the need for individual wire interconnections. This method is an example of flip-chip bonding, contrasted with solder bump. (Graf)

Beam rider A craft following a beam, particularly one which does so automatically, the beam providing the guidance.

Beam rider guidance A system for guiding aircraft or spacecraft in which a craft follows a radar beam, light beam, or other kind of beam along the desired path. Also called beam-climber guidance.

Beam switching tube A trochotron in which an electron beam can be formed and switched to any one of several (usually 10) positions.

Beam width A measure of the concentration of power of a directional antenna. It is the angle in degrees subtended at the antenna by arbitrary power-level points across the axis of the beam. This power level is usually the point where the power density is one-half that which is present in the axis of the beam at the same distance from the antenna (half-power points). Also called beam angle.

The beam width of a radar determines the minimum angular separation which two targets can have and still be resolved. Roughly speaking, two targets at the same range whose angular separations at the radar antenna exceeds one-half of the beam width between half-power points will be resolved or distinguishable as two individual targets. The smaller the beam width, the greater the annular resolving power. Beam width may be at different locations through the axis depending upon the shape of the antenna reflector.

Bearing The horizontal direction of an object or point, usually measured clockwise from a reference line or direction through 360°.

A bearing is often designated as true, magnetic, compass, grid, or relative as the reference direction is true, magnetic, compass, or grid north, or heading, respectively.

At one time bearing was restricted to reference to the direction of a terrestrial object or point as distinguished from azimuth which referred to the direction of a celestial body. This distinction has been blurred by usage.

Bearing angle Horizontal direction measured from 0° at the reference direction clockwise or counterclockwise through 90° or 180°. Compare bearing.

Bearing angle is labeled with the reference direction as a prefix and the direction of measurement from the reference direction as a suffix. Thus, bearing angle N 37° W is 37° west of north, or bearing 323°.

Beat 1. One complete cycle of the variations in the amplitude of two or more periodic phenomena of different frequency which mutually react.

Beat frequency Beat frequencies are periodic vibrations that result from the addition and subtraction of two or more sinusoids. For example, in the case of two turbine aircraft engines that are rotating at nearly the same frequency but not precisely at the same frequency; Four frequencies are generated:(f1) the rotational frequency of turbine one, (f2) the rotational frequency of turbine two, (f1 + f2) the sum of turbine rotational frequencies one and two, and (f1 - f2) which is the difference or "beat" frequency of turbines one and two. The difference of the two frequencies is the lower frequency and is the one that is "felt" as a beat or "wow" in this case.

Beating A wave phenomenon in which two or more periodic quantities of different frequencies produce a resultant having pulsations of amplitude.

This process may be controlled to produce a desired beat frequency.

Beavertail antenna A type of radar antenna which forms a beam having a greater beam width in azimuth than in elevation, or vice versa. In physical dimensions, its long axis lies in the plane of smaller beam width.

Behavior The way in which an organism, organ, body, or substance acts in an environment or responds to excitation, as the behavior of steel under stress , or the behavior of an animal in a test .

Behavioral design The design of how an overall system or CSCI will behave, from a user's point of view, in meeting its requirements, ignoring the internal implementation of the system or CSCI. This design contrasts with architectural design, which identifies the internal components of the system or CSCI, and with the detailed design of those components.

Belted radial tires A reinforcing bank, normally textile, fiberglass or steel, running around the circumference of a tire and strengthening the tread area.

Bench power supply A power source fitted with output controls, meters, terminals and displays for experimental bench-top use in a laboratory.

Benchmarking An improvement process in which a company measures the performance of its products or processes against that of best-in-class products or companies, determines how the product or company achieved their performance level, and uses the information to improve its own performance.

Bend 1. Pains in the extremities, abdomen, and chest caused by aeroemphysema and in some instances by aeroembolism resulting from the reduction of ambient air pressure.

2. Popularly used as synonymous with aeroembolism.

Berkeley software distribution (BSD) A version of the UNIX operating system that first included TCP/IP support. The UNIX operating systems that included TCP/IP are referred to as 4.2BSD or 4.3BSD.

Bernoulli law or Bernoulli theorem In aeronautics, a law or theorem stating that in a flow of incompressible fluid the sum of the static pressure and the dynamic pressure along a streamline is constant if gravity and frictional effects are disregarded.

From this law it follows that where there is a velocity increase in a fluid flow there must be a corresponding pressure decrease. Thus an airfoil, by increasing the velocity of the flow over its upper surface, derives lift from the decreased pressure.

Besselian star numbers Constants used in the reduction of a mean position of a star to an apparent position (used to account for short-term variations in the precession, nutation, aberration, and parallax).

Best practice Best Practice is a superior method or innovative practice that contributes to the improved performance of an organization, usually recognized as "best" by other peer organizations.

Beta decay Radioactive transformation of a nuclide in which the atomic number changes by ±1 with the emission of a beta particle and the mass number remains unchanged. Also called beta disintegration.

Increase of atomic number occurs with negative beta particle emission, decrease with positive beta particle (positron) emission or upon electron capture.

Beta particle A charged particle emitted from a nucleus during radioactive decay, with a mass equal to 1/1837 that of a proton. A negatively charged beta particle is identical to an electron. A positively charged beta particle is called a positron. Large amounts of beta radiation may cause skin burns, and beta emitters are harmful if they enter the body. Beta particles may be stopped by thin sheets of metal or plastic.

Betatron A particle accelerator in which magnetic induction is used to accelerate electrons.

Betatron refers to the accelerated particles, electrons, which are identical with beta particles.

Beyond design basis accidents This term is used as a technical way to discuss accident sequences that are possible but were not fully considered in the design process because they were judged to be too unlikely. As the regulatory process strives to be as thorough as possible, "beyond design-basis" accident sequences are analyzed to fully understand the capability of a design.

BIAS Current A very low-level DC current generated by the panel meter and superimposed on the signal. This current may introduce a measurable offset across a very high source impedance.

Bias error A measurement error that remains constant in magnitude for all observations. A kind of systematic error.

An example is an incorrectly set zero adjustment.

Bias ply tires A type of tire in which the plies or layers of cord in the tire casing are laid diagonally, criss-crossing one another at an angle of 30 to 40 degrees.

Bidirectional transducer A transducer device capable of measuring input in both a positive and a negative direction from a reference zero or rest position.

Bilateral transducer A transducer capable of transmission simultaneously in both directions.

Bilayer lipid membrane (BLM) The structure found in most biological membranes, in which two layers of lipid molecules are so arranged that their hydrophobic parts interpenetrate, whereas their hydrophilic parts form the two surfaces of the bilayer.

Bimetallic strip It is a strip that is made of two different metals joined together in such a way that any increase in temperature produces a bend on one side. Such a strip is used in thermostats and safety cut outs.

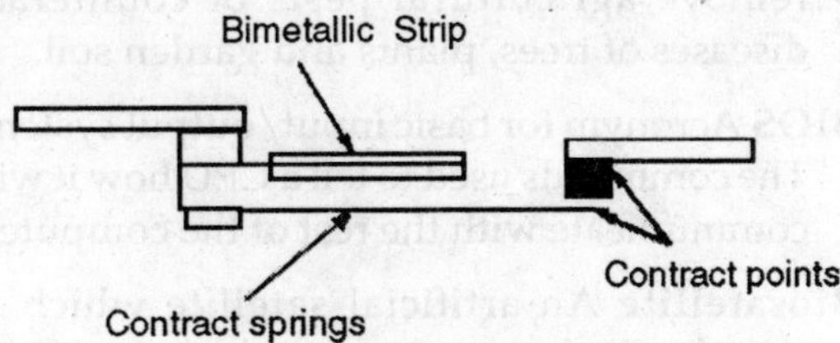

Fig. Bimetallic strip

Bimetallic strip gage A thermal conductivity vacuum gage in which deflection of a bimetallic strip with changing temperature indicates the changes in pressure.

Binary cell Any device or circuit that can be placed in either of two stable states to store a bit of binary information. Often called a binary .

Binary code A code composed of a combination of entities each of which can assume one of two possible states. Each entity must be identifiable in time or space.

Binary coded decimal (BCD) The representation of a decimal number (base 10, 0 through 9) by means of a 4 bit binary nibble.

Binary magnetic core A ferromagnetic material which can be caused to assume either of two stable magnetic states and thus can be used in a binary cell.

Binary star A system of two stars revolving about their barycenter.

The smaller star of the system is referred to as the companion or comes.

Visible binaries are those which can be resolved into two stars by a telescope.

Spectroscopic binaries are those which cannot be resolved by a telescope by show temporary displacement and doubling of the lines in their spectra.

Binder The nonvolatile portion of the vehicle of a coating which holds together the pigment particles.

Binding The process by which a molecule (or ligand) becomes bound, that is, confined in position (and often orientation) with respect to a receptor. Confinement occurs because structural features of the receptor create a potential well for the ligand; van der Waals and electrostatic interactions commonly contribute.

Binding energy The reduction in the free energy of a system that occurs when a ligand binds to a receptor. Generally used to describe the total energy required to remove something, or to take a system apart into its constituent particles—for example, to separate two atoms from one another, or to separate an atom into electrons and nuclei.

Bioassay The determination of kinds, quantities, or concentrations and, in some cases, locations of radioactive material in the human body, whether by direct measurement (in vivo counting) or by analysis and evaluation of materials excreted or removed (in vitro) from the human body.

Bioastronautics The study of biological, behavioral, and medical problems pertaining to astronautics. This includes systems functioning in the environments expected to be found in space, vehicles designed to travel in space, and the conditions on celestial bodies other than on earth.

Biochemical oxygen demand (BOD) The amount of oxygen required to oxidize any organic matter present in a water during a specified period of time, usually 5 days. It is an indirect measure of the amount of organic matter present in a water.

Biochemistry Chemistry dealing with the chemical processes and compounds of living organisms.

Bioclimatology The study of the relations of climate and life, especially the effects of climate on the health and activity of human beings (human bioclimatology) and on animals and plants.

Biodynamics The study of the effects of dynamic processes (motion, acceleration, weightlessness, etc) On living organisms.

Biofilm A film of microorganisms attached to a surface, such as that on a trickling filter, rotating biological contactor, or rocks in natural streams.

Biogeochemical cycle The cycle of elements through the biotic and abiotic environment.

Biological half life The time required for a biological system, such as that of a human, to eliminate, by natural processes, half of the amount of a substance (such as a radioactive material) that has entered it.

Biological shield A mass of absorbing material placed around a reactor or radioactive source to reduce the radiation to a level safe for humans.

Biomechanics The application of mechanical laws to the living systems, with specific regard to normal locomotion and includes the mechanical laws governing the structure, function, and position of the human body.

Biomedical engineering Biomedical engineering applies engineering principles and design to the biology and medical arena to improve health care and the lives of those with medical impairments. Bringing together knowledge from many engineering disciplines and technical fields, biomedical engineers design medical instruments, devices, and software; develop new procedures; conduct research; and solve clinical problems.

Bionics The study of systems, particularly electronic systems, which function after the manner of, or in a manner characteristic of, or resembling, living systems.

Biopak A container for housing a living organism in a habitable environment and for recording biological functions during space flight.

Bioremediation Use of living organisms to clean up oil spills or remove other pollutants from soil, water or wastewater; use of organisms such as non-harmful insects to remove agricultural pests of counteract diseases of trees, plants and garden soil.

BIOS Acronym for basic input/output system. The commands used to tell a CPU how it will communicate with the rest of the computer.

Biosatellite An artificial satellite which is specifically designed to contain and support man, animals, or other living material in a reasonably normal manner for an adequate period of time and which, particularly for man and animals, possesses the proper means for safe return to the earth.

Biosensor The term" biosensor" is a general designation that denotes either a sensor to detect a biological substance or a sensor which incorporates the use of biological molecules such as antibodies or enzymes. Biosensors are a subcategory of chemical sensors.

Biosphere That transition zone between earth and atmosphere within which most forms of terrestrial life are commonly found; the outer portion of the geosphere and inner or lower portion of the atmosphere.

Biosynthesis Catabolism, the production of new cellular materials from other organic or inorganic chemicals.

Biotechnology The application of engineering and technological principles to the life sciences.

Biotelemetry The remote measuring and evaluation of life functions, as, e.g. In spacecraft and artificial satellites.

Biotron A test chamber used for biological research within which the environmental conditions can be completely controlled, thus allowing observations of the effect of variations in environment on living organisms.

Bipolar junction transistor Transistor with n-type and p-type semiconductors having base-emitter and collector-base junctions.

Bipropellant A rocket propellant consisting of two unmixed or uncombined chemicals (fuel and oxidizer) fed to the combustion chamber separately.

Bipropellant rocket A rocket using two separate propellants which are kept separate until mixing in the combustion chamber.

Biquinary notation A numerical system in which each decimal digit is represented by a pair of digits consisting of a coefficient of five followed by a coefficient of one.

For example, the decimal digit 7 is represented in biquinary notation by 12 [(1 x 5) + (2 x 1)], and the decimal quantity 3648 is represented by 03 11 04 13.

The abacus is based on biquinary notation.

Bistatic reflectivity The characteristic of a reflector which reflects energy along a line, or lines, different from, or in addition to, that of the incident ray.

For example, any reflector that scatters the incident energy.

Bit 1. An abbreviation of binary digit.

2. A single character of a language employing only two distinct kinds of characters.

3. A quantity of intelligence which is carried by an identifiable entity and which can exist in either of two states.

4. A unit of storage capacity; the capacity in bits of a storage device is the logarithm to the base two of the number of possible states of the device.

5. A quantum of information.

6. Loosely, a mark.

Bitnet (Because It's Time Network) A non-TCP/IP network for small universities without Internet access. An electronic mail network connecting over 200 universities. It merged with the CSNET network to produce CREN.

Black body, blackbody (symbol b used as a subscript) 1. An ideal emitter which radiates energy at the maximum possible rate per unit area at each wavelength for any given temperature. A black body also absorbs all the radiant energy in the near visible spectrum incident upon it.

No actual substance behaves as a true black body, although platinum black and other soots rather closely approximate this ideal. However, one does speak of a black body with respect to a particular wavelength interval. This concept is fundamental to all the radiation laws, and is to be compared with the similarly idealized concepts of the white body and the gray body. In accordance with the Kirchhoff law, a black body not only absorbs all wavelengths, but emits at all wavelengths and does so with maximum possible intensity for any given temperature.

2. A laboratory device which simulates the characteristics of a black body.

Black box 1. In engineering design, a unit whose output is a specified function of the input, but for which the method of converting input to output is not necessarily specified.

2. Colloquially, any unit, usually an electronic device such as an amplifier, which can be mounted in, or removed from, a rocket, spacecraft, or the like as a single package.

Black-body radiation The electromagnetic radiation emitted by an ideal black body; it is the theoretical maximum amount of radiant energy of all wavelengths which can be emitted by a body at a given temperature.

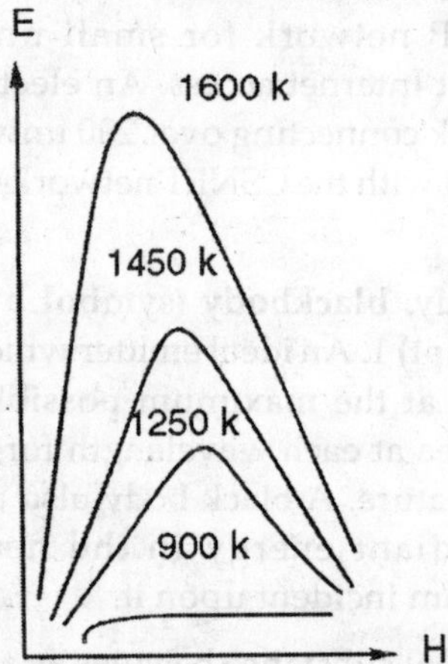

Fig. Black body temprature

The spectral distribution of black-body radiation is described by Planck law and the related radiation laws. If a very tiny opening is made into an otherwise completely enclosed space (hohlraum), the radiation passing out through this hole when the walls of the enclosure have come to thermal equilibrium at some temperature will closely approximate ideal black-body radiation for that temperature.

Blackout 1. A fadeout of radio communications due to ionospheric disturbances.

Blackouts are most common in, but are not restricted to, the arctic. An arctic blackout may last for days or even weeks during periods of intense auroral activity. Past experiments with high-altitude nuclear detonations have produced blackouts and artificial auroras over the subtropics.

2. A fadeout of radio and telemetry transmission between ground stations and vehicles traveling at high speeds in the atmosphere caused by signal attenuation in passing through ionized boundary layer (plasma sheath) and shock wave regions generated by the vehicle.

3. A vacuum tube characteristic which results from the formation of a dielectric film on the surface of the control grid.

A negative charge, accumulated on the film when the grid is driven positive with respect to the cathode, affects the operating characteristics of the tube.

4. A condition in which vision is temporarily obscured by a blackness, accompanied by a dullness of certain of the other senses, brought on by decreased blood pressure in the eye and a consequent lack of oxygen, as may occur, e.g., in pulling out of a high-speed dive in an airplane. Compare grayout, redout.

Blade 1. (a) An arm of a propeller; a rotating wing. (b) Specifically, restrictive, that part of a propeller arm or of a rotating wing from the shank outward, i.e., that part having an efficient airfoil shape and that cleaves the air.

2. A vane, such as a rotating vane or stationary vane in a rotary air compressor, or a vane of a turbine wheel.

Blast 1. The brief and rapid movement of air or other fluid away from a center of outward pressure, as in an explosion.

2. The characteristic instantaneous rise in pressure, followed by a sudden decrease, that results from this movement, differentiated from less rapid pressure changes.

3. To take off from a launching pad or stand. Said of a rocket in reference to the blast effects caused by rapid combustion of fuel as the rocket starts to move upward. (Popular).

This term is commonly used for explosion, but the two terms should be distinguished. In space, an explosion could take place, but no blast would follow.

Blast chamber A combustion chamber, especially a combustion chamber in a gas-turbine engine, jet engine, or rocket engine.

Blast cleaning The cleaning and roughing of a surface by the use of sand, artificial grit or fine metal shot which is projected at a surface by compressed air or mechanical means

Blast furnace A reaction vessel in which mixed charges of oxide ores, fluxes and fuels are blown with a continuous blast of hot air and oxygen-enriched air for the chemical reduc tion of metals to their metallic state. Iron ore is most commonly treated in this way, and so are some ores of copper, lead, etc.

Bleed off To take off a part or all of a fluid from a tank or line, normally through an escape valve or outlet, as in to bleed off excess oxygen from a tank.

Bleeder resistor A resistor that allows a small current drain on a power source to discharge filter capacitors or to stabilize an output.

BLEVE Boiling Liquid Expanding Vapour Explosion.

Blip A spot of light or deflection of the trace on a radarscope, loran indicator, or the like, caused by the received signal, as from a reflecting object. Also called a pip or echo.

Blister copper The product of the Bessemer converter furnace used in copper smelting. It is a crude form of copper, assaying about 99% copper, and requires further refining before being used for industrial purposes.

Blistering The formation of blisters in paint films by the local loss of adhesion and lifting of the film from the underlying substrate.

Blob A fairly small-scale temperature and moisture inhomogeneity produced by turbulence within the atmosphere.

The abnormal gradient of the index of refraction resulting from a blob can produce a radar echo of the type known as angels.

Block diagram Block Diagram is a diagram that shows the operation, interrelationships and interdependencies of components in a system. Boxes or blocks represent the components; connecting lines between the blocks represent interfaces.

BMP Best Manufacturing Practices

BOE Basis of Estimate

BOM Bill of Material

Block mode A string of data recorded or transmitted as a unit. Block mode transmission is usually used for high speed transmissions and in large, high speed networks.

Blockhouse, block house 1. A reinforced concrete structure, often built underground or half underground, and sometimes dome shaped, to provide protection against blast, heat, or explosion during rocket launchings or related activities; specifically, such a structure at a launch site that houses electronic control instruments used in launching a rocket.

2. The activity that works in such a structure.

Blocking oscillator A regenerative circuit which generates pulses of short duration.

Blocking oscillators are used in digital computers.

Blooming A haziness which develops on paint surfaces caused by the exudation of a component of the paint film.

Blowdown tunnel A type of wind tunnel in which stored compressed gas is allowed to expand through a test section to provide a stream of gas or air for model testing.

The downstream side may or may not be reduced in pressure to provide greater expansion potential.

Blowdown turbine A turbine attached to a reciprocating engine which receives exhaust gases separately from each cylinder, utilizing the kinetic energy of the gases.

Blowoff The action of applying an explosive force and separating a package section away from the remaining part of a rocket vehicle or reentry body, usually to retrieve an instrument or to obtain a record made during early flight.

Blushing A film defect which manifests itself as a milky appearance which is generally caused by rapid solvent evaporation or the presence of excessive moisture during the curing process.

Bobbing Fluctuation of the strength of a radar echo, or its indication on a radarscope, due to alternate interference and reinforcement of returning reflected waves.

Body 1. The main part or main central portion of an airplane, airship, rocket, or the like; a fuselage or hull.

2. In a general sense, any fabrication, structure, or other material form, especially one aerodynamically or ballistically designed, as, an airfoil is a body designed to produce an aerodynamic reaction.

Body angle The angle which the longitudinal axis of the airframe makes with some selected line.

Body axis Any one of a system of mutually perpendicular reference axes fixed in an aircraft or similar body and moving with it.

Body centered cubic (BCC) A common crystal structure found in some elemental metals. Within the cubic unit cell, atoms are located at corner and cell center positions.

Body of revolution A symmetrical body having the form described by rotating a plane curve about an axis in its plane.

Body on frame construction A type of automobile construction in which the body structure is attached to a separate frame.

Body planes For the convenience of anatomists and practitioners the human body is conveniently divided into sections or reference planes. These are used to describe relationships between different parts of the body with relation to the centre of the body. There are three mutually perpendicular axes about which movement of a joint could take place, although not all joints are capable of such complex movement. Body Planes are divided into Frontal Plane; Sagittal Plane; and Transverse Plane.

BoE (Barrel of Oil Equivalent) A measure of gas. A BoE has similar energy content as a barrel of Oil. Approx 167 MMscf depending on calorific value of the gas.

Bogie 1. A supporting and aligning wheel or roller on the inside of an endless track, used, e.g., in certain types of landing gear.

2. A type of landing-gear unit consisting of two sets of wheels in tandem with a central strut.

Boilerplate model A metal copy of a flight vehicle, the structure or components of which are heavier than the flight model.

Boiling point The temperature at which a substance in the liquid phase transforms to the gaseous phase; commonly refers to the boiling point of water which is 100°C (212°F) at sea level.

Boiling water reactor (BWR) A reactor in which water, used as both coolant and moderator, is allowed to boil in the core. The resulting steam can be used directly to drive a turbine and electrical generator, thereby producing electricity.

Boiloff The vaporization of a liquid, such as liquid oxygen or liquid hydrogen, as its temperature reaches its boiling point under conditions of exposure, as in the tank of rocket being readied for launch.

Bolometer An instrument which measures the intensity of radiant energy by employing a thermally sensitive electrical resistor; a type of actinometer. Also called actinic balance. Compare radiometer.

Two identical, blackened, thermally sensitive electrical resistors are used in a Wheatstone bridge circuit. Radiation is allowed to fall on one of the elements, causing a change in its resistance. The change is a measure of the intensity of the radiation.

Bolometric magnitude 1. The magnitude of a star for the entire electromagnetic spectrum without atmospheric absorption.

The magnitude measured within the earth's atmosphere by a bolometer is the radiometric magnitude.

2. Loosely = radiometric magnitude.

BOM (pronounced "bomb") Bill of Materials. A list of components to be included on an assembly such as a printed circuit board. For a PCB the BOM must include reference designators for the components used and descriptions which uniquely identify each component. A BOM is used for ordering parts and, along with an assembly drawing, directing which parts go where when the board is stuffed.

Bond 1. Adhesion between various components and materials, such as adhesion between mortar (or grout) and masonry units or reinforcement.

2. Patterns formed by exposed faces of units.

3. Tying various parts of a masonry wall by lapping units one over another or by connecting with metal ties.

4. Attachment between a (coating) film and the substrate material to which it is applied.

5. A form of insurance agreement under which a bonding company guarantees to pay an owner or developer within stated limits for any financial loss, or for failure of the binder (contractor) to perform in accordance with the terms of the contract, i.e., to follow specifications, to charge the agreed-upon price, or to otherwise be found in default of the contract.

6. Process of joining two structures together, i.e., to create an assembly by means of adhesive linkage

Bond albedo The ratio of the amount of light reflected from a sphere exposed to parallel light to the amount of light incident upon it. Sometimes shortened to albedo.

The Bond albedo is used in planetary astronomy.

Bond beam Course or courses of a masonry wall grouted and usually reinforced in the horizontal direction. Serves as horizontal tie of wall, bearing course for structural members or as a flexural member itself.

Bone seeker A radioisotope that tends to accumulate in the bones when it is introduced into the body. An example is strontium-90, which behaves chemically like calcium.

Boolean algebra The study of the manipulation of symbols representing operations according to the rules of logic.

Boolean algebra corresponds to an algebra using only the numbers 0 and 1, therefore can be used in programming digital computers which operate on the binary principle.

Boolean logic Logic dealing with True/False values (for example, the operators AND, OR, and NOT are Boolean operators).

Boom Telescoping, (usually) hydraulically powered steel arm used to deliver hoses (such as in concrete pumping), manbaskets or hydraulic hammers to the work location on a particular site.

Boost 1. Additional power, pressure, or force supplied by a booster, as, hydraulic boost, or extra propulsion given a flying vehicle during lift-off, climb, or other part of its flight as with a booster engine.

2. Boost pressure.

3. To supercharge.

4. To launch or to push along during a portion of flight, as to boost a ramjet to flight speed by means of a rocket, or a rocket boosted to altitude with another rocket.

Boost pressure Manifold pressure greater than the ambient atmospheric pressure, obtained by supercharging. Often called boost.

Boost regulator One of several basic families of switching power supply topologies. Energy is stored in an inductor during the pulse then released after the pulse.

Booster pump A pump in a fuel system, oil system, or the like, used to provide additional or auxiliary pressure when needed or to provide an initial pressure differential before entering a main pump, as in pumping hydrogen near the boiling point.

Booster rocket 1. A rocket motor, either solid or liquid, that assists the normal propulsive system or sustainer engine of a rocket or aeronautical vehicle in some phase of its flight.

2. A rocket used to set a vehicle in motion before another engine takes over.

Boostglide vehicle A vehicle designed to glide in the atmosphere following a rocket-powered phase. Portions of the flight may be ballistic, out of the atmosphere.

Bootstrap 1. Referring to a self-generating or self-sustaining process; specifically, the operation of liquid-propellant rocket engines in which, during main-stage operation, the gas generator is fed by the main propellants pumped by the turbopump, and the turbopump in turn is driven by hot gases from the gas generator system.

Such a system must be started in its operation by outside power or propellants. When its operation is no longer dependent on outside power or propellant the system is said to be in bootstrap operation.

2. In computer operations, the coded instructions at the beginning of an input tape which together with manually inserted instructions, initiate a routine.

3. = leap-frog.

BOP Blow Out Preventor. Hydraulically or mechanically actuated valve installed at the wellhead to control pressure within the well.

BOPD Barrels of Oil per Day. Unit of measurement of crude oil produced by a well or a field. The volume of a barrel is equivalent to 42 US gallons (0.16 metres cubed).

Border gateway protocol (BGP) A protocol that provides information about the devices that can be reached through a router (into an autonomous network). BGP is newer than EGP.

Boresight camera A camera mounted in the optical axis of a tracking radar to photograph rockets being tracked while in camera range and thus provide a correction for the alignment of the radar.

Bose Einstein statistics Formulas relating to equations of state and partition of kinetic energy when the wave functions are symmetric.

Bounce An e-mail message you receive that tells you that an e-mail message you sent wasn't delivered. Usually contains an error code and the contents of the message that wasn't delivered.

Bounce back The rebound of atomized paint, especially when applied by conventional air spray methods.

Bounce table A testing device which subjects devices and components to impacts such as might be encountered in accidental dropping.

Boundary conditions A set of mathematical conditions to be satisfied, in the solution of a differential equation, at the edges or physical boundaries (including fluid boundaries) of the region in which the solution is sought. The nature of these conditions usually is determined by the physical nature of the problem.

Boundary layer The layer of fluid in the immediate vicinity of a bounding surface; in fluid mechanics, the layer affected by viscosity of the fluid, referring ambiguously to the laminar boundary layer, turbulent boundary layer, planetary boundary layer, or surface boundary layer.

In aerodynamics the boundary-layer thickness is measured from the surface to an arbitrarily chosen point, e.g., where the velocity is 99 percent of the stream velocity. Thus, in aerodynamics, boundary layer by selection of the reference point, can include only the laminar boundary layer or the laminar boundary layer plus all, or a portion of, the turbulent boundary layer.

Boundary value problem A physical problem completely specified by a differential equation in an unknown, valid in a certain region of space, and certain information (boundary condition) about the unknown, given on the boundaries of that region. The information required to determine the solution depends completely and uniquely on the particular problem.

Boussinesq approximation The assumption (frequently used in the theory of convection) that the fluid is incompressible except insofar as the thermal expansion produces a buoyancy, represented by a term gat (a is equivalent to alpha), where g is the acceleration of gravity; a is the coefficient of thermal expansion; and T is the perturbationtemperature.

Boyle mariotte law The empirical generalization that for many so-called perfect gases, the product of pressure p and volume V is constant in an isothermal process: pv = F(T) where the function F the temperature T cannot be specified without reference to other laws (e.g., Charles-Gay-Lussac law). Also called Boyle law, Mariotte law.

Brake fade A condition brought about by repeated brake applications, resulting in build-up of heat that causes a temporary reduction or fading of braking effectiveness.

Brake horsepower (BHP) The actual horsepower of an engine, measured by a brake attached to the driving shaft and recorded by a dynamometer.

Brake linings The replaceable friction material which contacts the brake drum in a drum brake system to slow or stop the car.

Brake master cylinder A cylinder containing a movable piston activated by pressure on the brake pedal. The piston produces hydraulic pressure that pushes fluid through the lines and wheel cylinders. This forces the brake lining or pad against the drum or disc to slow or stop the car.

Brake pads In a disc system, they are the replaceable flat segments consisting of a rigid backing plate plus frictional lining that takes the place of the shoe and lining in a drum brake. Brake pads are sometimes referred to as brake pucks.

Brake shoe The arc-shaped carrier to which the brake linings are mounted in a drum brake. They also force the lining against the rotating drum during braking.

Brakes, disc A type of braking system in which brake shoes, in a vise-like caliper, grip a revolving disk mounted on a wheel to slow or stop disc and wheel rotation for braking.

Brakes, drum A type of braking system that utilizes a metal drum mounted on a wheel to form the outer shell of a brake. The brake shoes press against the drum to slow or stop drum and wheel rotation for braking.

Braking ellipses A series of ellipses, decreasing in size due to aerodynamic drag, followed by a spacecraft in entering a planetary atmosphere.

In theory, this maneuver will allow a spacecraft to dissipate the heat generated in entry without burning up.

Branch 1. In an electrical circuit, a portion of a network consisting of one or more two-terminal elements in series.

2. The point in a computer program at which the machine will proceed with one of two or more possible routines according to existing conditions and instructions.

Branched polymer A polymer having a molecular structure of secondary chains that extend from the primary main chains.

Brazing Method of joining metal pieces by using a nonferrous filler metal heated below the melting point of the base metals to be joined, but above 800° F (430° C).

The filler metal, also called hard solder, may be brass, bronze, or a silver or gold alloy; the filler metal distributes itself between the surfaces to be bonded by capillarity. Brazing is different from welding; in welding, partial melting of the surfaces is likely to occur, and the filler metal is not distributed by capillarity. Brazing differs from soldering only in the temperature of the operation: ordinary, or soft, solder melts at temperatures below 430° C.

Brazing requires surface preparation (thorough cleaning) and the use of a flux, such as borax, to reduce any oxide film on the surfaces. In manufacturing operations, furnaces are often used to heat the parts to be brazed, or the parts are brazed by dipping in baths of molten filler alloys. For construction site operations, the joint is usually preheated with a gas torch.

Breadboard 1. An assembly of preliminary circuits or parts used to prove the feasibility of a device, circuit, system, or principle without regard to the final configuration or packaging of the parts.

2. To prepare a breadboard.

Break even time Break Even Time - a metric that measures the time from the start of development through production and sales until the product has achieved financial breakeven considering the investment in development and other non-recurring expenses.

Break point In computer operations, a point at which a break-point instruction inserted in the routine will cause the machine to stop, upon a command from the operator, for a check of progress.

Breakaway The action of a boundary layer separating from a surface.

Breakdown Failure of a material resulting from an electrical overload. The resulting damage may be in the form of thermal damage (melting or burning) or electrical damage (loss of polarization in piezoelectric materials).

Breakdown voltage 1. The voltage level which causes insulation failure.

2. The reverse voltage at which a semiconductor device changes its conductance characteristics.

Breakdown voltage rating The dc or ac voltage which can be applied across insulation portions of a transducer without arcing or conduction above a specific current value.

Breaking joints Any arrangement of masonry units which prevents continuous vertical joints from occurring in adjacent courses.

Breakoff phenomenon The feeling which sometimes occurs during high-altitude flight of being totally separated and detached from the earth and human society. Also called the breakaway phonomonon.

Breeder A reactor that produces more nuclear fuel than it consumes. A fertile material, such as uranium-238, when bombarded by neutrons, is transformed into a fissile material, such as plutonium-239, which can be used as fuel.

Bremsstrahlung effect The emission of electromagnetic radiation as a consequence of the acceleration of charged elementary particles, such as electrons, under the influence of the attractive or repulsive force fields of atomic nuclei near which the charged particle moves.

In cosmic-ray shower production, bremsstrahlung effects give rise to emission of gamma rays as electrons encounter atmospheric nuclei. The emission of radiation in the bremsstrahlung effect is merely one instance of the general rule that electromagnetic radiation is emitted only when electric charges undergo acceleration.

Brennschluss (German, combustion termination). The cessation of burning in a rocket, resulting from consumption of the propellants, from deliberate shutoff, or from other cause; the time at which this cessation occurs.

Brewster point One of the three commonly detectable points of zero polarization of diffuse sky radiation, neutral points, along the vertical circle through the sun; the other two are the Arago point and Babinet point.

This neutral point, discovered by Brewster in 1840, is located about 15° to 20° directly below the sun; hence it is difficult to observe because of the glare of the sun.

Brick A solid masonry unit of clay or shale, formed into a rectangular prism while plastic and burned or fired in a kiln. 1. Acid-Resistant Brick: Brick suitable for use in contact with chemicals, usually in conjunction with acid-resistant mortars.

2. Adobe Brick: Large roughly-molded, sun-dried clay brick of varying size.

3. Angle Brick: Any brick shaped to an oblique angle to fit a salient corner.

4. Arch Brick: a: Wedge-shaped brick for special use in an arch; b:. Extremely hard-burned brick from an arch of a scove kiln.

5. Building Brick: Brick for building purposes not especially treated for texture or color. Formerly called common brick.

6. Clinker Brick: A very hard-burned brick whose shape is distorted or bloated due to nearly complete vitrification.

7. Common Brick:

8. Dry-Press Brick: Brick formed in molds under high pressures from relatively dry clay (5 to 7 percent moisture content).

9. Economy Brick: Brick whose nominal dimensions are 4 by 4 by 8 in.

10. Engineered Brick: Brick whose nominal dimensions are 4 by 3.2 by 8 in.

11. Facing Brick: Brick made especially for facing purposes, often treated to produce surface texture. They are made of selected clays, or treated, to produce desired color.

12. Fire Brick: Brick made of refractory ceramic material which will resist high temperatures.

13. Floor Brick: Smooth dense brick, highly resistant to abrasion, used as finished floor surfaces.

14. Gauged Brick: a. Brick which have been ground or otherwise produced to accurate dimensions. b. A tapered arch brick.

15. Hollow Brick: A masonry unit of clay or shale whose net cross-sectional area in any plane parallel to the bearing surface is not less than 60 percent of its gross cross-sectional area measured in the same plane.

16. Jumbo Brick: A generic term indicating a brick larger in size than the standard. Some

producers use this term to describe oversize brick of specific dimensions manufactured by them.

17. Norman Brick: A brick whose nominal dimensions are 4" by 2 2/3" by 12".

18. Paving Brick: Vitrified brick especially suitable for use in pavements where resistance to abrasion is important.

19. Roman Brick: Brick whose nominal dimensions are 4 by 2 by 12 in.

20. Salmon Brick: Generic term for under-burned brick which are more porous, slightly larger, and lighter colored than hard-burned brick. Usually pinkish-orange color.

21. "SCR Brick":

22. Sewer Brick: Low absorption, abrasive-resistant brick intended for use in drainage structures.

23. Soft-Mud Brick: Brick produced by molding relatively wet clay (20 to 30 percent moisture). Often a hand process. When insides of molds are sanded to prevent sticking of clay, the product is sand-struck brick. When molds are wetted to prevent sticking, the product is water-struck brick.

24. Stiff-Mud Brick: Brick produced by extruding a stiff but plastic clay (12 to 15 percent moisture) through a die

Brick grade Designation for durability of the unit expressed as SW for severe weathering, MW for moderate weathering, or NW for negligible weathering.

Brick type Designation for facing brick which controls tolerance, chippage and distortion. Expressed as FBS, FBX and FBA for solid brick, and HBS, HBX, HBA and HBB for hollow brick.

Brightness temperature 1. In astrophysics, the temperature of a black body radiating the same amount of energy per unit area at the wavelengths under consideration as the observed body. Compare effective temperature, antenna temperature.

2. The temperature of a nonblack body determined by measurement with an optical pyrometer.

Brinell hardness number (Bhn) A parameter describing the hardness of a material as the ratio of pressure on a standard sphere used to indent the material to be tested to the area of the indentation produced.

Brinell hardness test A test for determining the hardness of a material by forcing a hard steel or carbide ball of specified diameter into it under a specified load; the result is expressed as the Brinell hardness number.

British thermal units (BTU) A unit of heat energy defined as the amount of heat required to raise the temperature of one pound of water by one degree Fahrenheit. One Btu is equal to about 252 calories, or 778 foot pounds, or 1.055 kilojoules or 0.293 watt hours.

Brittle crack propagation A very sudden propagation of a crack with the absorption of no energy except that stored elastically in the body.

Brittle fracture Separation of a solid with little or no macroscopic plastic deformation; fracture occurs by rapid crack propagation with less expenditure of energy than for ductile fracture; brittle tensile fractures have a bright, granular appearance and exhibit little or no necking; typical fracture mode of a glass or ceramic.

Brittleness The tendency of a material to fracture without first undergoing significant plastic deformation.

Broadband A range of frequencies divided into several narrower bands. Each band can be used for different purposes.

Broadband signaling The type of signaling used in Local Area Networks that enables multiplexing of more than one transmission at a time.

Broadcast The simultaneous transmission of the same data to all nodes connected to the network.

Broadside array An antenna array whose direction of maximum radiation is perpendicular to the line or plane of the array according as the elements lie on a line or plane. A uniform broadside array is a linear array whose elements contribute fields of equal amplitude and phase.

Brouter A network device that is a combination of the functions of a bridge and a router. It can function as a bridge while filtering protocols and packets destined for nodes on different networks.

Brownfields Abandoned, idled or underused industrial and commercial facilities/sites where expansion or redevelopment is complicated by real or perceived environmental contamination. Brownfields can be in urban, suburban or rural areas.

Brownian assembly Brownian motion in a fluid brings molecules together in various positions and orientations. If molecules have suitable complementary surfaces, they can bind, assembling to form a specific structure.

Brownian motion Motion of a particle in a fluid owing to thermal agitation, observed in 1827 by Robert Brown. (Originally thought to be caused by a vital force, Brownian motion in fact plays a vital role in the assembly and activity of the molecular structures of life.)

Browser A utility that lets you look through collections of things. For example, a file browser lets you look through a file system. Applications that let you access the World Wide Web are called browsers.

B-scope A cathode-ray indicator in which a signal appears as a spot with bearing as the horizontal coordinate and distance as the vertical coordinate. Also called B-display .

B-station In loran, the designation applied to the transmitting station of a pair, the signal of which always occurs more than half a repetition period after the next succeeding signal and less than half a repetition period before the next preceding signal from the other station of the pair, designated an A-station.

BTU (British thermal units) The quantity of thermal energy required to raise one pound of water at its maximum density, 1 degree F. One BTU is equivalent to .293 watt hours, or 252 calories. One kilowatt hour is equivalent to 3412 BTU.

Bubbling A temporary or permanent film defect in which bubbles of air or solvent vapor are present in the applied film.

Buck regulator A switching regulator which incorporates a step down DC-DC converter. A transformerless design in which the lower output voltage is achieved by chopping the input voltage with a series connected switch (transistor) which applies pulses to an averaging inductor and capacitor.

Buckling 1. An unstable state of equilibrium of a thin-walled body stemming from compressive stresses in the walls.

2. The lateral deflection of a thin-walled body resulting from such instability.

Buffer In computers: 1. An isolating circuit used to avoid reaction of a driven circuit on the corresponding driving circuit.

2. A storage device used to compensate for a difference in rate of flow of information or time or occurrence of events when transmitting information from one device to another.

Buffer storage In computer operations, storage used to compensate for a difference in rate of flow or time of occurrence when transferring information from one device to another.

Build Of a radiant energy signal, to increase, often temporarily, in received signal strength without a change of receiver controls.

The opposite is fade.

Build and design standards · Baseline Build Standard Part of the Production Baseline, defining the manufacture drawing issues and equipment part numbers, including Hardware and Software identified by codification requirements. · Functional Design/Build Standard A system to define and retrieve information in respect of the functional characteristics of a system/item and its influence on other systems/items when it is changed. · Design Standard The definition of what is to be built, as defined by approved drawings, etc., and subsequent approved changes. · Build standard The incorporated Hardware/Software standards (product baseline) of the specific System, as confirmed by Quality Assurance. · Design Build Standard Incorporating both the Design and the build Standards, records the design requirements and the embodiment status.

Bulk sampling Removing mineral substances in substantial quantities (over 50 tonnes) in order to do mineral processing tests.

Bulkhead A wall, partition, or similar member in a rocket, spacecraft, airplane fuselage, or similar structure, at right angles to the longitudinal axis of the structure, and serving to strengthen, divide, or help give shape to the structure.

Burble A separation or breakdown of the laminar flow past a body; the eddying or turbulent flow resulting from this.

Burble occurs over an airfoil operating at an angle of attack greater than the angle of maximum lift, resulting in a loss of lift and an increase of drag.

Burn in The operation of a newly fabricated device or system prior to application with the intent to stabilize the device, detect defects, and expose infant mortality. In power supplies, a period during which a supply is energized and loaded to peak output, with the intent of finding potentially weak components. Typical burn-in tests can include temperature cycling, input cycling, and/or load cycling.

Burnout 1. An act or instance of fuel or oxidant depletion or, ideally, the simultaneous depletion of both; the time at which this occurs. Compare cutoff.

In the United Kingdom all burnt is preferred to burnout.

2. An act or instance of something burning out or of overheating; specifically, an act or instance of a rocket combustion chamber, nozzle, or other part overheating so as to result in damage or destruction.

Burnout velocity The velocity of a rocket, rocket-powered aircraft, or the like at the time the fuel or oxidant or both are depleted. Also called burnt velocity.

Burnup 1. In a reactor, the percentage of fissionable atoms that have been fissioned.

2. Depletion of reactor fuel by fission.

Burst 1. A single pulse of radio energy; specifically such a pulse at radar frequencies.

2. = solar radio burst.

3. = cosmic ray burst.

Burst disk A diaphragm designed to burst at a predetermined pressure differential; sometimes used as a valve, e.g., in a liquid-propellant line in a rocket. Also called a rupture disk.

Burst proportioning A fast-cycling output form on a time proportioning controller (typically adjustable from 2 to 4 seconds) used in conjunction with a solid state relay to prolong the life of heaters by minimizing thermal stress.

Bus Parallel lines used to transfer signals between devices or components. Computers are often described by their bus structure (i.e., S-100, IBM PC).

Busch lemniscate The locus in the sky, of all points at which the plane of polarization of diffuse sky radiation is inclined 45° to the vertical: Also called neutral line .

Business case Business Case refers to the results of market, technical and financial analyses used to justify the feasibility of a new product. Ideally defined just prior to the "go to development" decision (gate), the case defines the product and project, including the project justification and the action or business plan.

Business process reengineering Business Process Reengineering (BPR) is the analysis and redesign of workflow within and between enterprises. Authors Michael Hammer and James Champy promoted the idea of BPR as the radical redesign and reorganization of an enterprise to lower costs and increase quality of service.

CAD Computer-Aided Design

CAE Computer-Aided Engineering

CAGR Compound Annual Growth Rate

CAI Computer-Aided Inspection

CAIT Computer-Aided Inspection and Test

CAIV Cost as an Independent Variable (DoD initiative)

CAM Computer-Aided Manufacturing

Cost Account Manager.

Butler volmer equation Used by cell designers to predict the current which will flow in a battery. It is the sum of the anodic and cathodic contributions and is directly proportional to the surface area of the electrodes, increasing exponentially with temperature.

Butterfly diagram (Pedotti Diagram) This refers to plotted data which represents a tilting resultant force vector moving forwards. This data is collected from a force platform and reproduced by the software as a butterfly shaped set of vectors. Each vector is usually recorded at 20 ms intervals.

Buzz 1. In supersonic diffuser aerodynamics, a nonsteady shock motion and airflow associated with the shock system ahead of the inlet, very rapid pressure pulsations are produced which can affect downstream operation in the burner, nozzle, etc.

2. Sustained oscillation of an aerodynamic control surface caused by intermittent flow separation on the surface, or by a motion of shock waves across the surface, or by a combination of flow separation and shock-wave motion on the surface.

Byproduct Byproduct is 1. any radioactive material (except special nuclear material) yielded in, or made radioactive by, exposure to the radiation incident to the process of producing or using special nuclear material (as in a reactor); and

2. the tailings or wastes produced by the extraction or concentration of uranium or thorium from ore.

Byte Eight related bits of data or an eight-bit binary number. Also denotes the amount of memory required to store one byte of data.

C

C Variously, the abbreviation for capacitance, capacitor, Celsius, centigrade and coulomb.

C/10 rate = 500 Ah/10 = 50 A.

C Pillar The roof support between a vehicle's rearmost side window and its rear window. Also known as a C-Post. On a vehicle with four side pillars, the rearmost roof support may be called a D-pillar.

C Programming language The preferred programming language for embedded software used in many battery management applications. Robust, fast and powerful, it allows low level access to information and commands while still retaining the portability and syntax of a high level language.

C Rate C is a value which expresses the rated current capacity of a cell or battery. A cell discharging at the C rate will deliver its nominal rated capacity for 1 hour. Charging and discharging currents are generally expressed as multiples of C. The time to discharge a battery is inversely proportional to the discharge rate.

NC is a charge or discharge rate which is N times the rated current capacity of the battery where N is a number (fraction or multiple).

CN is the battery capacity in AmpHours which corresponds to complete discharge of the battery in N hours (N is usually a subscript). Also written as the N-Hour rate.

C/B RATIO The ratio of the weight of water absorbed by a masonry unit during immersion in cold water to weight absorbed during immersion in boiling water. An indication of the probable resistance of brick to freezing and thawing. Also called saturation coefficient.

C4 Controlled Collapsed Chip Connect. A type of flip-chip technology which is used in Intel's Pentium III and in Motorola's PowerPC 603 and PowerPC 604 RISC Microprocessors. Here is an Friday, February 07, 2003 introduction to the C4/CBGA interconnect technology by Kromann, Gerke and Huang of Motorola's Advanced Packaging Technology Division.

Cable A structural element formed from steel wire bound in strands; the suspending element in a bridge; the supporting element in some dome roofs.

Cache A memory location that keeps frequently requested material ready. Usually the cache is faster than a storage device. It is used to speed data and instruction transfer.

CAD (Computer Aided Design) A system where engineers create a design and see the proposed product in front of them on a graphics screen or in the form of a computer printout or plot. In electronics, the result would be a printed circuit layout.

CAFE The acronym for Corporate Average Fuel Economy. This single mileage figure is determined by taking a sales weighted average of the fuel consumption for all models produced by a manufacturer. The minimum required figure is an established U. S. government standard. Manufacturers which do not meet the minimum standard are fined.

Caisson disease An affliction developed by people moving in and out of caissons quickly; also called the bends and decompression sickness.

Calcaneal neutral stance position This captures the contours of the subtalar joint when it is in its anatomical neutral position. Using the three criteria of clinical observation, the neutral position of the subtalar joint should present the following:

- When observed from the anterior aspect of the ankle, the medial and lateral edges of the talus should not protrude.

- From the posterior aspect the lateral surface of the foot, immediately below the lateral malleolus, is parallel to the concavity on the lateral surface of the leg, immediately above the lateral malleolus.

- the lateral surface of the foot describes a straight line in the area of the calcaneal cuboid joint.

Calcaneal position Throughout stance phase, the relationship between the calcaneum and the talus alter as the subtalar joint goes through its range of motion . These changes can be observed and the trained clinician may distinguish the supinated calcaneal position from the neutral calcaneal position; and the supinated position from the pronated calcaneal position. There are three criteria used to distinguish calcaneal position:

- The congruity of the medial and lateral edges of the talus to the calcaneum. This is observed on the anterior aspect of the ankle.

- The asymmetry between concavities on the lateral surface of the foot and the lateral surface of the leg. This can be viewed from the posterior aspect of the leg.

- The lateral surface of the foot in the area of the calcaneal cuboid joint, again viewed from the posterior aspect of the leg.

Calcaneal relaxed position The relaxed position captures the subtalar joint in full range pronation. When observed in weightbearing, the outward signs are as follows:

- the medial head of the talus may protrude on the medial side of the anterior aspect of the foot.

- the cavity of the foot becomes deeper and the concavity above the malleolus more shallow. The two concavities are no longer parallel.

- The posterior aspect of the foot reveals a concavity on the lateral surface in the area of the calcaneal cuboid articulation.

Calibration The process of adjusting an instrument or compiling a deviation chart so that its reading can be correlated to the actual value being measured.

Caliper In a disk brake, a housing for cylinder, pistons and brake shoes, connected to the hydraulic system. The caliper holds the brake shoes so they straddle the brake disc.

Cam A component that translates or rotates to move a contoured surface past a follower; the contours impose a sequence of motions (potentially complex) on the follower.

CAM files CAM means Computer Aided Manufacturing. These are the data files used directly in the manufacture of printed wiring. . The types of CAM files are 1. Gerber file, which controls a photoplotter,

2. NC Drill file, which controls an NC Drill machine and 3. Fab and assembly drawings in soft form (pen-plotter files). CAM files represent the valuable final product of PCB design. They are handed off to the board house which further refines and manipulates CAM data in their processes, for example in step- and-repeat panelization. Some PCB design software companies refer to all plotter or printer files as CAM files , although some of the plots may be check plots which are not used in manufacture.

Camber angle The inward or outward angle which a front-wheel spindle makes with a vertical line, as viewed by either the front or the rear of the vehicle. Positive camber results when the top of the tire tilts out further than its bottom. The adjustment of this setting affects both tire wear and vehicle handling.

Camshaft The shaft in the engine which is driven by gears, belts or chain from the crankshaft. The camshaft has a series of cams that opens and closes intake and exhaust valves as it turns.

CAN Bus Controller Area Network The automotive industry standard for on-board vehicle communications. It is a two wire, serial communications bus which is used for networking intelligent sensors and actuators

Capability. A group of related requirements; the word capability may be replaced with 'function', 'subject', 'object' or other term useful for presenting the requirements. Capability is a measure of the ability of a system to perform within its specification limits. It uses a series of indices: Cp, Cpk, Cr, and Cpm.

Capability maturity model (CMM) A description of the stages through which software organizations evolve as they define, implement, measure, control and improve their software processes. The model is a guide for selecting the process improvement strategies by facilitating the determination of current process capabilities and identification of the issues most critical to software quality and process improvement.

Capacitance (C) The charge-storing ability of a capacitor, defined as the magnitude of charge stored on either plate divided by the applied voltage. A 1-F capacitor charged to 1 V contains C of charge and 1 C is an amount of charge equal to that of about 6.24×10^{18} electrons.

Capacitance, distributed The capacitance in a circuit resulting from adjacent turns on coils, parallel leads and connections.

Capacitive effect leakage current The current flow between segregated conductive metal parts; voltage and frequency dependent.

Capacitor A device that stores a charge. A simple capacitor consists of two conductors separated by a dielectric.

Capacity The ability of a component, battery or other device to store and discharge a given quantity of current (A) or power (W) over a specified period of time (Ah or Wh).

Capacity factor (gross) The ratio of the gross electricity generated, for the time considered, to the energy that could have been generated at continuous full-power operation during the same period.

Capacity offset A correction factor applied to the rating of a battery if discharged under different C-rates from the one rated.

Capitalized cost The price of the leased vehicle plus any other charges such as fees or taxes that become the cost basis for calculating the terms of a lease.

Capture To draw (schematics) with CAE software in such a way that data, especially connectivity, can be extracted electronically. The extracted data would minimally be a netlist and preferably also a BOM . The more useful data that is included (captured) in the schematic, the more useful will be the BOM and netlist extracted from it.

Carbanion A highly reactive anionic chemical species with an even number of electrons and an unshared pair of electrons on a tetravalent carbon atom.

Carbene A highly reactive chemical species containing an electrically neutral, divalent carbon atom with two nonbonding valence electrons; a prototype is CH_2.

Carbon steel Steel which owes its properties chiefly to various percentages of carbon without substantial amounts of other alloying elements; also known as ordinary, straight carbon, or plain carbon steel.

Carbonaceous biochemical oxygen demand (CBOD) The amount of oxygen required to oxidize any carbon containing matter present in a water.

Carbonium ion A highly reactive cationic chemical species with an even number of electrons and an unoccupied orbital on a carbon atom.

Carbonyl A chemical moiety consisting of O with a double bond to C. If the C is bonded to N, the resulting structure is termed an amide; if it is bonded to O, it is termed a carboxylic acid or an ester linkage.

Carboxylic acid A molecule that includes a C having a double bond to O and a single bond to OH.

Card Another name for a printed circuit board. card-edge connector A connector which is fabricated as an integral portion of a printed circuit board along part of its edge. Often employed to enable a daughter or add-on card to be plugged directly into another much larger printed board, the motherboard or backplane.

Casement A window sash that swings open on side hinges; in-swinging are French in origin; out-swinging are from England.

Casing Exposed molding or framing around a window or door, on either the inside or outside, to cover the space between the window frame or jamb and the wall.

Cask A heavily shielded container used to store and/or ship radioactive materials. Lead and steel are common materials used in the manufacture of casks.

Cast iron A brittle alloy with high carbon content; iron that has been melted, then poured into a form and cooled; can be made into any shape desired.

Caster angle The forward or backward tilt of the steering axis as viewed from the side. If the point of load is ahead of the point of contact, the caster angle is positive. The caster angle tends to keep wheels in a straight line. Proper caster adjustment improves both tire wear and fuel economy.

Casting Generic term referring to a process where a fluid material (usually a molten alloy) is made to flow into a shaped mold cavity where it solidifies; this method is used to produce complex component shapes and properties difficult to achieve otherwise. more

CAT(Carbon Abatement Technology) Includes improving the efficiency of and co-firing of power plant with low carbon alternatives such as biomass. Also CCS.

Cat Cracker A refinery unit used to break up large hydrocarbon molecules into smaller ones.

The conversion operation takes place at very high temperatures (500 degrees Celcius) in the presence of a catalyst.

Catabolism The production of energy by the degradation of organic compounds.

Catalyst A chemical agent which promotes or influences a chemical reaction without itself being permanently changed by the reaction. Used in recombinant cells and fuel cells

Catalytic converter Often simply called a "catalyst", this is a stainless steel canister that is part of a vehicle's exhaust system and contains a thin layer of catalytic material spread over a large area of inert supports. It induces chemical reactions that convert an engine's exhaust emissions into less harmful products prior to entering the environment.

Cathode 1. In an electron tube the electrode through which a primary source of electrons enters the interelectrode space.

2. General name for any negative electrode.

3. When a semiconductor diode is biased in the forward direction, that terminal of the diode which is negative with respect to the other terminal. 4. In electrolytic plating, the workpiece being plated. (Graf).

Cathodic protection The reduction or prevention of corrosion of a metal surface caused by making it cathodic. This is accomplished by using a sacrificial anode (such as in zinc rich coatings or galvanizing) or by using impressed current.

Cation Particles in the electrolyte of a galvanic cell carrying a positive charge and moving towards the cathode during operation of the cell.

Caulking A mastic compound for filling joints and sealing cracks to prevent leakage of water and air, commonly made of silicone, bituminous, acrylic, or rubber-based material

Cavitation A condition of liquid flow where a liquid vaporises and the vapour bubbles subsequently collapse.

Can produce surface damage to pumps; valves etc.

Cavity The cavity or air space between the brick and the positive water barrier at the exterior steel stud wall should not be less than two inches wide. Smaller cavities are permitted but often do not function well. The cavity acts to provide a buffer for wind driven rain and allows water that penetrates the brick veneer to flashing, the cavity should be kept clear of any obstructions that might allow water to bridge across. Mortar droppings should be prevented from falling into the cavity. When mortar droppings do enter the cavity they should be removed. Cavity spaces can be wider, but this will reduce the capacity of the brick ties. Construction tolerance on the cavity width should be limited to ± 1/2".

CCA(Cold cranking amperes) A measure used to specify the cold cranking capability of automotive SLI batteries. For Lead Acid batteries it is the constant current a battery can deliver during a continuous discharge over a period of 30 seconds at -18°C without the terminal voltage dropping below a minimum of 1.2 Volts/cell.

CCS (Carbon Capture and Storage) Three methods of capture used are prè combustion; post combustion and oxy-firing.

CE A mark applied to an end-use product certifying that the product meets "applicable directives" and can be sold in Europe. "Certified Europe" is the result of the "harmonization" or unification of European safety and other standards. Each type of end-use equipment has "applicable directives" which must be met in order to display the CE mark. Power supplies intended for use within an end-use system are not required to have a CE mark. However, many power supply manufacturers offer CE-certified power supplies which have been tested to the applicable directives which the power supply manufacturer believes will be necessary for the anticipated end-use equipment.

CELL (Corrosion Eng./Electrochem.) Reactor used to assess the corrosion of metals, e.g. polarization cell.

Cell 1. The basic electrochemical unit used to generate or store electrical energy. A cell consists of two electrodes of dissimilar material isolated from one another electronically, in a common ionically conductive electrolyte.

2. An electrochemical system which converts chemical energy into electrical energy and also the reverse for rechargeable units.

Cell balancing The process used during charging to ensure that every cell is charged to the same state of charge. Also called "Equalisation".

Cell chemistry The active materials used in the energy cell.

Cell reversal A condition which may occur multi cell series chains in which an over discharge of the battery can cause one or more cells to become completely discharged. The subsequent volt drop across the discharged cell effectively reverses its normal polarity.

Cell voltage The dc voltage potential between the individual positive and negative terminals of a cell in a battery.

Cellosolve Proprietary name for ethylene glycol monoethyl ether. A slow evaporating, water miscible, relatively strong solvent often used in epoxy coatings.

Celsius (centrigrade) A temperature scale defined by 0°C at the ice point and 100°C at boiling point of water at sea level.

CEM-1 A NEMA grade of industrial laminate having a substrate of woven glass surfaces over a cellulose paper core and a resin binder of epoxy. It has good electrical and mechanical properties, somewhat surpassed by those of FR-4 .

Cement 1. Powder of alumina, silica, lime, iron oxide, and magnesium oxide burned together in a kiln and finely ground; used as an ingredient of mortar, grout and concrete; also: ather mixtures used for similar purposes.

Cementitious material Having the property of or acting like cement (ie binding substances together when reacting with water), such as certain limestones and tuffs when used in the surfacing of roads, but also Portland cement and fly ash in concrete, mortar or grout mixes.

Cement-modified soil The addition relatively small amounts of cement (1% to 2%) to fine-grained soils to reduce the liquid limit, plasticity index, volume change and water-retaining tendency. The effect of the cement is to bring individual soil particles into aggregations, thus artificially adjusting the grading of the soil, and icreasing its load bearing capacity and shearing strength. This is different from **cement** which contains more cement Also called "soil stabilization".

Center of gravity (Mass Center) The center of gravity of a body is that point in the body through which passes the resultant of weights of its component particles for all orientations of the body with respect to a uniform gravitational field.

Centipoise One hundredth of a poise which is a unit of measurement for viscosity. Water at room temperature has a viscosity of 1.0 Centipoise.

Centre of gravity The centre of gravity (Cof G) represents the centre of concentration of body mass. It is the point at which all parts of the body exactly balance each other. It is the point in the body at which all three reference planes intersect. The theoretical C of G, in the anatomical position, sits in front of the lumbrosacral junction in males but lower in females. Any movement in the body will alter its Cof G and in some movement e.g. bending to touch toes, will require the Cof G to sit outside the body. Control of the Cof G is essential to balance.

Centripetal force A force exerted on an object moving in a circular path which is exerted inward toward the center of rotation.

Ceramic Polycrystalline ferroelectric materials which are used as the sensing units in piezoelectric accelerometers. There are many different grades, all of which can be made in various configurations to satisfy different design requirements.

Ceramic color glaze An opaque colored glaze of satin or gloss finish obtained by spraying the clay body with a compound of metallic oxides, chemicals and clays. It is burned at high temperatures, fusing glaze to body making them inseparable.

Ceramic insulation High-temperature compositions of metal oxides used to insulate a pair of thermocouple wires The most common are Alumina (Al_2O_3), Beryllia (BeO), and Magnesia (MgO). Their application depends upon temperature and type of thermocouple. High-purity alumina is required for platinum alloy thermocouples. Ceramic insulators are available as single and multihole tubes or as beads.

Cerebral palsy A neurological disorder with spacticity and inco-ordination, caused by brain damage in and around the time of birth.

Cermet A composite material consisting of a combination of ceramic and metallic meterials. The most common cermets are the cemented carbides, composed of an extremely hard ceramic (e.g. WC, TiC), bonded together by a ductile metal such as cobalt or nickel.

Cern The European Laboratory for Particle Physics, where the World Wide Web was first conceived of and implemented.

Certification A process, which may be incremental, by which a contractor provides evidence to the acquirer that a product meets contractual or otherwise specified requirements.

CESI Testing institute and Notified Body based in Italy.

Also provide power system studies and consultancy to power producers; electrical utilities; large-scale users of electricity and Financial Institutions.

Chain reaction A reaction that initiates its own repetition. In a fission chain reaction, a fissionable nucleus absorbs a neutron and fissions spontaneously, releasing additional neutrons. These, in turn, can be absorbed by other fissionable nuclei, releasing still more neutrons. A fission chain reaction is self-sustaining when the number of neutrons released in a given time equals or exceeds the number of neutrons lost by absorption in nonfissionable material or by escape from the system.

Chaku Chaku A Japanese word that means "load-load." It is a method of conducting single-piece flow in which the operator proceeds from machine to machine, taking a part from the previous operation and loading it in the next machine, then taking the part just removed from that machine and loading it in the following machine. Chaku-chaku lines allow different parts of a production process to be completed by one operator, eliminating the need to move around large batches of work-in-progress inventory.

Chalking The formation of a friable powdery coating on the surface of a paint film, generally caused by exposure to ultraviolet radiation resulting in a loss of gloss.

Change request The formal documentation that is prepared for a request to change a specification in accordance with the CMP Change Procedure.

Character A letter, digit or other symbol that is used as the representation of data. A connected sequence of characters is called a character string.

Charge 1. The conversion of electrical energy, provided in the form of a current from an external source, into chemical energy within a cell or battery.

2. The potential energy stored in a capacitive electrical device.

Charge acceptance The ability of a secondary cell to convert the active material to a dischargeable form. A charge acceptance of 90% means that only 90% of the energy can become available for useful output. Also called Coulombic Efficiency or Charge Efficiency.

Charge carriers The particle carrying the electrical charge during the flow of electrical current . In metallic conductors the charge carriers are electrons , while ions carry the charges in electrolyte solutions .

Charge efficiency The ratio (expressed as a percentage) between the energy removed from a battery during discharge compared with the energy used during charging to restore the original capacity. Also called the Coulombic Efficiency or Charge Acceptance.

Charge pump A power supply which uses capacitors instead of inductors to store and transfer energy to the output. A voltage doubler or tripler.

Charge rate The current applied to a secondary cell or battery to restore its capacity. This rate is commonly expressed as a multiple of the rated capacity of the cell or battery.

Charge retention The ability of a battery to retain its charge in zero current conditions. Charge retention is much poorer at high temperatures.

Charge sensitivity For accelerometers that are rated in terms of charge sensitivity, the output voltage (V)is proportional to the charge (Q) divided by the shunt capacitance (C). This type of accelerometer is characterized by a high output impedance. The sensitivity is given in terms of charge; picocoulombs per unit of acceleration (g).

Charge transport The movement of electrical charge from one part of the system to another, occurring through the drift of ions under the influence of electrical potential difference. Also called Electromigration.

Charpy test An impact test in which a V-notched, keyhole-notched, or U-notched specimen, supported at both ends horizontally, is struck behind the notch by a striker mounted at the lower end of a pendulum; the energy that is absorbed in fracture is calculated from the height to which the striker would have risen had there been no specimen and the height to which it actually rises after fracture of the specimen.

Check plots Pen plots that are suitable for checking only. Pads are represented as circles and thick traces as rectangular outlines instead of filled-in artwork. This technique is used to enhance transparency of multiple layers.

Check rail The bottom horizontal member of the upper sash and the top horizontal member of the lower sash which meet at the middle of a double-hung window.

Checks (Corrosion Eng.) Slight microscopic breaks in a coating which do not penetrate to the underlying metal.

Chemical engineering Chemical engineering applies principles of chemistry and physics to the design and production of materials that undergo chemical changes during their manufacture. Chemical engineers also participate in efforts to maintain a clean environment and to create substitutes for or find ways to preserve our natural resources.

Chemical fixation (or stabilization/ solidification) A term for several different methods of chemically immobilizing hazardous materials into a cement, plastic, or other matrix.

Chemical oxygen demand (COD) The amount of oxygen required to oxidize any organic matter in the water using harsh chemical conditions.

Chemical recombination Following an ionization event, the positively and negatively charged ion pairs may or may not realign themselves to form the same chemical substance they formed before ionization. Thus, chemical recombination could change the chemical composition of the material bombarded by ionizing radiation.

Chemoautotrophic Organisms which utilize inorganic carbon (carbon dioxide or carbonates) for synthesis and inorganic chemicals for energy.

Chemotroph Organisms which obtain energy from the metabolism of chemicals, either organic or inorganic.

Chip 1. An integrated circuit manufactured on a semiconductor substrate and then cut or etched away from the silicon wafer . (Also called a die.) A chip is not ready for use until packaged and provided with external connections.

2. Commonly used to mean a packaged semiconductor device.

Chip on board In this technology integrated circuits are glued and wire-bonded directly to printed circuit boards instead of first being packaged. The electronics for many mass-produced toys are embedded by this system, which can be identified by the black glob of plastic sitting on the board. Underneath that glob (technical term: glob top), is a chip with fine wires bonded to both it and the landing pads on the board.

Chip scale package A chip package in which the total package size is no more than 20% greater than the size of the die within. Eg: Micro BGA.

Chipping Small pieces of paint removed from the surface, typically a sign of physical damage incurred in shipping or handling. Use of a surface tolerant primer for touch up followed by the same finish coat generally solves the problem.

Chlorinated hydrocarbon A class of strong, fast evaporating, nonflammable solvents such as carbon tetrachloride, methylene chloride or trichloroethylene.

Chlorinated rubber A coating resin formed by the reaction of rubber with chlorine gas. Often used for chemical or water resistant properties.

Chromatography The general name for a series of methods for separating mixtures by employing a system with a mobile phase and a stationary phase.

CIM (Computer Integrated Manufacturing) Used by an assembly house, this software inputs assembly data from a PCB CAM/CAD package, such as Gerber and BOM, as input and, using a pre-defined factory modeling system, outputs routing of components to machine programming points and assembly and inspection documentation. In higher end systems, CIM can integrate multiple factories with customers and suppliers.

Circumduction This type of movement is used to delay ground contact, by swinging the leg in an arc to increase ground clearance, rather than straight forwards. The swing phase of the other leg will usually be normal.

Civil engineering Civil engineering involves planning, designing, and building a wide variety of structures and facilities. These include bridges, roads and highways, dams, high-rise buildings, airports, water treatment centers, industrial manufacturing and processing facilities, and sanitation plants. Many civil engineers hold managerial and supervisory positions in government, industry, construction, and private practice. Others may work in research and teaching.

Cladding The thin-walled metal tube that forms the outer jacket of a nuclear fuel rod. It prevents corrosion of the fuel by the coolant and the release of fission products into the coolant. Aluminum, stainless steel, and zirconium alloys are common cladding materials.

Clarifier (sedimentation basin) A tank in which quiescent settling occurs, allowing solid particles suspended in the water to agglomerate and settle to the bottom of the tank. The solids resulting from the settling being removed as a sludge.

Classical mechanics Classical mechanics describes a mechanical system as a set of particles (which in a limiting case can form continuous media) having a well-defined geometry at any given time, and undergoing motions determined by applied forces and by the initial positions and velocities of the particles. The forces themselves may have electromagnetic or quantum mechanical origins. Classical statistical mechanics uses the same physical model, but treats the geometry and velocities as uncertain, statistical quantities subject to random thermally-induced fluctuations. Classical mechanics and classical statistical mechanics give a good account of many mechanical properties and behaviors of molecules; but for describing the electronic properties and behaviors of molecules, they are often useless.

Clay A natural, mineral aggregate consisting essentially of hydrous aluminum silicate; it is plastic when sufficiently wetted, rigid when dried and vitrified when fired to a sufficiently high temperature.

Clay mortar-mix Finely ground clay used as a plasticizer for masonry mortars.

Clean and dry Rather than a method, the requirement for Clean and Dry describes the condition of the surface prior to painting.The surface shall be clean, dry, and free of oil, grease, wax, form oils, and any other contaminant that may effect the adhesion of the coating. For best results and high performance requirements remove latencies and contaminants from precast and cast-in-place concrete by abrasive blasting or high pressure water blasting.Dry means that the substrate contains less then 15% moisture.Concrete should be cured at least 28 days and mortar joints at least 15 days @ 75 F and 50% RH.

Clean room A special decontaminated facility for constructing sensitive items, such as satellites. All dust and stray particles are diligently removed and controlled to avoid damaging delicate parts. Team members must wear sterile suits to avoid bringing outside impurities into the room. In addition, special care is taken to avoid electric shocks that might harm intricate electronics. There is a clean room in the fabrication area of the UT Satellite Design Lab.

Cleanup Actions taken to deal with a release or threat of release of a hazardous substance that could affect humans and/or the environment. This term is sometimes used interchangeably with the terms "remedial action," "response action" or "corrective action."

Clear ceramic glaze Same as Ceramic Color Glaze except that it is translucent or slightly tinted, with a gloss finish.

Clearance distance The shortest path separating two conductors or two circuit components.

Cleavage Transcrystalline fracture along specific crystallographic planes; usually associated with low-energy fracture; may exhibit river patterns and/or tongues.

Clerestory A window in the upper part of a lofty room that admits light to the center of the room.

Client/server architecture A catch where one program can initiate a session and another program can answer its requests. The origin of client/server designs is closely allied with the TCP/IP protocol suite.

Climatology The study of the climate, how the earth's atmosphere performs over long periods of time.

Clinical diagnosis A working hypothesis(es) based on the collection of symptom data, both subjective and objective, which have been used to consider the potential cause and effect relationship(s).

Clipping The term applied to the phenomenon which occurs when an output signal is limited in some way by the full range of an amplifier, ADC or other device. When this occurs, the signal is flattened at the peak values, the signal approaches the shape of a square wave, and high frequency components are introduced. Clipping may be hard, as is the case when the signal is strictly limited at some level; or it may be soft, in which case the clipping signal continues to follow the input at some reduced gain.

Clock An oscillator producing timing pulses to synchronize various elements of a system. In switching mode power supplies, a clock is used to produce the power pulses that are modulated to control power transfer. In digital interfaces that communicate on a bus (such as the IEEE-488) a clock is used to synchronize the data transfer and commands.

Closed chain motion This describes motion of the joints when a non-constant external force is applied to the body e.g. (in gait the GRF's)

Closed chain pronation This describes the movement of the subtalar joint during weightbearing. Subtalar pronation occurs by eversion of the calcaneus, with adduction and plantarflexion of the talus. During closed chain pronation, the leg internally rotates and the talus functions as an extension of the leg in the frontal and transverse plane.

Closed crankcase ventilation (CCV) A system in which crankcase vapors are discharged into the engine intake system (usually via the intake manifold) where they are burned during the combustion process rather than being discharged into the atmosphere.

Closed end lease A lease in which the lessee is not responsible for the value of a vehicle when the lease is done. Under this arrangement the lessee may return the vehicle at the end of the term and have no further obligation for the car. Also known as a "walk-away" lease.

Closed loop buck converter A circuit where energy is pulsed to an averaging LIC filter and delivered to a load.

Closed loop gain In a feedback control circuit, the increase in value of an output signal due to the effects on it of various other components or signals in the circuit.

Closeness of control Total temperature variation from a desired set point of system. Expressed as "closeness of control" is ±2°C or a system bandwidth with 4°C, also referred to as amplitude of deviation.

Closure The act of preparing a landfill for long term inactivity, including placement of a cover over the landfill to prevent infiltration of surface water.

CMR (Common-Mode Rejection) The ability of a panel meter to eliminate the effect of AC or DC noise between signal and ground. Normally expressed in dB at dc to 60 Hz. One type of CMR is specified between SIG LO and PWR GND. In differential meters, a second type of CMR is specified between SIG LO and ANA GND (METER GND).

CMV (Common-Mode Voltage) The AC or DC voltage which is tolerable between signal and ground. One type of CMV is specified between SIG LO and PWR GND. In differential meters, a second type of CMV is specified between SIG HI or LO and ANA GND (METER GND).

Coalescence The formation of resinous or polymeric material when water evaporates from an emulsion or a latex system, permitting contact and fusion of adjacent particles; fusing or flowing together of liquid particles.

Coastdown An action that permits the reactor power level to decrease gradually as the fuel in the core is depleted.

Coating asphalt Weather-resistant layer of asphalt applied to a roofing material surface during manufacture. Surfacing material such as aggregate is embeded in this asphalt.

Coating system A number of coats separately applied, in a predetermined order, at suitable intervals to allow for drying and curing, resulting in a completed job.

Cobwebbing Premature drying of a coating during spraying causing a spider web effect.

Coherence function. A frequency domain function computed to show the degree of a linear, noise-free relationship between a system's input and output. The value of the coherence function ranges between zero and one, where a value of zero indicates there is no causal relationship between the input and the output. A value of one indicates the existence of linear noise-free frequency response between the input and the output.

Col In describing landforms, a pass between two valleys is sometimes termed a col. In describing molecular potential energy functions, this term is commonly used to describe analogous features of the PES; a col is the region around a saddle point having negative curvature along one axis and positive curvature along all orthogonal axes.

Cold cranking amps (CCA) A rating, measured in amperes. Used for comparing cranking strength of automotive batteries during extremely cold (0 F or lower) weather.

Cold rolled steel Low carbon, cold-reduced, sheet steel. Differs from hot rolled steel by the absence of mill scale.

Cold shutdown The term used to define a reactor coolant system at atmospheric pressure and at a temperature below 200 degrees Fahrenheit following a reactor cooldown.

Cold working Plastic deformation of a metal at a temperature below that at which it recrystallizes; increasing the amount of cold work causes the dislocation density to rise in the material, making it more difficult to plastically deform the material and eventually cause brittle fracture.

Collar joint The vertical, longitudinal joint between wythes of masonry.

Collective dose The sum of the individual doses received in a given period by a specified population from exposure to a specified source of radiation.

Collector 1. An electrode in a transistor that collects electrons or holes.

2. In certain electron tubes, an electrode to which electrons or ions flow after they have completed their function.

Collision An event that occurs when two or more nodes broadcast packets at the same time the packets collide.

Collision detection A device's capability to detect whether a collision has occurred.

Colloids Small particles which have a negligible settling velocity. These particles have a very small mass so gravitational force is low compared to surface frictional forces. Typical colloidal sizes range from 10-3 mm to 1 mm.

Color code The ANSI established color code for thermocouple wires in the negative lead is always red. Color Code for base metal thermocouples is yellow for Type K, black for Type J, purple for Type E and blue for Type T.

Column A vertical member whose horizontal dimension measured at right angles to the thickness does not exceed three times its thickness.

Columnar structure Coarse structure of parallel columns of grains caused by highly directional solidification of molten metal resulting from sharp thermal gradients.

Combustible liquid Any liquid having a flash point at or above 100 F (37.8 C)

Combustion chamber The volume of space at the top of the cylinder where burning of the air/fuel mixture begins.

Combution engine An engine that uses combution to produce energy which can be converted into mechanical energy. Example car engine.

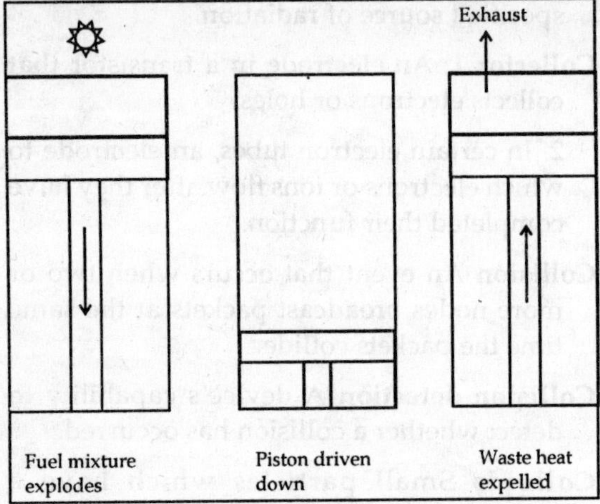

Fig. Combution engine

Command and control regulations Specific requirements prescribing how to comply with specific standards defining acceptable levels of pollution.

Command line Line on a terminal-based interface where you enter commands to the operating system.

Commercial off the shelf (COTS) software Commercially available applications sold by Vendors through public catalogue listings, COTS software is not intended to be customerised or enhanced. Contract-negotiated software developed for a specific application is not COTS software.

Commercial quarry Term that includes open mining for aggregate or limestone for industrial and agricultural purposes.

Commercial sampling of aggregates or coal Laboratory procedures intended to produce an accuracy such that if a large number of samples are taken from a single lot of aggregates or coal, 95 out of 100 test results will be within + or - 10% of the average of these samples.

Comminution The reduction of a rock to progressively smaller particles by natural methods (weathering, erosion, or tectonic movements) or manufacturing processes (breaking, crushing, or grinding by mechanical means).

Committed dose equivalent This is the dose to some specific organ or tissue that is received from an intake of radioactive material by an individual during the 50-year period following the intake.

Committed effective dose equivalent The committed dose equivalent for a given organ multiplied by a weighting factor.

Common management information protocol (CMIP) A network management protocol usually associated with OSI. When used with TCP/IP, CMIP is called CMOT.

Common mode The output form or type of control action used by a temperature controller to control temperature, i.e. on/off, time proportioning, PID.

Common mode noise The component of noise voltage that appears equally and in phase on conductors relative to a common reference.

Common mode rejection ratio The ability of an instrument to reject interference from a common voltage at its input terminals with relation to ground. Usually expressed in db (decibels).

Common point With respect to operationally programmable power supplies one output/ sense terminal is designated "common to which load, reference and external programming signal all return.

Communication Transmission and reception of data among data processing equipment and related peripherals.

Communications port A standard communications interface, such as an IEEE 488 or RS-232, that provides information flow from a processor to a peripheral device, such as a power supply.

Compact A group of two or more states formed to dispose of low-level radioactive waste on a regional basis. Forty-two states have formed nine compacts.

Compact rock A rock so closely grained that no component particles or crystals can be recognized by the eye.

Compaction curve The curve showing the relationship between the density (dry unit weight) and the water content of a soil for a given compactive effort.

Compaction equipment Machines, such as rollers, tc expel air from a soil mass and so achieve a high density. Smooth-wheel rollers are best for gravels, sands, and gravels-and-clay soils with reasonably high moisture contents. Pneumatic-tired rollers are best for clays with reasonably high moisture content, and sheepsfoot rollers are the best for clays with low moisture content.

Compaction test A basic laboratory compacting procedure to determine the optimum water content at which a soil can be compacted so as to yield the maximum density (dry unit weight). The method involves placing (in a specified manner) a soil sample at a known water content in a mold of given dimensions, subjecting it to a compactive effort of controlled magnitude, and determining the resulting unit weight (ASCE, 1958, term 74). The procedure is repeated for various water contents sufficient to establish a relation between water content and unit weight. The maximum dry density for a given compactive effort will usually produce a sample whose saturated strength is near maximum.

Comparison amplifier A dc amplifier which compares one signal to a stable reference, and amplifies the difference to regulate the power supply power-control elements.

Compatibility The ability to mix with or adhere properly to other coatings without detriment.

Compensated connector A connector made of thermocouple alloys used to connect thermocouple probes and wires.

Compensating alloys Alloys used to connect thermocouples to instrumentation. These alloys are selected to have similar thermal electric properties as the thermocouple alloys (however, only over a very limited temperature range).

Compensating loop Lead wire resistance compensation for RTD elements where an extra length of wire is run from the instrument to the RTD and back to the instrument, with no connection to the RTD.

Compensation A change of structure, position or function of a part in an attempt by the body to adjust to or neutralise the abnormal force of a deviation of structure, position or function of another part.

Compiler A program that translates a high-level language, such as Basic, into machine language.

Complex wave The resultant form of a number of sinusoidal waves that are summed together forming a periodic wave. Such waves may be analyzed in the frequency domain to readily determine their component parts.

Complexation The ionic bonding of one or more central ions or molecules by one or more surrounding ions or molecules.

Component Components are the named "pieces" of design and/or actual entities of the system/subsystem. In system/subsystem architectures, components consist of subsystems (or other variations), and manual operations.

Component library A representation of components as decals, stored in a computer data file which can be accessed by a PCB CAD program.

Components/equipment Reparable assemblies which currently require repair parts support or will require it when introduced into an inventory.

Composite headlamps Usually manufactured with replaceable halogen bulbs and separate hard acrylic or glass lenses. This type of lamp provides superior illumination compared to the long-conventional sealed beam unit.

Composting The controlled aerobic degradation of organic wastes into a material which can be used for landscaping, landfill cover, or soil conditioning.

Compound A chemical combination of two or more elements combined in a fixed and definite proportion by weight.

Compress A program that compacts a file so it fits into a smaller space. Also can refer to the technique of reducing the amount of space a file takes up.

Compressed air chamber The space at the bottom of a caisson into which air is introduced under pressure to exclude water so that excavation can take place.

Compression ratio The volume of the combustion chamber and cylinder when the piston is at the bottom of its stroke, divided by the volume of the combustion chamber and cylinder when the piston is at the top of its stroke. Higher compression ratios tend to increase engine efficiency.

Compression settling Settling which occurs in the lower reaches of clarifiers where particle concentrations are highest. Particles can settle only by compressing the mass of particles below.

Compressor (Air Conditioning) The machinism is an air conditioner that pumps vaporized refrigerant out of the evaporator, compresses it to a relatively high pressure, and then delivers it to the condenser.

Compuserve A commercial online service that gives its subscribers access to the Internet in addition to its other features.

Computer engineering Computer engineering involves the design, construction, and operation of computer systems. Computer engineers work on both computer hardware and software (programming) problems.

Computer hardware Devices capable of accepting and storing computer data, executing a systematic sequence of operations on computer data, or producing control outputs. Such devices can perform substantial interpretation, computation, communication, control, or other logical functions.

Computer program A combination of computer instructions and data definitions that enable computer hardware to perform computational or control functions.

Computer software component (CSC) A functionally or logically distinct part of a computer software configuration item. Computer software components may be top-level or low-level. (DOD-STD-2167A) - now referred to as software units by MIL-STD-498.

Computer software configuration item (CSCI) A software item which is identified for configuration management. (MIL-STD-973) or, An aggregation of software that satisfies an end use function and designated for separate configuration management. (MIL-STD-498).

Concatenate To join two strings.

Concentric contraction Muscles are known to exhibit two forms of control i.e. concentric contraction where muscles undergo shortening and eccentric contraction where they resist elongation.

Concept of execution Represents the dynamic relationships of the components. It can include such descriptions as flow of execution control, data flow, dynamically controlled sequencing, state transition diagrams, timing diagrams, priorities among units, handling of interrupts, etc.

Concrete A mixture of water, sand, small stones, and a gray powder called cement.

Concurrent engineering A systematic approach to the integrated, concurrent design of products and their related processes, including manufacture and support. This approach is intended to cause the developers, from the outset, to consider all elements of the product life cycle from conception through disposal, including quality, cost, schedule and user requirements.

Condensation The deposit of water vapor from the air on any cold surface whose temperature is below the dew point, such as a cold window glass or frame that is exposed to humid indoor air.

Condensation polymerization The formation of polymer macromolecules by an intermolecular reaction involving at least two monomer species, usually with the production of a by-product of low molecular weight, such as water.

Condenser A large heat exchanger designed to cool exhaust steam from a turbine below the boiling point so that it can be returned to the heat source as water. In a pressurized water reactor, the water is returned to the steam generator. In a boiling water reactor, it returns to the reactor core. The heat removed from the steam by the condenser is transferred to a circulating water system and is exhausted to the environment, either through a cooling tower or directly into a body of water.

Conditioning Cycle charging and discharging to ensure that formation is complete when a cell enters service or returns to service after a period of inactivity

Conductance Strictly speaking the Conductance applies to resistive circuits and is the reciprocal of the Resistance. Battery manufacturers have their own definition which applies to the frequency dependent elements of the circuit, that is - C=I/E where C is the conductance, I is the test current applied to a component (the cell) and E is the in phase component of the ac voltage E producing it.(Compare with Ohm's Law R=E/I) Measuring the conductance of a battery gives a good indication of its state of health.

Conducting polymer Plastic materials which have some of the properties of metals. Used as solid electrolytes in batteries. Also used in the construction of fuel cell membranes, capacitor electrodes and in applications requiring anti-static plastics.

Conduction Heat transfer through a solid material by contact of one molecule to the next. Heat flows from a higher-temperature area to a lower-temperature one

Conductivity, electrical The proportionality constant between current density and applied electric field; a measure of the ease with which a material is capable of conducting an electric current.

Confidence level The range (with a specified value of uncertainty, usually expressed in percent) within which the true value of a measured quantity exists.

Configuration The functional and/or physical characteristics of hardware/software as set forth in technical documentation and achieved in a product. (MIL-STD-973).

Configuration control The systematic evaluation, co-ordination, approval or disapproval and dissemination of proposed changes and implementation of all approved changes in the configuration of any item after formal establishment of its configuration Baseline.

Configuration control board A board composed of technical and administrative representatives who approve or disapprove proposed engineering changes to an approved baseline.

Configuration documentation Configuration documentation is the sum of all the documents that define the physical and functional characteristics of a System, subsystem, CSCI, HWCI or designated equipment, for example, specifications, design documents, engineering drawings, source code listings.

Configuration identification The current approved or conditionally approved technical documentation for a configuration item as set forth in specifications, drawings, and associated lists, and documents referenced therein. (MIL-STD-973).

Configuration item (CI) A configuration item is an aggregation of hardware or software that satisfies an end use function and is designated by the acquirer for separate configuration management. (MIL-STD-973).

Configuration management (CM) A discipline applying technical and administrative direction and surveillance to: · identify and document the functional and physical characteristics of CIs; · audit the CIs to verify conformance to specifications, interface control documents and other contract requirements; · control changes to CIs and their related documentation; and · record and report information needed to manage CIs effectively, including the status of proposed changes and the implementation status of approved changes.

Configuration space A mathematical space describing the three-dimensional configuration of a system of particles (e.g., atoms in a nanomechanical structure) as a single point; the configuration space for an N particle system has 3N dimensions.

Configuration status accounting The recording and reporting of information needed to manage configuration effectively, including: · a listing of the approved configuration identification; · the status of proposed changes, deviations, and waivers to the configuration; · the implementation status of approved changes, and; ·the configuration all units of the CI in the operational inventory.

Conformal coating A thin nonconducting coating that is either plastic or inorganic; it is applied to a circuit for environmental and mechanical protection.

Conformation A molecular geometry that differs from other geometries chiefly by rotation about single or triple bonds; distinct conformations (termed con formers) are associated with distinct potential wells. Typical biomolecules and products of organic synthesis can interconvert among many conformations; typical diamondoid structures are locked into a single potential well, and thus lack conformational flexibility.

Conformity error For thermocouples and RTDs, the difference between the actual reading and the temperature shown in published tables for a specific voltage input.

Congenital dislocation of the hip A congenital condition in which the head of the femur is not properly contained within the acetabulum.

Conjugated A conjugated pi system is one in which pi bonds alternate with single bonds. The resulting electron distribution gives the intervening single bonds partial double-bond character, the pi electrons become delocalized, and the energy of the system is reduced.

Connection A link between two or more processes, applications, machines, network, and so forth. Connections can be logical, physical, or both.

Connection head An enclosure attached to the end of a thermocouple which can be cast iron, aluminum or plastic within which the electrical connections are made.

Connectionless A type of network service that does not send acknowledgments to the sender upon receipt of data. UDP is a connectionless protocol.

Connectivity The intelligence inherent in PCB CAD software which maintains the correct connections between pins of components as defined by the schematic.

Connector A plug or receptacle which can be easily joined to or separated from its mate. Multiple-contact connectors join two or more conductors with others in one mechanical assembly.

Conservative In design and analysis, a conservative model or a conservative assumption is one that departs from accuracy in such a way that it reduces the chances of a false-positive assessment of the feasibility of the system in question. Conservative assumptions overestimate problems and underestimate capabilities.

Consortium for research and education network (CREN) The name for the body arising from the combination of CSNET and BITNET.

Constant current limiting circuit Current-limiting circuit that holds output current at some maximum value whenever an overload of any magnitude is experienced.

Constant current load An electronic load with a control loop to regulate the current drawn from the power supply.

Constant current power supply A power supply that regulates its output current, within specified limits, against changes in line, load, ambient temperature and time. Constant current supplies are sometimes used to drive certain types of loads (such as large incandescent lamps) or in certain applications where very long output leads can result in significant induced noise and resistive losses (such as industrial environments).

Constant voltage power supply A power supply that regulates its output voltage within specified limits, against changes in line, load, ambient temperature and time.

Constant voltage transformer Maintains approximately constant voltage ratio over the range from zero to rated output.

Construction manager A person who coordinates the entire construction process — from initial planning and foundation work through the structure's completion.

Construction recapture The maximum number of years that could be added to the license expiration date to recover the period from the construction permit to the date when the operating license was granted. A licensee is required to submit an application for such a change.

Consumers Organisms which consume protoplasm produced from photosynthesis or consume organisms from higher levels which indirectly consume protoplasm from photosynthesis.

Containment structure A gaslight shell or other enclosure around a nuclear reactor to confine fission products that otherwise might be released to the atmosphere in the event of an accident.

Contamination The deposit, absorption, or adsorption of biological or chemical agents (including the growth of bacteria or fungi)on or by materials, structures, areas, personnel, or objects

Contention A condition occurring in some LANs where the Media Access Control (MAC) sublayer allows more than one node to transmit at the same time, risking collisions.

Context Many functions return either array values or scalar values depending on the context, that is, whether returning an array or a scalar value is appropriate for the place where the call was made.

Continuity Means more than being without holes. Because the component that performs the role of the air barrier changes from the wall to the window to the roof, continuity means that all these assemblies must be connected together so as to ensure that there is no break in the airtightness of the envelope

Continuous flow A concept where items are processed and moved directly from one processing step to the next, one piece at a time. Also referred to as "one piece flow" and "single piece flow."

Continuous span beam bridge Simple bridge made by linking one beam bridge to another; some of the longest bridges in the world are continuous span beam bridges.

Continuous spectrum A frequency spectrum that is characterized by non-periodic data The spectrum is continuous in the frequency domain and is characterized by an infinite number of frequency components.

Contract A mutually binding legal relationship obligating the seller to furnish the supplies or services (including construction) and the buyer to pay for them. It includes all types of commitments that obligate the acquirer to an expenditure of appropriate funds and that, except otherwise authorized, are in writing. In addition to bilateral agreements, contracts include, but are not limited to, awards and notices of awards; job orders or task letter issued under purchase orders, under which the contract becomes effective by written acceptance or performance; and bilateral modifications.

Contract data requirements list That portion of a contract that identifies deliverable data products.

Contractor An individual, partnership, company, corporation, association or other service, having a contract with an acquirer for the design, development, manufacture, maintenance, modification, or supply of items under the terms of a contract.

Control arm A suspension element that has one joint at one end and two joints at the other end, typically on the chassis side. Also known as a wishbone or an A-arm.

Control character A character whose occurrence in a particular context starts, modifies or stops an operation that effects the recording, processing, transmission or interpretation of data.

Control code A non-printing character which is input or output to cause some special action rather than to appear as part of the data. Control codes are generated by holding down the <Ctrl> key on your computer keyboard while pressing one of the letter keys (e.g. < CTRL-G>. Sometimes called "control characters."

Control mode The output form or type of control action used by a temperature controller to control temperature, i.e., on/off, time proportioning, PID.

Control point The temperature at which a system is to be maintained.

Control rod A rod, plate, or tube containing a material such as hafnium, boron, etc., used to control the power of a nuclear reactor. By absorbing neutrons, a control rod prevents the neutrons from causing further fissions.

Control room The area in a nuclear power plant from which most of the plant power production and emergency safety equipment can be operated by remote control.

Control systems engineering Control systems engineering involves the design and manufacture of instrumentation and ways to control dynamic processes automatically. Such engineers draw on a variety of disciplines, including elements of electrical, mechanical, and chemical, and focus on the technologies needed for feedback and feedforward control of dynamic systems. These engineers ensure safe and efficient system performance.

Controlled area At a nuclear facility, an area outside a restricted area but within the site boundary, access to which the licensee can limit for any reason.

Convection A heat transfer process involving motion in a fluid (such as air) caused by the difference in density of the fluid and the action of gravity. Convection affects heat transfer from the glass surface to room air, and between two panes of glass.

Convection cooled power supply A power supply cooled exclusively from the natural motion of a gas or a liquid over the surfaces of heat dissipating elements.

Converter A device that changes the value of a signal or quantity. Examples: DC-DC; a device that delivers dc power when energized from a dc source. Fly-Back; a type of switching power supply circuit.

Coolant A substance circulated through a nuclear reactor to remove or transfer heat. The most commonly used coolant in the United States is water. Other coolants include heavy water, air, carbon dioxide, helium, liquid sodium, and a sodium-potassium alloy.

Cooldown The gradual decrease in reactor fuel rod temperature caused by the removal of heat from the reactor coolant system after the reactor has been shutdown.

Cooling The process of removing heat dissipated by a power supply during transformation and regulation.

Cooling system The system that removes heat from the engine by the forced circulation of coolant and thereby prevents engine overheating. In a liquid-cooled engine, it includes the water jackets, water pump, radiator, and thermostat.

Cooling tower A heat exchanger designed to aid in the cooling of water that was used to cool exhaust steam exiting the turbines of a power plant. Cooling towers transfer exhaust heat into the air instead of into a body of water.

Cooling: forced air Forced air cooling ratings may be specified in either linear feet per minute (LFM), cubic feet per minute (CFM), or pound mass per minute (lb/mm). Linear feet per minute ratings are a more valid measure of the air flow because they specify exactly where the air flows and directly relate to heat transfer. However, CFM values are usually specified. Flow measured in lb/mm is most useful when dealing with varying air densities or high altitude, and is calculated by CFM and density (lb/w̃). This application note explains the relationship between forced air specifications in linear feet per minute and in cubic feet per minute. This application note applies to all Astec power supplies requiring forced air cooling. To Convert from CFM to LFM and LFM to CFM given the cross sectional area: To convert from CFM to LFM and vice-versa, we need the cross section area of the power supply. This would be the actual area that the air is being blown through. The formula is CFM LFM x cross sectional area. A popular selection is 2.5" x 5.0". This corresponds to 0.0868 square feet. For any fixed cross sectional area, the relationship is linear and can be shown with a graph.

Coordination number The number of atomic or ionic nearest neighbors.

Coping The material or masonry units forming a cap or finish on top of a wall, pier, pilaster, chimney, etc. It protects masonry below from penetration of water from above.

Copolymer A polymer that consists of two or more dissimilar monomer units in combination in its molecular chains. Also a polymer formed from the polymerization of more than one type of monomer.

Corbel A shelf or ledge formed by projecting successive courses of masonry out from the face of the wall.

Cord set An assembly of a suitable length of flexible cord provided with an attachment plug at one end and a second connector at the other.

Core damage frequency An expression of the likelihood that, given the way a reactor is désigned and operated, an accident could cause the fuel in the reactor to be damaged.

Correction (Balancing) plane A plane perpendicular to the shaft axis of a rotor in which correction for unbalance is made.

Corrosion A surface electrochemical phenomenon common to all base metals in aqueous or humid environments whereby metal ions are developed at an anodic site and the electrons associated with this dissolution accepted at a cathodic site.

Corrosive waste A waste that is outside the pH range of 2 to 12.5 or a waste that corrodes steel at a rate greater than 6.35 mm (0.25 in) per year. One of EPA's four hazardous waste properties.

Cosmic radiation Penetrating ionizing radiation, both particulate and electromagnetic, originating in outer space. Secondary cosmic rays, formed by interactions in the Earth's atmosphere, account for about 45 to 50 millirem of the 360 millirem background radiation that an average individual receives in a year.

Cost escalation Change in estimated cost of a project due to inflation, aggravated conditions, etc.

Cost of ownership Several items enter into the calculation of how much it actually costs to own a vehicle. Understanding these items can help determine what is the best purchase for each person. Ownership costs are divided into fixed costs and running costs. Fixed costs are the same whether the car is used or just sits. These costs usually decrease each year. Running costs are variable and are incurred when the vehicle is used. Running costs could increase as the vehicle ages.

Coulomb A unit of electric charge. One coulomb (1C) is equal to the charge transferred by a current of one ampere in one second.

Coulomb counting A method of determining the state of charge of a battery by integrating the ingoing and outgoing discharge currents of a battery over time.

Coulomb sensitivity Charge/unit acceleration, expressed in Pc/g (charge sensitivity).

Coulombic efficiency The ratio (expressed as a percentage) between the energy removed from a battery during discharge compared with the energy used during charging to restore the original capacity. Also called Charge Efficiency or Charge Acceptance.

Counter A general designation applied to radiation detection instruments or survey meters that detect and measure radiation. The signal that announces an ionization event is called a count.

Counts The number of time intervals counted by the dual-slope A/D converter and displayed as the reading of the panel meter, before addition of the decimal point.

Coup de fouet (Whiplash) A dramatic initial voltage drop when a battery is suddenly called upon to supply a heavy load. The voltage recovers after a short time once the electro-chemical discharge process stabilises.

Coupling The characteristic of isolated circuit elements to interact with one another.

Covalent bond A primary interatomic bond that is formed by the sharing of electrons between neighboring atoms.

Covalent radius Given a set of N elements that can form covalent single bonds in molecules, with N(N - 1) possible elemental pairings, it has proved possible to define a covalent radius for each element such that the actual bond length between any two elements that form a covalent single bond is roughly equal to the sums of their covalent radii.

Covers .A product "covers" a given set of items if every item in the set has been dealt with in the specific product. The set could comprise of system, subsystem, hardware, or software items or a combination. For example, a plan covers the SOW if every provision in the SOW is dealt with in the 'Systems Engineering Management Plan' and supporting specialty plans; a design covers a set of requirements of every requirement has been dealt with in the design; a test plan covers a set of requirements if every requirement is the subject of one or more tests.

Coxa valga The reverse of Coxa Vara, where the angle between the neck and the shaft of the femur is increased above 140 degrees.

CPAN (Central Perl Archive Network) A series of machines on the Internet that act as central repositories for Perl distributions, documentation, libraries, and modules.

CPD (Continued Professional Development) The planned acquisition of knowledge; experience; skills and the development of personal qualities necessary for the execution of professional duties throughout the working life.

Crack 1. A : to break so that fissures appear on the surface or through the element b: fracture caused by the effects of stress(es) on weak or weakened parts of a member or material.

2. To impair seriously or irreparably

3. To subject (hydrocarbons) to cracking (to break a large, complex compound into simpler compounds)

4. To break up (chemical compounds) into simpler compounds by means of heat

Crack length Total outside perimeter of window vent. Used in figuring air infiltration during AAMA certification testing.

Crankcase A shaft with one or more cranks, or "throws," that are coupled by connecting rods to the engine's pistons. The combustion process creates reciprocating motion in the rods and pistons which in turn is converted to a rotating motion by the crankshaft.

Creep The time-dependent permanent deformation that occurs under stress; for most materials it is important only at elevated temperatures.

Creepage The movement of electrolyte onto surfaces of electrodes or other components of a cell with which it is not normally in contact.

Creepage distance The shortest distance separating two conductors as measured along a surface touching both conductors.

CRF Condensation Resistance Factor. An indication of a window's ability to resist condensation. The higher the CRF, the less likely condensation is to occur

CRI Cuttings Re Injection. Name given to the process of injecting drilling debris back in to the well.

Critical damping Critical damping is the smallest amount of damping at which a given system is able to respond to a step function without overshoot.

Critical mass The smallest mass of fissionable material that will support a self-sustaining chain reaction.

Critical organ That part of the body that is most susceptible to radiation damage under the specific conditions under consideration.

Critical resolved shear stress The shear stress, resolved within a slip plane and direction, which is required to initiate slip.

Critical speed The rotational speed of the rotor or rotating element at which resonance occurs in the system. The shaft speed at which at least one of the "critical" or natural frequencies of a shaft is excited.

Critical temperature (Superconductor) The temperature below which a superconducting material must be cooled in order to exhibit the property of superconductivity.

Criticality A term used in reactor physics to describe the state when the number of neutrons released by fission is exactly balanced by the neutrons being absorbed (by the fuel and poisons) and escaping the reactor core. A reactor is said to be "critical" when it achieves a self-sustaining nuclear chain reaction, as when the reactor is operating.

Cross contamination The movement of underground contaminants from one level or area to another due to invasive subsurface activities.

Cross sensitivity The influence of one measurand on the sensitivity of a sensor, another measurand.

Crosslink A pathway for communication between the two FASTRAC satellites. Crosslinks are used for satellite-to-satellite communication, while downlinks and uplinks are used for satellite-to-groundstation and groundstation-to-satellite communications.

Crosslinked polymer A polymer in which adjacent linear molecular chains are joined at various positions by covalent bonds.

Crossmember One of several horizontal members in a vehicle frame which join the side members and add to overall strength and stability.

Crosstalk Electromagnetic noise transmitted between leads or circuits in close proximity to each other.

Crowbar An overvoltage protection circuit which rapidly places a low resistance shunt across the power supply output terminals if a predetermined voltage is exceeded.

Crud A colloquial term for corrosion and wear products (rust particles, etc.) that become radioactive (i.e., activated) when exposed to radiation. Because the activated deposits were first discovered at Chalk River, a Canadian nuclear plant, "crud" has been used as shorthand for Chalk River Unidentified Deposits.

Crude unit The refinery processing unit where initial crude oil distillation takes place. The crude unit makes the first rough distillation cut.

The lighter products produced in this process are further refined in the cat cracker or the reforming unit. Heavier products which cannot be vaporised and separated in this process are distilled still further in the vacuum distillation unit or the coker.

Cryogenics Measurement of temperature at extremely low values, i.e., below -200°C.

Crystal structure For crystalline materials, the manner in which atoms or ions are arrayed in space. It is defined in terms of the unit cell geometry and the atom positions within the unit cell.

Crystalline The state of a solid material characterized by a periodic and repeating three-dimensional array of atoms, ions, or molecules.

Crystallization Act or process of forming crystals or bodies by elements or compounds solidifying so they are bounded by plane surfaces, symmetrically arranged, and are external expressions of definite internal structure.

Culls Something rejected, especially as being inferior or worthless, i.e. masonry units which do not meet the standards or specifications and have been rejected.

Cumulative dose The total dose resulting from repeated exposures of ionizing radiation to an occupationally exposed worker to the same portion of the body, or to the whole body, over time.

Cure point The temperature at which a normally magnetic material goes through a magnetic transformation and becomes non-magnetic.

Curie (Ci) The basic unit used to describe the intensity of radioactivity in a sample of material. The curie is equal to 37 billion (3.7 x 10^{10}) disintegrations per second, which is approximately the activity of 1 gram of radium. A curie is also a quantity of any radionuclide that decays at a rate of 37 billion disintegrations per second. It is named for Marie and Pierre Curie, who discovered radium in 1898.

Curie point or **Curie temperature** The temperature above which a ferromagnets and some other materials undergo a sharp change in their magnetic properties losing their ability to possess a net spontaneous or remanent magnetization in the absence of an external magnetic field.

Current The rate of flow of electricity. The unit of the ampere (A) defined as 1 ampere = 1 coulomb per second.

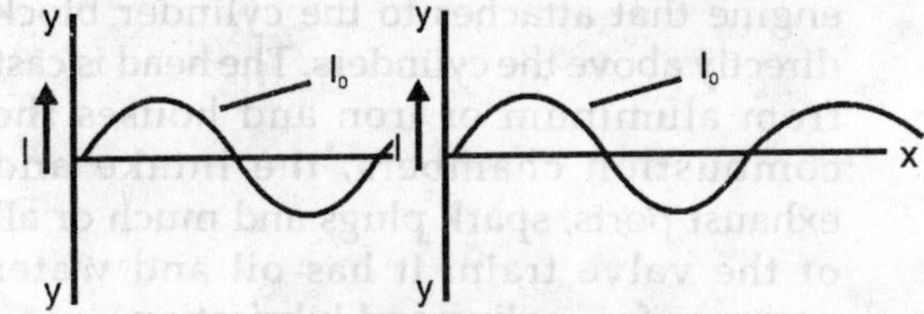

Fig. Alternating current

Current (I) The rate of transfer of electrical energy measured in amperes. (One 'international" ampere will deposit silver from a specified silver nitrate solution at the rate of 0.00111800 grams per second. An "international" ampere, in turn, is defined as 0.99985 "absolute" amperes, one coulomb per second.)

Current density The current per unit active area of the surface of an electrode.

Current fed supply A buck regulator with a choke in the primary.

Current limit The maximum current drain under which the particular battery will perform adequately under a continuous drain.

Current limit knee The point on the plot of current vs. voltage of a supply at which the current starts to foldback, or limit.

Current proportioning An output form of a temperature controller which provides a current proportional to the amount of control required. Normally is a 4 to 20 milliamp current proportioning band.

Current transformer 1. Instrument Transformer: Intended to have its primary winding connected in series with the conductor carrying the current to be measured or controlled.

2. Metering: Designed for use in the measurement or control of current. Its primary winding may be single turn or bus bar, and is connected in series with the load.

3. Power and Distribution Transformer: Intended to have its primary winding connected in series with the conductor carrying the current to be measured or controlled. (In window-type current transformers, the primary winding is provided by the line conductor and is not an integral part of the transformer.)

Curtain wall as Exterior non-loadbearing wall not wholly supported at each story. Such walls may be anchored to columns, spandrel beams, floors or bearing back-up walls, but not necessarily built between structural elements. Curtain walls must be capable of supporting their own weight for the height of the wall. In contrast, panel walls are required to be self-supporting between stories. Both walls resist lateral forces such as wind pressures and must transfer these forces to adjacent structural members.

Curve fitting Curve fitting is the process of computing the coefficients of a function to approximate the values of a given data set within that function. The approximation is called a "fit". A mathematical function, such as a least squares regression, is used to judge the accuracy of the fit.

Cut and cover A method of tunnel construction that involves digging a trench, building a tunnel, and then covering it with fill.

Cutoff Condition in a diode or bipolar junction transistor in which the potential across a p-n junction prevents current flow.

Cutoff voltage The cell or battery voltage at which the discharge is terminated. The cutoff voltage is specified by the cell manufacturer and is generally a function of discharge rate.

CV valve flow coefficient. A measurement of valve capacity.

The number of gallons per minute of room temperature water that will flow through the valve with a pressure drop of 1 psi across the valve.

Cybernetic Systems which change in response to feedback.

Cyberspace A term used to refer to the entire collection of sites accessible electronically. If your computer is attached to the Internet or another large network, it exists in cyberspace.

Cycle time The time required to complete one cycle of an operation. If cycle time for every operation in a complete process can be reduced to equal takt time, products can be made in single-piece flow.

Cyclic redundancy check (CRC) A mathematical function performed on the contents of an entity that is then included to enable a receiving system to recalculate the value and compare to the original. If the values are different, corruption of the contents has occurred.

Cyclotron A type of particle accelerator in which an ion introduced at the center is accelerated in an expanding spiral path by use of alternating electrical fields in the presence of a magnetic field.

Cylinder block The basic part of the engine to which other engine parts are attached. It is usually a casting and includes engine cylinders and the upper part of the crankcase.

Cylinder head The removable part of the engine that attaches to the cylinder block directly above the cylinders. The head is cast from aluminum or iron and houses the combustion chambers, the intake and exhaust ports, spark plugs and much or all of the valve train. It has oil and water passages for cooling and lubrication.

D code Draft code. 1. A datum in a Gerber file which acts as a command to a photoplotter . A D code in a Gerber file takes the form of a number prefixed by the letter D, e.g. "D20". However, in some aperture lists the D is dropped. In aperture lists of Cadstar, the column heading "Position" actually refers to D code, and the D prefix is dropped.

2. D codes have multiple purposes. The first is to control the state of the light being on or off. Valid codes for light state are D01, D02, and D03.

D01 - Light on for next move.

D02 - Light off for next move.

D03 - Flash (Light On, Light Off) after move (effect is limited to block in which appears, i.e. non-modal). You can also think of a D03 as D02, D01, D02 series of commands linked together.

D codes with values of 10 or greater represent the aperture's position on the list or wheel. It is very important to understand that there is no universal "D10" or "D30". Unlike the D01 , D02, and D03 counterparts which have a fixed meaning (draw , move, flash), D10 and higher values have aperture shapes and dimensions assigned to them by each individual user. Hence, one job's D10 could be a 10 mil Round, when another job's D10 could be a 45 mil Square.

There are two distinct ways to number an aperture list. The traditional 24 aperture system started with D10 - D19, jumping suddenly to D70 - D71, then back to D20 - D29, ending with D72 -D73. This is still a common format for output for CAD packages, and is still mandatory for old 24 aperture Gerber vector Photoplotters.

It is now common to start with D10, then increase numerically in steps of 1 (D10, D11, etc.) continuing up to D70 and beyond, rarely beyond 1000 individual apertures.

D Pillar or D Post The vertical or sometimes diagonal roof supporting member located at the extreme rear of the roof or greenhouse structure on station wagons and some sedan models.

D&ID Duct and Instrument Diagram - the HVAC equivalent of a P&ID.

An air distribution diagram which typically includes HVAC equipment; instrumentation; connections to other systems; duct sizes etc.

Daemon A UNIX process that operates continuously and unattended to perform a service. TCP/IP uses several daemons to establish communications processes and provide server facilities.

Damp course A course or layer of impervious material which prevents capillary entrance of moisture from the ground or a lower course. Often called damp check.

Damp proofing or dampproofing Prevention of moisture penetration by capillary action. It is achieved by application of materials or by the treatment of surfaces in order to stop (rarely) or to retard the passage of moisture. It allows evaporation from the inside of the building.

Damping The reduction of vibratory movement through dissipation of energy. Types include viscous, coulomb, and solid.

DARPA (Defense Advanced Research Project Agency, originally ARPA) The governmental body that created the ARPANET for widespread communications. ARPANET eventually became the Internet.

Data Recorded information, regardless of medium or characteristics, of any nature, including administrative, managerial, financial, and technical.

Database An structured way of storing data in an organized way, often described in terms of a number of tables each made up of a series of records, each record being made of a number of fields.

Database management system An integrated set of computer programs that provide the capabilities needed to establish, modify, make available, and maintain the integrity of a database.

Data circuit Terminating Equipment (DCE) Terminal Equipment (DTE) to a network or serial line. A modem is a DCE device. Also called Data Communications Equipment and Data Circuit Equipment.

Data item description (DID) A completed document that defines the data required of a supplier (contractor). The document specifically defines the data content, format, and intended use.

Data link The part of a node controlled by a data link protocol. It is the logical connection between two nodes.

Data link protocol (DLP) A method of handling the establishment, maintenance, and termination of a logical link between nodes. Ethernet is one example of a Data Link Protocol.

Data product Information which is inherently generated as the result of work tasks cited in a Statement of Work (SOW) or in a Source Document invoked in the contract. Such information is as a separate entity (for example, drawing, specification, manual, report, records, and parts list).

Data recorded information regardless of medium or characteristics, of any nature, including administrative, managerial, financial, and technical.

Data terminal equipment (DTE) The source or destination of data, usually attached to a network by DCE devices. A terminal or computer acting as a node on a network is usually a DTE device.

Datagram A basic unit of data used with TCP/IP.

Dative bond A dipolar bond.

Daughter products Isotopes that are formed by the radioactive decay of some other isotope. In the case of radium-226, for example, there are 10 successive daughter products, ending in the stable isotope lead-206.

DB (Decibel) 20 times the log to the base 10 of the ratio of two voltages. Every 20 dBs correspond to a voltage ratio of 10, every 10 dBs to a voltage ratio of 3.162. For instance, a CMR of 120 dB provides voltage noise rejection of 1,000,000/1. An NMR of 70 dB provides voltage noise rejection of 3,162/1.

DBM A UNIX database format.

DC (Direct current) An electric current flowing in one direction only and substantially constant in value.

DC Component The dc value of an ac wave that has an axis other than zero.

DC/DC Converter A circuit or device that changes a dc input signal value to a different dc output signal value.

Dead band 1. For chart records: the minimum change of input signal required to cause a deflection in the pen position.

2. For temperature controllers: the temperature band where heat is turned off upon rising temperature and turned on upon falling temperature expressed in degrees. The area where no heating (or cooling) takes place.

Debugging The process of tracking down errors in a program, often aided by examining or outputting extra information designed to help this process.

Debye H¸ckel equation Used in relating the activity coefficient (fi) to ion strength:where I is the ionic strength, A and B the temperature-dependent constants, Zi the valence of the ion (i), and Â the ion-size parameter in angstroms.

Debye Shielding The Debye length in front of a sensing electrode depends on the ionic strength of the electrolyte used. In a 0.001N NaCl the Debye length measures 96.5 Å, while for a 1.0 N solution it is reduced to 3.0 Å. An adsorbed protein can stick out from the surface for as much as 50 to 100 Å. As a result, the charges which could contribute to the surface potential will be shielded in a 1.0 N solution. To make more sensitive measurements a solution of low ionic strength should be used.

Decal A graphic software representation of a component, so named because hand tape-up of printed circuit boards employed the use of pull-off and paste decals to represent components. Also called a part, footprint or package . On a manufactured board the body of a footprint is an epoxy-ink outline.

Decay heat The heat produced by the decay of radioactive fission products after a reactor has been shut down.

Decay, radioactive The decrease in the amount of any radioactive material with the passage of time due to the spontaneous emission from the atomic nuclei of either alpha or beta particles, often accompanied by gamma radiation.

Decibel The numerical expression of the relative loudness of two signals, such as sound. The difference in decibels between two signals is ten times the common logarithm of the ratio of their powers.

Declared pregnant woman A woman who is an occupational radiation worker and has voluntarily informed her employer, in writing, of her pregnancy and the estimated date of conception.

Decommissioning The process of closing down a facility followed by reducing residual radioactivity to a level that permits the release of the property for unrestricted use.

Decomposition Decomposition is the process of dividing the system into its smallest, coherent, self-contained elements. Decomposition is used in systems engineering, software engineering, process mapping and functional analysis system technique.

DECON A method of decommissioning in which the equipment, structures, and portions of a facility and site containing radioactive contaminants are removed and safety buried in a low-level radioactive waste landfill or decontaminated to a level that permits the property to be released for unrestricted use shortly after cessation of operations.

Decontamination The reduction or removal of contaminating radioactive material from a structure, area, object, or person. Decontamination may be accomplished by

1. Treating the surface to remove or decrease the contamination,

2. Letting the material stand so that the radioactivity is decreased as a result of natural radioactive decay, or

3. Covering the contamination to shield or attenuate the radiation emitted .

Deep cycle battery A battery designed to be discharged to below 80% Depth of Discharge. Used in marine, traction and EV applications.

Defense communications agency (DCA) The governmental agency responsible for the Defense Data Network (DDN).

Defense data network (DDN) Refers to military networks such as MILNET and ARPANET and the communications protocols (including TCP/IP) that they employ.

Defense in depth A design and operational philosophy with regard to nuclear facilities that calls for multiple layers of protection to prevent and mitigate accidents. It includes the use of controls, multiple physical barriers to prevent release of radiation, redundant and diverse key safety functions, and emergency response measures.

Deficiencies Deficiencies consist of two types: · conditions or characteristics in any hardware or software which are not in compliance with the specified configuration identification; or · inadequate (or erroneous) configuration identification which has resulted, or may result, in CIs that do not fulfil approved operational requirements. Deficiencies are corrected by Class II methods of change control or revisions with "change bars".

Degradation A term used to describe the deteriorative processes that occur with polymeric materials, including swelling, dissolution, and chain scission.

Degree An incremental value in the temperature scale, i.e., there are 100 degrees between the ice point and the boiling point of water in the Celsius scale and 180°F between the same two points in the Fahrenheit scale.

Dendritic growth The formation from small crystals in the electrolyte of tree like structures which degrade the performance of the cell.

Denitrification The anoxic biological conversion of nitrate to nitrogen gas. It occurs naturally in surface waters low in oxygen, and it can be engineered in wastewater treatment systems.

Deoxygenation The consumption of oxygen by the different aquatic organisms as they oxidized materials in the aquatic environment.

Deoxyribonucleic acid (DNA) A huge nucleotide polymer having a double helical structure with complementary bases on two strands. Its major functions are protein synthesis and the storage and transport of genetic information.

Departure from nucleate boiling (DNB) The point at which the heat transfer from a fuel rod rapidly decreases due to the insulating effect of a steam blanket that forms on the rod surface when the temperature continues to increase.

Depleted uranium Uranium having a percentage of uranium-235 smaller than the 0.7 percent found in natural uranium. It is obtained from spent (used) fuel elements or as byproduct tails, or residues, from uranium isotope separation.

Depth of discharge DOD The ratio of the quantity of electricity or charge removed from a cell on discharge to its rated capacity.

Derivative The derivative function senses the rate of rise or fall of the system temperature and automatically adjusts the cycle time of the controller to minimize overshoot or undershoot.

Derived air concentration (DAC) The concentration of radioactive material in air and the time of exposure to that radionuclide in hours. An NRC licensee may take 2000 hours to represent one ALI, equivalent to a committed effective dose equivalent of 5 rems (0.05 sievert).

Design Those characteristics of a system or CSCI that are selected by the developer in response to the requirements, such as definitions of all error messages, others will be implementation related, such as decisions about what software units and logic to use to satisfy the requirements. For a more detailed discussion Definition of Design.

Design authority An organization responsible for the detailed design of materiel to approved specifications and authorized to sign certificates of design or certified sealed drawings in accordance with procedures identified in the Program 'Configuration Management Plan'.

Design basis accident A postulated accident that a nuclear facility must be designed and built to withstand without loss to the systems, structures, and components necessary to assure public health and safety.

Design basis phenomena Earthquakes, tornadoes, hurricanes, floods, etc., that a nuclear facility must be designed and built to withstand without loss of systems, structures, and components necessary to assure public health and safety.

Design completeness The design shall be complete from a total system element viewpoint (hardware, facilities, personnel, computer software, procedural data).

Design for Assembly. Design for Assembly (DFA) refers to the principles of designing assemblies so that they are more manufacturable. DFA principles address general part size and geometry for handling and orientation, features to facilitate insertion, assembly orientation for part insertion and fastening, fastening principles, etc. The objective of DFA is to reduce manufacturing effort and cost related to assembly processes.

Design for environment Design for Environment is a process for the systematic consideration during design of issues associated with environmental safety and health over the entire product life cycle. DFE can be thought of as the migration of traditional pollution prevention concepts upstream into the development phase of products before production and use.

Design for manufacturability/assembly Design for Manufacturability / Assembly (DFM/A) is the broad definition of optimizing a product's design for make it's parts more manufacturable (fabrication) and easier to assemble. DFM/A includes: understanding the organization's process capabilities, obtaining early manufacturing involvement, using formalized DFM/A guidelines, using DFM/A analysis tools, and addressing DFM/A as part of formal design reviews.

Design for six sigma. Design for Six Sigma (DFSS) is a systematic methodology or quality framework utilizing tools, processes and measurements to enable the design of products and processes that meet customer expectations and can be produced at Six Sigma quality levels. DFSS is built around five connected phases of Define, Measure, Analyze, Design, Verify or DMADV. The tools and methods to support DFSS include voice of the customer (VOC), quality function deployment (QFD), design for manufacturability and assembly, design of experiments (DOE), failure mode and effects analysis (FMEA), process capability studies, design reviews, control plans, etc.

Design life Is an important quantitative measure that defines the quality of the project. Buildings will not last forever. The owner and designer should establish a reasonable design life for each project. This requires consideration of the economic factors such as initial cost and maintenance costs. The design life will have an impact on the selection of materials, maintenance procedures, and the selected factors of safety.

The expected performance is also an important qualitative measure for the design of the project. The minimum performance level is set by the building code, however, there are aspects of a Structural Brick Veneer System performance that are not explicitly covered by the code and require judgment. It has been useful to define two distinct levels of expected life and performance:

Level 1 (institutional) is intended to signify a high level of quality and long life. Buildings of this type might include public or institutional buildings. Specifically, these are buildings where the additional costs associated with higher quality are judged to be necessary in meeting the overall project requirements.

Level 2 (commercial) is intended to signify a good level of quality and an average design life. Buildings of this type might include general office, industrial, and residential buildings. These are buildings where the additional cost of Level 1 (institutional) quality is not economically justified or necessary. The primary difference in design life is obtained by increasing the quality of the connectors, improving the weather resistance of the materials and expanding on the amount of inspection and testing.

Design of experiments A statistical methodology for designing, conducting and analyzing experiments or tests to evaluate product and process design parameters or factors that affect the achievement of a product performance characteristic. The response of interest is evaluated under the various conditions to: 1. Identify the influential variables among the ones tested,

2. Quantify the effects across the range represented by the levels of the variables,

3. Gain a better understanding of the nature of the causal system at work in the process, and

4. Compare the effects and interactions. These experiments lead to setting parameter or factor levels (values) that can optimize the product performance characteristic under study and minimize the affects of variation. There are several techniques including Taguchi Methods, fractional factorial, and Plackett-Burman. DOE's are often classified in one of three categories: Screening Designs, which are intended to identify which main effects (factors) are the vital few important factors that require further study (usually a fractional factorial); Characterization Designs, which are used to gain some quantitive understanding of the relationships among the factors, including interactions, on the response variable (usually a factorial); and Optimization Designs, which are used to gain a precise understanding of the mathematical relationships that is sufficient to allow prediction and optimization throughput the experimental region.

Design optimization 1. Design Optimization in the broad sense refers to optimizing a design to meet its functional, environmental and lifecycle requirements at a minimum of cost.

2. Design Optimization in a narrower sense refers to the use of computer-aided engineering applications which analyze a design and, given constraints and objectives, seek to improve or optimize the design to meet the stated objectives within the stated constraints. These applications will typically use an iterative, goal-seeking cycle to seek design optimization.

Design reviews Design reviews are formal technical reviews conducted during the development of a product to assure that the requirements, concept, product or process satisfies the requirements of that stage of development, the design is sound, the issues are understood, the risks are being managed, any problems are identified, and needed solutions proposed. Typical design reviews include: requirements review, concept/ preliminary design review, final design review, and a production readiness/launch review.

Design structure matrix Design Structure Matrix - matrix used to represent and analyze task dependencies in a product development project / process.

Design to cost A development methodology that treats cost as an independent design parameter. A realistic cost objective is established based on customer affordability, trade-off's are made between the cost objective and other product functions/ parameters, cost models are used to project the cost early in the development cycle, and a variety of techniques such as function analysis and DFM are used to proactively achieve the cost objective.

Design to life cycle cost This represents the totality of design-to-cost addressing all costs related to acquisition, operation, support and disposal.

Design to unit production costs The average unit production costs for producing the specified hardware lot(s) which generally includes recurring material, labor and overhead costs; engineering change costs; program management costs; and production support costs.

Design validation Testing to assure that the product conforms to defined user needs and requirements. This normally occurs toward the end of the Design Phase following successful design verification and prior to pilot production, beta/market testing, and product launch. Design validation is normally performed on the final product under defined, operating conditions. Multiple validations may be performed if there are different intended uses.

Design verification Design verification is the process of ensuring the design conforms to specification (design outputs meet design input requirements). Design verification may include: alternate calculations, design reviews, comparison to similar designs, inspection, and system or product testing.

Destructive testing Sectioning a portion of printed circuit panel and examining the sections with a microscope. This is performed on coupons , not the funtional part of the PCB.

Detailed design The conversion of product specifications into designs and their associated process and/or code-to documentation. Detailed design includes design capture, modeling, analysis, developmental testing, documentation, process design, producibility analysis, test plan development, coding, and design verification and validation.

Detector A material or device that is sensitive to radiation and can produce a response signal suitable for measurement or analysis. A radiation detection instrument.

Deterioration Deterioration is the length of time during which an item or element, subject to degradation or having a limited life which cannot be renewed, is considered serviceable while stored. Otherwise, deterioration implies the impairment of usefulness or value.

Determinants of gait There are six factors which limit the movement of the centre of body mass during gait. They are rotation of the pelvis about the vertical axis ; the way the pelvis tilts from side to side, so that when the hip of the stance phase leg is at its highest point, the pelvis is inclined, to lower the hip of the swing phase leg; Knee flexion and ankle movement are coordinated to adjusting the effective length of the leg during stance phase; Foot mechanism helps lengthen the leg at the end of stance phase; Lateral displacement of the body and keeping the base of gait narrow helps reduce the use of muscular energy.

Deterministic (probabilistic) Consistent with the principles of "determinism," which hold that specific causes completely and certainly determine effects of all sorts. As applied in nuclear technology, it generally deals with evaluating the safety of a nuclear power plant in terms of the consequences of a predetermined bounding subset of accident sequences. The term "probabilistic" is associated with an evaluation that explicitly accounts for the likelihood and consequences of possible accident sequences in an integrated fashion.

Deterministic effect The health effectsof radiation, the severity of which varies with the dose and for which a threshold is believed to exist. Radiation-induced cataract formation is an example of a deterministic effect (also called a non-stochastic effect).

Developer An organization that develops products ("develops" may include new development, modification, reuse, reengineering, maintenance, or any other activity that results in products) for itself or another organization.

Development strategy There are three basic development strategies and a generic 'other' which defines the variations, combinations of the other three. · Grand design - is essentially a right-first-time approach; · Incremental - identifies user needs, defines system requirements and performs the remaining software development in a sequence of builds. The first build contains part of the planned capabilities and so on until the system is complete. · Evolutionary - Development in builds but differs from incremental in that the system requirements need not be defined up front, but refined in each succeeding build.

Developmental configuration The contractor's design and associated technical documentation that defines the evolving configuration of a configuration item under development. It is under the developing contractor's configuration control and describes the design definition and implementation. The developmental configuration for a configuration item consists of the contractors released hardware and software designs and associated technical engineering documentation until establishment of the formal product baseline. (MIL-STD-973).

Deviations A specific written authorization, granted prior to the manufacture of a Configuration Item, to depart from a particular performance or design requirement of a specification, drawing or other document for a specific number of units or a specific period of time. A deviation differs from an engineering change in that an approved change requires a corresponding revision of the documentation defining, the affected item, whereas a deviation does not contemplate revision of the applicable specification or drawing.

Device Any type of electrical component on a PC board. It will have functions and properties unique to its type. In a schematic (and the extracted BOM) , it will be labeled with a value or device number. There are two main classes of devices, passive and active. .

Dialysis A phenomenon in which a semipermeable membrane allows transfer of both solvent molecules and small solute molecules and ions.

Diamagnetism The property of a substance which is repelled instead of attracted by a magnet. A diamagnetic material will be repelled from a magnet no matter what pole it is near. It is exhibited by all common materials, but is very weak and often swamped by stronger paramagnetic or ferromagnetic effects. Metals such as bismuth, copper, gold, silver and lead, as well as many nonmetals such as graphite, water and most organic compounds are diamagnetic.

Diamondoid As used in this volume, this term describes structures that resemble diamond in a broad sense: strong, stiff structures containing dense, three-dimensional networks of covalent bonds, formed chiefly from first and second row atoms with a valence of three or more. Many of the most useful diamondoid structures will in fact be rich in tetrahedrally coordinated carbon. Diamondoid is used more narrowly elsewhere in the literature.

Diaphragm The sensing element consisting of a membrane which is deformed by the pressure differential applied across it.

DICY Dicyandiamide, the most common cross-linking agent used in FR-4 . (Erik J. Bergum, "CAF Resistance of NON- DICY FR-4," PC FAB, 9/2002)

Die 1. A metal form used as a permanent mold for die casting or for a wax pattern in investment casting;

2. An integrated circuit chip as diced or cut from the finished wafer.

Dielectric A nonconductor of electricity, such as an insulator, or a substance in which an electric field can be maintained with a minimum loss of power. The material used between two conducting plates to form a capacitor.

Dielectric constant A relative measurement of the degree of polarization (shift of positive charge toward the negative electrode and negative charge toward the positive electrode) that occurs when a material is placed in an electric field.

Diesel engine A diesel engine uses heavier weight components than gas engines to handle higher compression ratios. Typically, diesel engines run with greater efficiency and higher torque than similar size gas engines. These attributes lead to better fuel economy and towing performance. Diesel engines do not have spark plugs or carburetors. Instead glow plugs are used to preheat air in the cylinders to ensure easy starts. Once the engine is started, compression heats the fuel in the cylinders for combustion.

Dieseling A condition in which gasoline continues to fire after the ignition has been shut off. In late-model engines, dieseling , or run-on, is caused by heat and the unusually high manifold pressure that result from retarding the spark at idle. In fuel-injected cars when the engine is turned off, fuel is automatically shut off, eliminating dieseling.

Differential For an on/off controller, it refers to the temperature difference between the temperature at which the controller turns heat off and the temperature at which the heat is turned back on. It is expressed in degrees.

Differential input A signal-input circuit where SIG LO and SIG HI are electrically floating with respect to ANALOG GND (METER GND, which is normally tied to DIG GND). This allows the measurement of the voltage difference between two signals tied to the same ground and provides superior common-mode noise rejection.

Differential mode noise The component of noise, excluding common-mode noise, that is measured between two lines with respect to a common reference point. The value is the difference of the noise components on the two lines.

Differential pressure The difference in static pressure between two identical pressure taps at the same elevation located in two different locations in a primary device.

Differential signaling A method of signal transmission through two wires which always have opposite states. The signal data is the polarity difference between the wires: Whenever either is high, the other is low. Neither wire is grounded.

Differential, locking The same attributes of a standard differential, except that when one wheel is slipping, the most torque is supplied to the wheel with best traction. A locking differential reduces the possibility of a vehicle becoming immobile when one driving wheel loses traction.

Diffusion coefficient (D) The constant of proportionality between the diffusion flux and the concentration gradient in Fick's first law; its magnitude is indicative of the rate of atomic diffusion.

Diffusion flux (J) The quantity of mass diffusing through and perpendicular to a unit cross-sectional area of material per unit time.

Digit A measure of the display span of a panel meter. By convention, a full digit can assume any value from 0 through 9, a 1/2-digit will display a 1 and overload at 2, a 3/4-digit will display digits up to 3 and overload at 4, etc. For example, a meter with a display span of ±3999 counts is said to be a 3-3/4 digit meter.

Digital Refers to systems employing only quantized (discrete) states to convey information.

Digital angle The angle formed between the longitudinal bisection of the digits and the longitundinal bisection of the metatarsus. The angle is normally an adductus angle. The digital angle is usually equal and opposite in direction to the forefoot angle. Any irregularity or alteration to the forefoot angle will proportionally alter the digital angle in the opposite direction.

Digital circuit A circuit which operates like a switch (it is either "on" or "off"), and can make logical decisions. It is used in computers or similar decision making equipment.

Digital output An output signal which represents the size of an input in the form of a series of discrete quantities.

DIN (Deutsche Industrial Norm) A set of German standards recognized throughout the world. The 1/8 DIN standard for panel meters specifies an outer bezel dimension of 96 x 48 mm and a panel cutout of 92 x 45 mm.

DIN 43760 The standard that defines the characteristics of a 100 ohm platinum RTD having a resistance vs. temperature curve specified by a = 0.00385 ohms per degree.

Diode 1. A device, as a two-element electron tube or a semiconductor, through which current can pass freely in only one direction. (Random House)

2. A semiconductor device with two terminals and a single junction, exhibiting varying conduction properties depending on the polarity of the applied voltage. (Graf)

DIP A type of housing for integrated circuits. The standard form is a molded plastic container of varying lengths and 0.3 inch wide (although there are other standard widths), with two rows of through-hole pins spaced 0.1 inch between centers of adjacent pins.

Dipolar bond A covalent bond in which one atom supplies both bonding electrons, and the other atom supplies an empty orbital in which to share them. Also termed a dative bond.

Direction of motion Most joints have at least two directions and some in more. The three factors which determine the direction of joint motion are the type of anatomical joint ; the axial position; and the equity of movement from the neutral position.

Directory In most computer file systems files are grouped into a hierarchical tree structure with a number of files in each directory (the files are like the leaves and the directories are like the branches of this tree).

Disc brakes Properly called caliper disc brakes, a type of brake that consists of a rotor that rotates at wheel speed, straddled by a caliper that can squeeze the surfaces of the rotor with brake pads near its edge. Disc brakes provide a more linear response and operate more efficiently at high temperatures and during wet weather than drum brakes.

Discharge capacity The amount of energy taken from the battery when discharged at the rated current and ambient temperature until the discharge end voltage is reached. Generally expressed in units of Watt hours (or Ampere hours for batteries with the same voltage).

Discharge time constant The time required for the output-voltage from a sensor or system to discharge 37% of its original value in response to a zero rise time step function input. This parameter determines a low frequency response.

Discharge tube 1. A tube biased to cut-off, and therefore non-conducting except when triggered by a positive pulse A condenser, connected in the plate circuit of the tube, charges when the tube is non-conducting and discharges when the tube is triggered and forced into conduction. The tube may be either vacuum or gaseous.

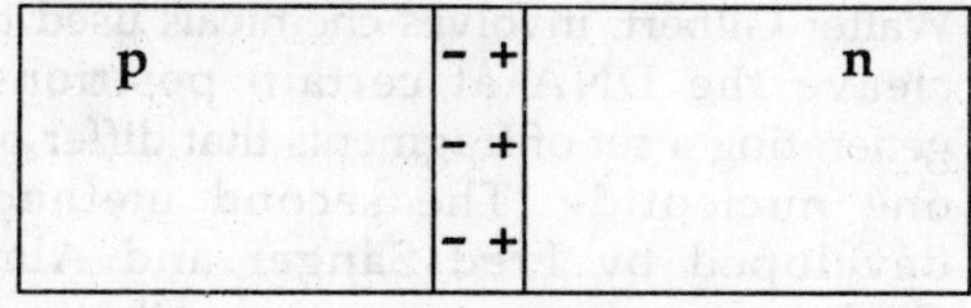

Fig. Diode

2. Any tube containing a gas or vapor at low pressure and capable of showing a gaseous discharge.

Discharge voltage The voltage between the terminals of a cell or battery under load, during discharge.

Discrete settling Settling in which individual particles settle independently, neither agglomerating or interfering with the settling of the other particles present. This occurs in waters with a low concentration of particles.

Dislocation A linear crystalline defect around which there is atomic misalignment. Plastic deformation corresponds to the motion of dislocations in response to an applied shear stress. Edge, screw, and mixed dislocations are possible.

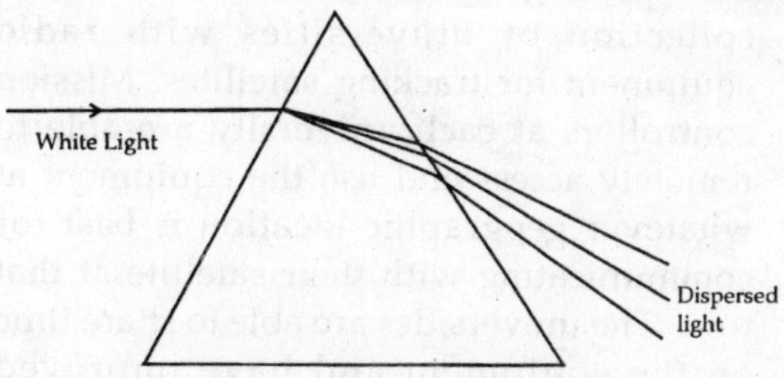

Fig. Dispersion of light beam

Dislocation line The line that extends along the end of the extra half-plane of atoms for an edge dislocation, and along the center of the spiral of a screw dislocation.

Displacement In an engine, the total volume of air or air-fuel mixture an engine is theoretically capable of drawing into all cylinders during one operating cycle. Generally expressed in liters or cubic inches. Engine displacement is equal to (bore) x (bore) x (stroke) x (number of pistons) x (.785).

Dissociation constant For systems in which ligands of a particular kind bind to a receptor in a solvent there will be a characteristic frequency with which existing ligand-receptor complexes dissociate as a result of thermal excitation, and a characteristic frequency with which empty receptors bind ligands as a result of Brownian encounters, forming new complexes. The frequency of binding is proportional to the concentration of the ligand in solution. The dissociation constant is the magnitude of the ligand concentration at which the probability that the receptor will be found occupied is 1/2.

Dissolved oxygen (DO) The amount of molecular oxygen dissolved in water.

Distributed computing environment (DCE) A set of technologies developed by the Open Software Foundation (0SF) supporting distributed computing.

Distributed file service (DFS) An Open Software Foundation (OSF) fileserver technology sometimes used with TCP/IP.

Distributed ground station network A collection of universities with radio equipment for tracking satellites. Mission controllers at each university are able to remotely access and use the equipment at whatever geographic location is best for communicating with their satellite at that time. The universities are able to share time on the equipment and have improved capabilities then any could have individually.

Distributed management environment (DME) A system and network management technology developed by the Open Software Foundation (OSF).

Distributed processing When a process is spread over two or more devices, it is distributed. It is usually used to spread CPU loads among a network of machines.

Distribution statement A statement used in marking a technical document to denote the extent of its availability for distribution release, and disclosure without the need for additional approvals and authorizations from the controlling agency.

Distributor A component of the ignition system, usually driven by the camshaft that directs high-voltage surges to the spark plugs in the proper sequence.

Diversion channel A bypass created to divert water around a dam so that construction can take place.

DMA Acronym direct memory access. A high speed data storage mode of the IBM PC.

DNA Probes A DNA or nucleic acid probe is a short strand of DNA that locates and binds to its complementary sequence in samples containing single strands of DNA or RNA enabling identification of specific sequences. Nucleic acid probe assays exploit the fundamental hybridization reaction that occurs spontaneously between two complementary DNA:DNA or DNA:RNA strands. As in immunoassays, detection of the hybrid requires that the probe be labeled. Various direct and indirect methods have been devised for the detection of the hybrid. Direct labeling involves attaching the label directly to the probe sequence; indirect labeling binds an antibody to the DNA:DNA or DNA:RNA hybrid. As in immunoassays, non-isotopically-labeled probes are preferred over radio-labeled probes primarily because of radiation hazards, disposal problems, and short reagent shelf life. In addition, the factors determining the detection limits of hybridization assays based on labeled probes are similar to those in immunoassays. Therefore, the development of a simple, inexpensive and sensitive direct detection system which eliminates the use of labels is highly desirable.

DNA Sequencing There are two main classical methods for sequencing DNA: The first method, developed by Allan Maxam and Walter Gilbert, involves chemicals used to cleave the DNA at certain positions, generating a set of fragments that differ by one nucleotide. The second method, developed by Fred Sanger and Alan Coulson, involves enzymatic synthesis of DNA strands that terminate in a modified nucleotide. Analysis of fragments is similar for both methods and involves gel electrophoresis and autoradiography or fluorescence. The enzymatic method has largely replaced the chemical method as the technique of choice, although there are some situations where chemical sequencing can provide data more easily than the enzymatic method.

DOL Direct On Line - A type of motor starter for induction motors.

This is the simplest form of starter comprising a switch (typically an electromagnetic contactor) and an overload protection relay.

Domain name system (DNS) A service that converts symbolic node names to IP addresses. DNS is frequently used with TCP/IP. DNS uses a distributed database.

Dome A curved roof enclosing a circular space; a three-dimensional arch.

Doping The intentional alloying of semiconducting materials with controlled concentrations of donor or acceptor impurities.

Doppler coefficient Another name used for the fuel temperature coefficient of reactivity.

Doppler effect (CJ Doppler) Waves emitted by a moving object as received by an observer will be blueshifted (compressed) if approaching, redshifted (elongated) if receding. It occurs both in sound as well as electromagnetic phenomena, although it takes on different forms in each.

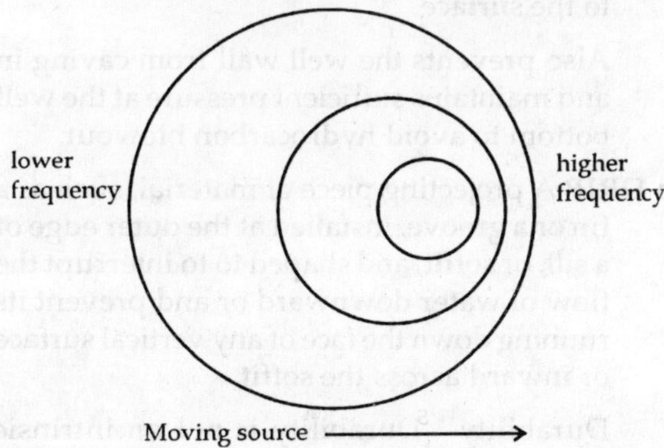

Fig. Doppler effect

Dorsiflexed A foot, or any part of a foot, is in a dorsiflexed position when the distal aspect of the foot or part is angulated in a sagittal plane toward the tibia and away from a transverse plane passing through the proximal aspect of the foot or part, or away from some other specific reference point.

DOS Disk Operating System. A program that controls the computer's transfer of data to and from a hard or floppy disk. Personal computers that are IBM-compatible run DOS rather than other early varieties of operating systems.

Dose The absorbed dose, given in rads (or in SI units, grays), that represents the energy absorbed from the radiation in a gram of any material. Furthermore, the biological dose or dose equivalent, given in rem or sieverts, is a measure of the biological damage to living tissue from radiation exposure.

Dose, absorbed The amount of energy deposited in any substance by ionizing radiation per unit mass of the substance. It is expressed numerically in rads or grays.

Dose equivalent The product of absorbed dose in tissue multiplied by a quality factor and then sometimes multiplied by other necessary modifying factors at the location of interest. It is expressed numerically in rems or sieverts .

Dose rate The ionizing radiation dose delivered per unit time. For example, rem or sieverts per hour.

Dosimeter A small portable instrument (such as a film badge or thermoluminescent or pocket dosimeter) for measuring and recording the total accumulated personal dose of ionizing radiation.

Dot address A unique number assigned to identify a host on the Internet (also called IP address or dot address). This address is usually represented as four numbers between 1 and 254 and separated by periods, for example, 192.58.107.230.

Dotted decimal notation A representation of IP addresses. Also called "dotted quad notation" because it uses four sets of numbers separated by decimals (for example, 255.255.255.255).

Double bond Two atoms sharing electrons as in a single bond (that is, a sigma bond) may also share electrons in an orbital with a node passing through the two atoms. This adds a second, weaker bonding interaction (a pi bond); the combination is termed a double bond. A twisting motion that forces the nodal plane at one atom to become perpendicular to the nodal plane on the other atom eliminates the (signed) overlap between the atomic orbitals, destroying the pi bond. The energy required to do this creates a large barrier to rotation about the bond.

Double byte character set A character set where alphanumeric characters are represented by two bytes, instead of one byte as with ASCII. Double byte characters are often necessary for Asian languages, which have more than 255 symbols.

Double peak curves During normal walking, within the stance phase there are two peaks of shock, where ground reaction exceeds body action. Graphic representation of force versus time illustrates a twin peak. Peak forces occur just after heel strike and again, but to a lesser extend, at heel off.

Double precision The degree of accuracy that requires two computer words to represent a number. Numbers are stored with 17 digits of accuracy and printed with up to 16 digits.

Double wishbone suspension ("A" Arm Suspension) A system of independent suspension in which each wheel is located on a "knuckle" that is connected by ball joints to an upper A arm and a lower A arm. Usually, the lower A arms are longer. This system provides minimal changes in track and camber when the suspension is under load, as when going over bumps or in hard cornering.

Doublet The electronic state of a molecule having one unpaired spin is termed a doublet. This term is derived from spectroscopy: an unpaired spin can be either up or down with respect to a magnetic field, and these states have different energy, resulting in field-dependent pairs, or doublets, of spectral lines.

Downstream Generic term that includes oil refining; petrochemicals; synthetic gas and fertilizer production facilities.

Downshore activities are oshore based.

Drag coefficient A measure of the aerodynamic sleekness of an object. Drag coefficient is signified by "dc.: The lower the number, the greater the aerodynamic efficiency. The higher the drag coefficient, the more a car's engine must work to keep a given road speed. Also known as "CD" for coefficient of drag."

Drift A change in output over a period of time independent of input, environment or load. For power supplies, the change in DC output as a function of time at constant line voltage, load, and ambient temperature. Normally specified for an eight hour period after a half hour warm-up.

Drilling mud Mixture of water and special additives circulating within the well for the purpose of cooling the drill-bit; removing rock cuttings and transporting them back up to the surface.

Also prevents the well wall from caving in and maintains sufficient pressure at the well bottom to avoid hydrocarbon blowout.

DRIP A projecting piece of material, such as a fin or a groove, installed at the outer edge of a sill, or soffit, and shaped to to interrupt the flow of water downward or and prevent its running down the face of any vertical surface or inward across the soffit.

Durability Durability is not an intrinsic property of a material but depends largely on how a material reacts to a specific environment such as moisture, temperature, ultra-violet radiation, and to the presence of other materials (incompatibility). A material also often sees two different environments during its service life, one during construction and another after the building has been completed. The environment the air barrier sees only briefly during construction may nonetheless be detrimental to the long term performance of some of the materials that compose it. These materials should always therefore be adequately protected from rain, heat, ultra-violet radiation, cold, and mechanical damage during construction.

Drivetrain The power-transmitting components in a car, including clutch, gearbox (or automatic transmission), driveshaft, universal joints, differential and axle shafts.

Droop A common occurrence in time-proportional controllers. It refers to the difference in temperature between the set point and where the system temperature actually stabilizes due to the time-proportioning action of the controller.

Drop cable In Ethernet networks it refers to the cable connecting the device to the network, sometimes through a transceiver.

Dropout In a voltage regulator, the lower limit of the AC input voltage where the power supply just begins to experience insufficient input to maintain regulation. The dropout voltage for linears is quite load dependent. For most switchers it is largely design dependent, and to a smaller degree load dependent.

Dry film solder mask A solder mask film applied to a printed board with photographic methods. This method can manage the higher resolution required for fine line design and surface mount. It is more expensive than liquid photoimageable solder mask.

Drywell The containment structure enclosing a boiling water reactor vessel and its recirculation system. The drywell provides both a pressure suppression system and a fission product barrier under accident conditions.

DT/dt The rate of change of temperature with time. The rapid rate of temperature rise is used to detect the end of the charging cycle in NiMH batteries.

Dual overhead camshafts (DOHC) A DOHC engine has two camshafts in each cylinder head; one camshaft actuates intake valves and the other actuates exhaust valves. The camshafts act directly on the valves, eliminating pushrods and rocker arms. This reduced reciprocating mass of the valve train enables the engine to build RPM more quickly. DOHC designs are typically high-performance, four valve per cylinder engines. (A four valve per cylinder two intake and two exhaust design helps the engine "breathe" more freely for increased performance.)

Dual slope A/D Converter An analog-to-digital converter which integrates the signal for a specific time, then counts time intervals for a reference voltage to bring the integrated signal back to zero. Such converters provide high resolution at low cost, excellent normal-mode noise rejection, and minimal dependence on circuit elements.

Ductile to brittle transition The transition from ductile to brittle behavior with a decrease in temperature exhibited by BCC alloys; the temperature range over which the transition occurs is determined by Charpy and Izod impact tests.

Ductility A measure of a material's ability to undergo appreciable plastic deformation before fracture; it may be expressed as percent elongation or percent area reduction from a tensile test.

Dumb terminal A terminal with no significant processing capability of its own, usually with no graphics capabilities beyond the ASCII set.

Dump An illegal and uncontrolled area where wastes have been placed on or in the ground.

Duplex Name given to a family of stainless steels which have a near equal mix of austenite and ferrite.

Duplex stainless steels exhibit high strength and excellent corrosion resistance.

Duplex wire A pair of wires insulated from each other and with an outer jacket of insulation around the inner insulated pair.

Duration The time interval between the first and last instants at which the instantaneous amplitude reaches a stated fraction of the peak pulse amplitude.

DUT Device Under Test. A DUT board (probe card) is used in automated testing of integrated circuits. It is part of the interface between the chip and a test head, which in turn attaches to computerized test equipment. The specific test equipment used will determine the value of the controlled impedance required for the chip tester boards. Depending on which system it is designed for, one type of DUT board is used in testing individual integrated circuits in a silicon wafer before they are cut free and packaged, and another type is used for testing packaged IC's.

Duty cycle 1. The ratio of time on to time off in a recurring event.

2. The operating regime of a cell or battery including factors such as charge and discharge rates, depth of discharge, cycle length a length of time in the standby mode.

Dynamic (Two-Plane) Balancing Machine A dynamic balancing machine is a centrifugal balancing machine that furnishes information for performing two-plane balancing.

Dynamic calibration Calibration in which the input varies over a specific length of time and the output is recorded vs. time.

Dynamic characteristics A description of an instrument's behavior between the time a measured quantity changes value and the time the instrument obtains a steady response.

Dynamic error The error that occurs when the output does not precisely follow the transient response of the measured quantity.

Dynamic load A load that rapidly changes from one level to another. To be properly specified, both the total change and the rate of change must be stated.

Dynamic pressure The difference in pressure levels from static pressure to stagnation pressure caused by an increase in velocity. Dynamic pressure increases by the square of the velocity.

Dynamic range The ratio of the largest to the smallest values of a range, often expressed in decibels.

Dynamic unbalance Dynamic unbalance is that condition in which the central principal axis is not coincident with the shaft axis.

E pad "Engineering-pad." A plated-through hole or surface mount pad on a PCB placed on the board for the purpose of attaching a wire by soldering. These are usually labeled with silkscreen. E-pads are used to facilitate proto-typing, or simply because wires are used for interconnections instead of headers or terminal blocks .

E Rate Discharge or charge power, in watts, expressed as a multiple of the rated capacity of a cell or battery which is expressed in watt-hours. For example, the E/10 rate for a cell or battery rated at 23.4 watt-hours is 2.34 watts. (This is similar to the method for calculating C-Rate.)

E&C Engineering & Construction. Usually in reference to a contract e.g. an E&C contract.

EA Environment Agency. The leading public organisation for protecting and improving the environment in England and Wales.

Earthquake, operating basis An earthquake that could be expected to affect the reactor plant site, but for which the plant power production equipment is designed to remain functional without undue risk to public health and safety.

Eccentric contraction Muscles are known to exhibit two forms of control i.e. eccentric contraction where muscles resist elongation and concentric contraction where they undergo shortening.

Echo To reflect received data to the sender. For example, keys depressed on a keyboard are usually echoed as characters displayed on the screen.

ECL Emitter Coupled Logic. A type of unsaturated logic performed by emitter-coupled transistors. Higher speeds may be achieved with ECL than are obtainable with standard logic circuits. ECL is costly, power hungry, and difficult to use, but it is four times faster than TTL. (Graf)

Ecosystem An organism or group of organisms and their surroundings. The boundary of an ecosystem may be arbitrarily chosen to suit the area of interest or study.

ED Explosive Decompression. Phenomenon whereby seals of a valve (or similar) absorb high pressure gas and when the system is depressurised the seals release their pressure causing failure.

Edge dislocation A linear crystalline defect associated with the lattice distortion produced in the vicinity of the end of an extra half-plane of atoms within a crystal; the Burgers vector is perpendicular to the dislocation line.

Effective gid The group identifier of the current process, which may have been changed from the original GID by various means.

Effective half life The time required for the amount of a radioactive element deposited in a living organism to be diminished 50 percent as a result of the combined action of radioactive decay and biological elimination.

Effective mass In a vibrating system, a particular vibrational mode can be described as a harmonic oscillator with some mass and stiffness. Given some measure of vibrational amplitude, there exists a unique choice of mass and stiffness that yields the correct values for both frequency and energy; these are the effective mass and effective stiffness.

Effective uid The user identifier of the current process, which may have been changed from the original UID by various means.

Efficiency 1. The ratio of total output power to total input power, expressed as a percentage, under specified conditions. For power supplies efficiency is generally measured at full load with nominal line conditions.

2. The ratio of the output of a secondary cell or battery on discharge to the input required to restore it to the initial state of charge under specified conditions.

Efficiency, plant The percentage of the total energy content of a power plant's fuel that is converted into electricity. The remaining energy is lost to the environment as heat.

Efflorescence A powder or stain sometimes found on the surface of masonry, Because mortar must be wet to render it plastic for bricklaying, brickwork becomes damp in the course of construction. As it subsequently dries, the moisture, in which the various salts derived from mortar and bricks have been dissolved, moves to the wall surface to evaporate, leaving a deposit of salts, usually as a white coating on the bricks resulting from deposition of water-soluble salts.

Effluent The fluid exiting a system, process, tank, etc. An effluent from one process can be an influent to another process.

Effluent based standards Standards which set concentration or mass per time limits on the effluent being discharged to a receiving water.

Egrep A UNIX pattern matching utility that finds matching patterns in text files.

Elastic An object behaves elastically if it returns to its original shape after a force is applied and then removed. (If an applied force causes a permanent deformation, the behavior is termed plastic.) In an elastic system, the internal potential energy is a function of shape alone, independent of past forces and deformations.

Elastomer Elastic or plastic materials that resemble rubber which resume their original shape when a deforming force is removed.

Electret The electrostatic equivalent of the permanent magnet. Dielectric materials that have been permanently electrically charged or polarised.

Electric field (V/m) In simplest form, the potential difference between two points divided by the distance between the two.

Electrical & electronics engineering Electrical and electronics engineering is the practical application of electricity. Electrical engineers are concerned with electrical devices and systems and with the use of electrical energy.

Electrical breakdown Condition in which, particularly with high electric field, a nominal insulator becomes electrically conducting.

Electrical engineer An engineer concerned with electrical devices and systems and with the use of electrical energy.

Electrical generator An electromagnetic device that converts mechanical (rotational) energy into electrical energy. Most large electrical generators are driven by steam or water turbine systems.

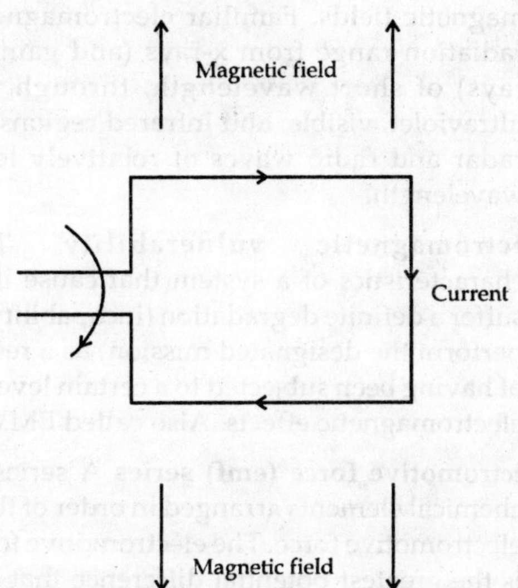

Fig. Electric motor/electric generator

Electrical object (Protel) A graphical object (in a PCB or schematic database) to which an electrical connection can be made, such as a component pin or a wire.

Electricity Property of fundamental particles of matter that have a force field associated with them to gain or lose electrons.

Electrode These are used to record the electrical activity of contracting muscles. There are several types used to collect electromyogrphic data e.g. surface electrodes, fine wire and needle electrodes.

Electrode potential (E) The difference in potential established between an electrode and a solution when the electrode is immersed in the solution.

Electrodeposition 1. The deposition of a conductive material from a plating solution by the application of electrical current;

2. The deposition of a substance on an electrode by passing electric current through an electrolyte; electroplating, electroforming, electrorefining, and electrotwinning result from electrodeposition.

Electrogoniometers This is a device for measuring the angle of a joint during walking. There are several types of electrogoniometers including the potentiometer, the flexible strain gauge, and the resistance of a rubber tube filled with mercury. An associated device, though not strictly an electrogoniometer, is the polarized light goniometer.

Electroluminescence In electrical engineering: the emission of visible light by a p-n junction across which a forward-biased voltage is applied. In electrochemistry: emission of light by a molecule which is being reduced or oxidized on a biased electrode. If the exciting cause is a photon, rather than an electron, the process is called photoluminescence.

Electrolysis Chemical modifications, oxidation and reduction produced by passing an electric current through an electrolyte.

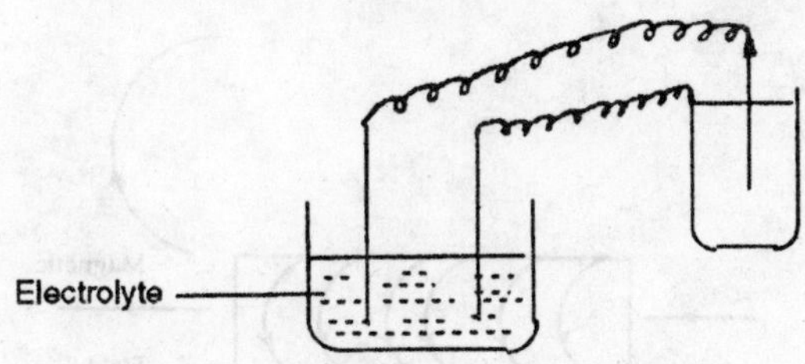

Fig. Electrolysis

Electrolyte A substance which dissociates into ions (charged particles) when in aqueous solution or molten form and is thus able to conduct electricity. It is the medium which transports the ions carrying the charge between the electrodes during the electrochemical reaction in a battery.

Electrolyte/Insulator/Silicon (EIS) Structures at the heart of a broad family of potentiometric silicon sensors. The best-known member of the family is the ion-sensitive field effect transistor, known as the ISFET or CHEMFET and the light-addressable potentiometric sensor LAPS . The principle of operation of devices using such structures is as follows. A potential with respect to a reference electrode is generated at the interface between the liquid solution and the insulator. The surface potential (ø0) is determined by that ionic species which has the fastest exchange rate (io) with the membrane covering the insulator. If no intentional membrane is deposited on an oxide covered insulator that species will be H+. Surface potential changes in turn change the Si flat-band voltage VFB. The flat-band voltage is the potential one needs to apply to the Si in order to have the bands flat throughout the semiconductor. For a given EIS sensor, these inaccurately known quantities are constant, and variations in flat-band voltage can be equated to variations of the surface potential.

Electromagnet A device consisting of a ferromagnetic core and a coil that produces appreciable magnetic effects only when an electric current exists in the coil.

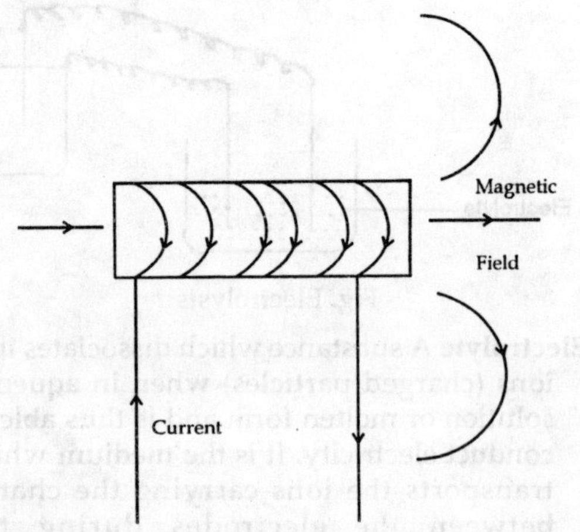

Fig. Electromagnet

Electromagnetic interference (EMI) Any electronic disturbance that interrupts, obstructs, or otherwise impairs the performance of electronic equipment. For power supplies, unwanted energy, generally emitted from switching power supplies, which may be conducted or radiated.

Electromagnetic radiation A traveling wave motion resulting from changing electric or magnetic fields. Familiar electromagnetic radiation range from x-rays (and gamma rays) of short wavelength, through the ultraviolet, visible, and infrared regions, to radar and radio waves of relatively long wavelength.

Electromagnetic vulnerability The characteristics of a system that cause it to suffer a definite degradation (incapability to perform the designated mission) as a result of having been subjected to a certain level of electromagnetic effects. Also called EMV

Electromotive force (emf) series A series of chemical elements arranged in order of their electromotive force. The electromotive force is the greatest potential difference that can be generated by a particular source of electric current. In practice this potential may be observable only when the source is not supplying current, because of its internal resistance.

Electromyography This is the measurement of the electrical activity of a contracting muscle i.e. the muscle action potential. The electromyogram can measure electrical activity of muscle contraction but not mechanical activity. It cannot be used, therefore, to distinguish between concentric, isometric and eccentric contractions. The relationship between EMG activity and the force of contraction remains unclear.

Electron An elementary particle with a negative charge and a mass 1/1837 that of the proton. Electrons surround the positively charged nucleus and determine the chemical properties of the atom.

Electron density The location of an electron is not fixed, but is instead described by a probability density function. The sum of the probability densities of all the electrons in a region is the electron density in that region.

Electron state (level) One of a set of discrete, quantized energies that are allowed for electrons. In the atomic case each state is specified by 4 quantum numbers.

Electronegative Describing elements that tend to gain electrons and form negative ions. The halogens are typical electronegative elements.

Electronegativity A measure of the tendency of an atom (or moiety) to withdraw electrons from structures to which it is bonded. In most circumstances, for example, sodium tends to donate electron density (it has a low electronegativity) and fluorine tends to withdraw electron density (it has a high electronegativity).

Electronic Pertaining to the energies, distributions, and behaviors of electrons;

Electronic fuel injection system A system that injects fuel into the engine and includes an electronic control unit to time and meter the fuel flow.

Electronic ignition system An ignition system that uses transistors and other semiconductor devices as an electronic switch to turn the primary current on and off.

Electronic industries association (EIA) A standards organization specializing in the electrical and functional characteristics of interface equipment.

Electrostatic discharge (ESD) The flow of current that results when objects having a static charge come into a close enough proximity to discharge.

Electrostatic precipitator A device which uses an electric field to trap particulate pollutants.

Element One of the 103 known chemical substances that cannot be broken down further without changing its chemical properties. Some examples include hydrogen, nitrogen, gold, lead, and uranium.

E-mail address An address used to send e-mail to a user on the Internet, consisting of the user name and host name (and any other necessary information, such as a gateway machine). An Internet e-mail address is usually of the form username@hostname.

Embankment dam A dam composed of a mound of earth and rock; the simplest type of gravity dam.

Embedded (Of a micro-processor(s), or system controlled by such) Dedicated to doing one job or supporting one device and built into the product.

Embedded system A special-purpose computer system, which is completely encapsulated within the device it controls, usually performing a limited range of specific pre-determined tasks. This allows the use of simpler or cheaper dedicated microprocessors providing only the minimum functionality required by the application, or alternatively the entire processing power of the microprocessor can be focused on a single task. Battery Management Systems will normally be implemented with an embedded system.

EMC Electromagnetic compatibility (EMC) is the ability of electronic and electrical equipment and systems to operate without adversely affecting other electrical or electronic equipment or being affected by other sources of electromagnetic interference.

Emergency classifications Response by an offsite organization is required to protect local citizens near the site. A request for assistance from offsite emergency response organizations may be required.

Emergency core cooling systems (ECCS) Reactor system components (pumps, valves, heat exchangers, tanks, and piping) that are specifically designed to remove residual heat from the reactor fuel rods should the normal core cooling system (reactor coolant system) fail.

EMI Filter A circuit composed of reactive and resistive components for the attenuation of radio frequency components being emitted from a power supply. **Emissivity** The ratio of energy emitted by an object to the energy emitted by a blackbody at the same temperature. The emissivity of an object depends upon its material and surface texture; a polished metal surface can have an emissivity around 0.2 and a piece of wood can have an emissivity around 0.95.

Emitter An electrode on a transistor from which a flow of electrons or holes enters the region between the electrodes. (Random House).

Emoticon An ASCII drawing such as :-) (look at it sideways) used to help indicate an emotion in a message. Also called emoticon.

Empowerment Act by which an employee, contractor or consultant is given the necessary freedom to make full use of his knowledge, energies and judgement to provide better service. It necessarily involves freedom of and responsibility for decision-making.

Emulation A program that simulates another device. For example, a 3270 emulator emulates an IBM 3270 terminal, sending the same codes as the real device would.

Encapsulation Including an incoming message into a larger message by adding information at the front, back, or both. Encapsulation is used by layered network protocols. With each layer, new headers and trailers are added.

Enclosure A housing for an electronic device designed to separate and protect both the internal componentry and the outside environment.

Encryption The process of scrambling a message so that it can be read only by someone who knows how to unscramble it.

End item A deliverable item which is formally accepted by the acquirer in accordance with requirements of a detail specification.

End point (Potentiometric) The apparent equivalence point of a titration at which a relatively large potential change is observed.

End to end design A version of CADCAM CAE in which the software packages used and their inputs and outputs are integrated with each other and allow design to flow smoothly with no manual intervention necessary (other than a few keystrokes or menu selections) to get from one step to the other. Flow can occur in both directions. In the field of PCB design, end-to-end design sometimes refers to only the electronic schematic/pcb layout interface, but this is a narrow view of the potentialities of the concept. For example, end-to-end systems can also implement electronic circuit simulation, parts procurement and beyond.

Endoergic A transformation is termed endoergic if it absorbs energy; such a reaction increases molecular potential energy. (Sometimes wrongly equated to the narrower term endothermic.)

Endothermic A transformation is termed endothermic if it absorbs energy in the form of heat. A typical endothermic reaction increases both entropy and molecular potential energy (and is thus analogous to a gas expanding while absorbing heat and compressing a spring).

Energy A conserved quantity that can be interconverted among many forms, including kinetic energy, potential energy, and electromagnetic energy. Sometimes defined as "the capacity to do work," but in an environment at a uniform nonzero temperature, thermal energy does not provide this capacity. (Note, however, that all energy has mass, and thus can be used to do work by virtue of its gravitational potential energy; this caveat, however, is of no practical significance unless a really deep gravity well is available.)

Energy consumption This is an attempt to measure the energy expenditure and is most commonly based on measuring the body's oxygen consumption. There are less direct methods, using either mechanical calculations or the measurement of the heart rate.

Energy content The absolute amount of energy stored in a battery expressed in Wh or Joules

Energy density The amount of energy stored in a battery. It is expressed as the amount of energy stored per unit volume or per unit weight (Wh/L or Wh/kg).

Engineered wood Relatively new term that means just what it says: wood products that are engineered. Plywood is considered to be the original engineered wood product and has been used for structural applications since the 1940s. Since then, other products have been developed which fit into the engineered wood family: glulam, OSB, and wood I-joists are some that are represented by APA and EWS.

Glued engineered wood products, including plywood, oriented strand board (OSB), glued laminated timber (glulam), wood I-joists and structural composite lumber, are manufactured by bonding together wood strands, veneers, lumber or other wood fiber to form a larger, more efficient composite structural unit. These products are extremely resource efficient because they utilize more of the available resource with minimal waste. In addition, in many cases, they are produced using faster growing and often underutilized wood species from managed forests and tree farms, thus reducing the industry's reliance on so-called "old growth" forests.

At the same time, glued engineered wood products meet marketplace needs for products with superior and consistent performance characteristics. For example, glulam beams and wood I-joists, "engineered" to exacting standards, can carry greater loads over longer spans than is possible with an equivalent size of solid sawn wood. Definitions vary, but engineered wood is generally understood to include a wide range of products manufactured by bonding together wood strands, veneers, lumber or other forms of wood fiber to produce a larger and integral composite unit.

Structural engineered wood products are "engineered" by virtue of possessing design values that are confirmed by methods other than simple visual grading.

APA divides structural engineered wood products into four general categories: 1. Wood structural panels, including plywood, oriented strand board, and composite panels;

2. Glued laminated timber (glulam);

3. Structural composite lumber (SCL), including primarily laminated veneer lumber (LVL) but also parallel strand lumber and oriented strand lumber; and

4. Wood I-joists.

A number of factors have contributed to what is emerging as a new era for engineered wood products, including changing resource supplies, manufacturing innovations, environmental considerations, superior performance, and industry standardization. All of these considerations suggest that engineered wood products will occupy an increasingly important place in the building materials market of the future.

Engineering A profession in which a knowledge of math and natural science is applied to develop ways to utilize the materials and forces of nature for the benefit of all human beings.

Engineering decision traceability Significant engineering decisions shall be traceable to the system engineering process activities on which they were based.

Engineering management The management of the engineering and technical effort to transform a conceptual requirement into an operational system. It includes the system performance parameters and preferred system configuration to satisfy the requirement, the planning and control of technical program tasks, integration of the engineering specialties, and the management of a totally integrated effort of design engineering, computer software engineering, test engineering, safety engineering, security engineering, logistics engineering, production engineering, and specialty engineering (EMC, environmental, etc.,) to meet cost, technical performance and schedule objectives.

Engineering specialty integration The timely and appropriate intermeshing of engineering efforts and disciplines such as reliability, maintainability, logistics engineering, human factors, safety, value engineering, computer software, standardization, transportability, etc., to insure their influence on system design.

Enthalpy The enthalpy of a system is its actual energy (termed the internal energy) plus the product of its volume and the external pressure. Though sometimes termed "heat content," the enthalpy in fact includes energy not contained in the system. Enthalpy proves convenient for describing processes in gases and liquids in laboratory environments, if one does not wish to account explicitly for energy stored in the atmosphere by work done when a system expands. It is of little use, however, in describing processes in nanomechanical systems, where work can take many forms: internal energy is then more convenient. Enthalpy is to energy what the Gibbs free energy is to the Helmholtz free energy.

ENTOMB A method of decommissioning in which radioactive contaminants are encased in a structurally long-lived material, such as concrete. The entombment structure is appropriately maintained and continued surveillance is carried out until the radioactivity decays to a level permitting decommissioning and ultimate unrestricted release of the property.

Entropy A measure of uncertainty regarding the state of a system: for example, a gas molecule at an unknown location in a large volume has a higher entropy than one known to be confined to a smaller volume. Free energy can be extracted in converting a low-entropy state to a high-entropy state: the (time-average) pressure exerted by a gas molecule can do useful work as a small volume is expanded to a larger volume. In the classical configuration space picture, any molecular system can be viewed as a single-particle gas in a high-dimensional space. In the quantum mechanical picture, entropy is described as a function of the probabilities of occupancy of different members of a set of alternative quantum states. Increased information regarding the state of a system reduces its entropy and thereby increases its free energy, as shown by the resulting ability to extract more work from it.

An illustrative contradiction in the simple textbook view of entropy as a local property of a material (defining an entropy per mole, and so forth) can be shown as follows: The third law of thermodynamics states that a perfect crystal at absolute zero has zero entropy*; this is true regardless of its size. A piece of disordered material, such as a glass, has some finite entropy G0 > 0 at absolute zero. In the local-property view, N pieces of glass, even (or especially) if all are atomically identical, must have an entropy of NG0. If these N pieces of glass are arranged in a regular three-dimensional lattice, however, the resulting structure constitutes a perfect crystal (with a large unit cell); at absolute zero, the third law states that this crystal has zero entropy, not NG0. To understand the informational perspective on entropy, it is a useful exercise to consider 1. What the actual entropy of such crystal is as a function of N, with and without information describing the structure of the unit cell,

2. How the third law can be phrased more precisely, and

3. What this more precise statement implies for the entropy of well-defined aperiodic structures. Note that any one unit cell in the crystal can be regarded as a description of all the rest.

Environmental audit An independent assessment of the current status of a party's compliance with applicable environmental requirements or of a party's environmental compliance policies, practices and controls.

Environmental conditions All conditions in which a transducer may be exposed during shipping, storage, handling, and operation.

Environmental engineering Environmental engineering is the development of processes and infrastructure for the supply of water, the disposal of waste, and the control of all kinds of pollution. The work of environmental engineers protects public health by preventing disease transmission and preserves the quality of the environment by averting the contamination and degradation of air, water, and land resources.

Environmental qualification A process for ensuring that equipment will be capable of withstanding the ambient conditions that could exist when the specific function to be performed by the equipment is actually called upon to be performed under accident conditions.

Environmental services Various combinations of scientific, technical, and advisory activities (including modification processes, i.e., the influence of manmade and natural factors) required to acquire, produce, and supply information on the past, present, and future states of space, atmospheric, oceanographic, and terrestrial surroundings for use in planning and decisionmaking processes, or to modify those surroundings.

Environmental stress screening Environmental Stress Screening (ESS) is a process which applies specific kinds of environmental stresses to products on an accelerated basis, but within their design parameters and limits to cause latent and intermittent flaws to become detectable failures.

Enzyme A protein molecule that acts as a specific catalyst, binding to other molecules in a manner that facilitates a particular chemical reaction.

Enzyme immunoassay/EIA In an enzyme immunoassay (EIA), an enzyme-labeled antibody or antigen is used for the detection and quantification of the antigen-antibody reaction. In an electrochemical EIA, the enzyme-catalyzed reaction is monitored electrochemically (amperometric, potentiometric, voltametric or conductometric). In EIA, the antibody-antigen reaction furnishes the needed specificity. The enzyme label provides the sensitivity via chemical amplification.

Epicycle In the Ptolemaic system of astronomy, the epicycle (literally: on the cycle in Greek) was a geometric model to explain the variations in speed and direction of the apparent motion of the Moon, Sun, and planets. It was designed by Apollonius of Perga at the end of the 3rd century BC. In particular it explained retrograde motion.

In the Ptolemaic system, the planets are assumed to move in a small circle, called an epicycle, which in turn moves along a larger circle called a deferent. Both circles rotate counterclockwise and are roughly parallel to the Earth's plane of orbit (ecliptic). The orbits of planets in this system are epitrochoids, and the point around which an epicyclic path revolves is the equant.

The deferent would be considered to be centered on the Earth.As viewed from Earth, the planets were seen as mostly moving eastward along the deferent. Half of the time, the added motion along the epicycle was eastward, in parallel with the eastward movement of the epicycle on the deference. However, at times the planet would move along the epicycle in an opposite direction to the motion of the epicycle along the deferent. This would cause the planet to slow down and reverse course, ie. retrogradation.

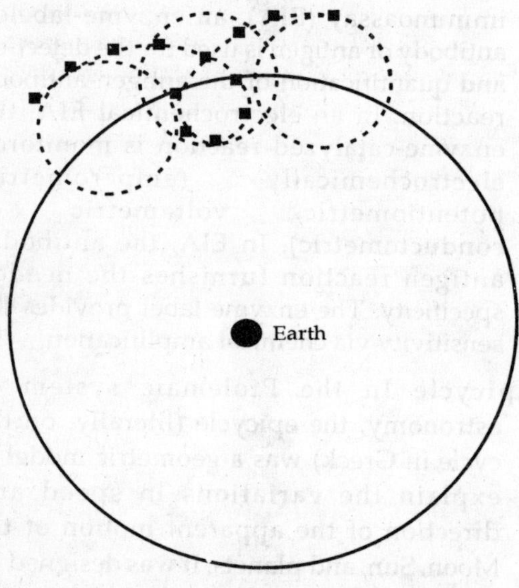

Fig. Epicycle

EPROM Electronically Erasable Programmable Read-Only Memory. Re-writable memory that does not lose data if power is lost to the system (non-volatile). Available in three types:

OTP One Time Programmable non-erasable.

Windowed (ultraviolet light erasable) used for prototyping and development work.

EEPROM Electronically Erasable Programmable Read-Only memory.

Equilibrium A system is said to be at equilibrium (with respect to some set of feasible transformations) if it has minimal free energy. A system containing objects at different temperatures is in disequilibrium, because heat flow can reduce the free energy. Springs have equilibrium lengths, reactants and products in solution have equilibrium concentrations, thermally excited systems have equilibrium probabilities of occupying various states, and so forth.

Equilibrium constant The product of the concentrations (or activities) of the substances produced at equilibrium in a chemical reaction divided by the product of concentrations of the reacting substances, each concentration raised to that power which is the coefficient of the substance in the chemical equation.

Equinus Equinus describes a fixation of the foot, or part of a foot, in the position it would assume if plantarflexed. The distal end of the part is farther away from the tibia. Talipes Equinus is the fixed position of plantar flexed attitude of the ankle joint. Forefoot Equinus is the fixed position of plantar flexed attitude of the forefoot in relation to the rearfoot.

Equitransference Equal diffusion rates of the positively and negatively charged ions of an electrolyte across a liquid junction without charge separation.

Equivalent conductance (l) Equivalent conductance of an electrolyte is defined as the conductance of a volume of solution containing one equivalent weight of dissolved substances when placed between two parallel electrodes 1 cm apart, and large enough to contain between them all of the solution. l is never determined directly, but is calculated from the specific conductance (Ls). If C is the concentration of a solution in gram equivalents per liter, then the concentration of a solution in gram equivalents per liter, then the concentration per cubic centimeter is C/1000, and the volume containing one equivalent of the solute, is, therefore, 1000/C.

Equivalent series inductance (ESL) The amount of inductance in series with an ideal capacitor which exactly duplicates the performance of a real capacitor.

Equivalent The mass of the compound which will produce one mole of available reacting substance. Thus, for an acid, this would be the mass of acid which will produce one mole of H+, for a base, one mole of OH-.

Error The difference between the value indicated by the transducer and the true value of the measurand being sensed. Usually expressed in percent of full scale output.

Error amplifier An operational amplifier, or differential amplifier, in a control loop that produces an error signal whenever a sensed output differs from a reference voltage.

Error band The allowable deviations to output from a specific reference norm. Usually expressed as a percentage of full scale.

Error proofing A process used to prevent errors from occurring or to immediately point out a defect as it occurs. If defects don't get passed down an assembly line, throughput and quality improve.

Error signal The output voltage of an error amplifier produced by the difference between the reference and the input signal times the gain of the amplifier.

Etching Chemical surface corrosion, usually conducted in a controlled fashion on a polished surface of a material sample to reveal details of the microstructure.

Ether An organic compound which has two hydrocarbon groups bound by an interior oxygen atom. The general formula is R'-O-R".

Ethernet A data link level protocol comprising the OSI model's bottom two layers. It is a broadcast networking technology that can use several different physical media, including twisted pair cable and coaxial cable. Ethernet usually uses CSMA/CD. TCP/IP is commonly used with Ethernet networks.

Ethernet address A 48 that uniquely identifies the Ethernet Network Interface Card (NIC) and hence the device the card resides in.

Ethernet meltdown A slang term for a situation where an Ethernet network becomes saturated. The condition usually persists for only a short time and is usually caused by a misrouted or invalid packet.

Eucaryotic organisms Organisms which possess a nuclear membrane. This includes all known organisms except viruses and bacteria.

Eutactic Characterized by precise molecular order, like that of a perfect crystal, the interior of a protein molecule, or a machine-phase system; contrasted to the disorder of bulk materials, solution environments, or biological structures on a cellular scale. Borderline cases can be identified, but perfection is not necessary. As a crystal with sparse defects is best described as a crystal (rather than as amorphous), so a eutactic structure with sparse defects is best described as (imperfectly) eutactic, rather than as disordered.

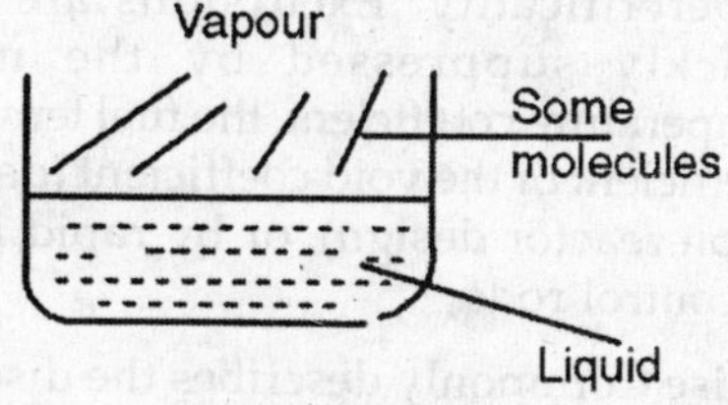

Fig. Eutectic

Eutectoid Upon cooling, one solid phase transforms isothermally and reversibly into two new solid phases that are intimately mixed.

Evaporation The conversion of liquid water to water vapor. It occurs on the surface of water bodies such as lakes and rivers and immediately after precipitation events in small depressions and other storage areas.

Evapotranspiration The sum of evaporation and transpiration. Since it is difficult to measure the two terms independently, they are often grouped as one value.

Eversion Motion occurring on the frontal plane during which the plantar aspect of the foot is tilted away from the midline of the body. The axis lies on the sagittal and transverse planes.

Everted A foot, or part of a foot is in an everted position when the foot or part of the foot is tilted in a frontal plane, so the plantar surface of the foot or part of the foot faces away from the midline of the body and away from the transverse plane, or away from some other specified reference point.

Excluded volume The presence of one molecule (or moiety) reduces the volume available for other molecules (or moieties); resulting reductions in their entropy are termed excluded volume effects.

Exclusion area The area surrounding the reactor where the reactor licensee has the authority to determine all activities, including exclusion or removal of personnel and property.

Excursion A sudden, very rapid rise in the power level of a reactor caused by supercriticality. Excursions are usually quickly suppressed by the negative temperature coefficient, the fuel temperature coefficient or the void coefficient (depending upon reactor design), or by rapid insertion of control rods.

Exercise Commonly describes the discharging to one volt per cell and subsequent charging. Used to maintain or condition NiCad and NiMH cells.

Exhaust gas recirculation (EGR) An exhaust-emission control system in which a portion of the exhaust gas is picked up from the exhaust manifold and sent back to the intake manifold t be reburned in the engine. Mixing exhaust gases with the fresh air / fuel mixture lowers the combustion temperature and reduces the formation of oxides of nitrogen in the exhaust.

Exhaust manifold The network of passages that gathers the exhaust gases from the various exhaust ports and routes them toward the catalyst, the muffler and the exhaust system.

Exia Explosion protection by means of the concept of Intrinsic Safety where protection is maintained with up to 2 component or other faults.

Exothermic The opposite of endothermic; describes an exoergic transformation in which energy is released as heat. Exoergic reactions in solution are commonly exothermic.

Explosion proof enclosure An enclosure that can withstand an explosion of gases within it and prevent the explosion of gases surrounding it due to sparks, flashes or the explosion of the container itself, and maintain an external temperature which will not ignite the surrounding gases.

Exposed junction A form of construction of a thermocouple probe where the hot or measuring junction protrudes beyond the sheath material so as to be fully exposed to the medium being measured. This form of construction usually gives the fastest response time.

Extended binary coded decimal interchange code (EBCDIC) An alternative to ASCII used extensively in IBM machinery. Some other vendors use it for mainframes. EBCDIC and ASCII are not compatible but are easy to convert between.

Exterior gateway protocol (EGP) A protocol used by gateways to transfer information about devices that can be reached within their autonomous systems.

Fab For "fabrication", a term referring to the making of semiconductor devices such as microprocessors.

Fabrication drawing A drawing used to aid the construction of a printed board. It shows all of the locations of the holes to be drilled, their sizes and tolerances, dimensions of the board edges, and notes on the materials and methods to be used. Called "fab drawing" for short. It relates the board edge to at least on hole location as a reference point so that the NC Drill file can be properly lined up.

Face 1. The exposed surface of a wall or masonry unit.

2. The surface of a unit designed to be exposed in the finished masonry.

Face centered cubic (FCC) A crystal structure found in some of the common elemental metals; within the cubic unit cell atoms are located at all corner and face-centered positions.

Faceshell bedding In concrete block masonry construction - application of mortar to all vertical and horizontal edges of the face shells of hollow masonry units.

Factory equipment In used vehicles: the combination of original standard equipment and production options that make up the equipment of a used vehicle. May also be referred to as "base" equipment.

Facultative A group of microorganisms which prefer or preferentially use molecular oxygen when available, but are capable of suing other pathways for energy and synthesis if molecular oxygen is not available.

Fahrenheit A temperature scale defined by 32° at the ice point and 212° at the boiling point of water at sea level.

Fail-stop Describes a component or subsystem that, in the event of a failure, produces no output (e.g., of material or data) rather than producing a damaged or incorrect output.

Failure analysis Failure Analysis is a collection of techniques to determine the root cause of a component or process defect or failure.

Failure modes and effects analysis. Failure Modes and Effects Analysis (FMEA) is a procedure in which each potential failure mode in every sub-item of an item is analyzed to determine its effect on other sub-items and on the required function of the item. It is used to identify potential failure modes and their associated causes/ mechanisms, consider risks of these failure modes, and identify mitigating actions to reduce the probability or impact of the failure.

Failure reporting and corrective action system Failure Reporting and Corrective Action System (FRACAS) is a closed-loop system to capture reports of failure from customers or service technicians in the field or from the factory, analyze these reports, detect trends or problems, and use this analysis to take corrective action in the design, component selection, supplier selection, manufacturing process, or operating manual of the product. Features of a FRACAS system include a database manager, tracking system for document controls, user definable reports which allow selection of data elements and sort options, and search functions.

Fan pressurization. Most buildings rely on some form of ventilation system to exhaust contaminated air. This system may run continuously, or may be operated automatically or manually. Ventilation systems can be of three types: a) exhaust only, resulting in a lower air pressure inside a building relative to the outside and increased infiltration; b) supply only, raising the pressure inside a building and increasing exfiltration, and c) balanced, with both supply and exhaust being operated by fans (Figure 3). Even in a balanced system, the amount of air may be increased or decreased relative to the amount being exhausted, increasing or decreasing the inside pressure accordingly.

Farad Unit of measurement of capacitance. A capacitor has a capacitance of one farad when a charge of one coulomb raises its potential one volt: C = QIE.

Faraday One gram equivalent weight of matter is chemically altered at each electrode of a cell for (approximately) each 96,500 international coulombs, or one Faraday, of electricity passed through the electrolyte.

Faraday cage An enclosure with no apertures (holes, slits, windows or doors) made of a perfectly conducting material. No electric fields are produced within the Faraday cage by the incidence of external fields upon it or by currents flowing on the perfect conductor such that the perfectly conducting enclosure is a perfect electromagnetic shield.

Fast fission Fission of a heavy atom (such as uranium-238) when it absorbs a high energy (fast) neutron. Most fissionable materials need thermal (slow) neutrons in order to fission.

Fast neutron A neutron with kinetic energy greater than its surroundings when released during fission.

FAT Factory Acceptance Test. A test of equipment carried out at supplier's factory prior to shipment of equipment - usually witnessed by purchaser.

Fault tree analysis Fault Tree Analysis is a top-down, hierarchical analysis of faults to identify the various fault mechanisms and their cause. It graphically describes the cause and effect relationships that result in major failures. The fault or major failure being analyzed is identified as the "top event." All of the possible causes of the top event are identified in a tree using "or" nodes for independent causes and "and" nodes for multiple causes that must exist concurrently for a failure to occur.

FEA Finite element analysis.

Feasibility Capable of being completed to meet goals. The feasibility of a proposed new product has two dimensions: a) Technical Feasibility and b) Financial Feasibility.

Feature Features are elements of the product that provide a distinctive benefit to the customer and are often highlighted in describing the benefits of the product to the customer. Features are the differentiating functionality of a product. This functionality may not be available in other products, or it may not be available with the same quality characteristics.

Feed forward A control technique whereby the line regulation of a power supply is improved by directly sensing the input voltage.

FEED(Front End Engineering and Design) A study used to analyse the various technical options for new developments with the objective to define the facilities required.

Feedback The process of returning part of the output signal of a system to its input.

Feed through A plated-through hole in a printed circuit board which electrically connects a trace a top of the beard with a trace on the bottom side.

Feeder lines A series of special assembly lines that allow assemblers to perform preassembly tasks off the main production line. Performing certain processes off the main production line means fewer parts in the main assembly area, the availability of service-ready components and assemblies in the main production area, improved quality and less lead time to build a product.

Feeder Part of the gating system that forms the reservoir of molten metal necessary to compensate for losses due to shrinkage as the metal solidifies; sometimes referred to as a riser.

Feedwater Water supplied to the reactor pressure vessel (in a BWR) or the steam generator (in a PWR) that removes heat from the reactor fuel rods by boiling and becoming steam. The steam becomes the driving force for the plant turbine generator.

Femoral position This refers to soft tissues abnormalities which cause the entire leg to rotate internally or externally when the hip is in its neutral transverse plane position.

Femoral torsion This refers to osseous deformities which cause the entire lower limb to rotate, either internally or externally, when the hip is in its neutral transverse position. Internal torsion causes the leg, ankle and foot to assume an internally rotated position as the individual walks. An adducted gait is evident when the deformity is uncompensated.

Fermentation Energy production without the benefit of oxygen as a terminal electron acceptor, i.e. oxidation in which the net effect is one organic compound oxidizing another.

Fermi energy The energy level in a solid at which the probability of finding an electron is 1/2. For a metal, the energy corresponding to the highest filled electron state in the valence band at 0 K.

Ferrimagnetism Permanent and large magnetizations found in some ceramic materials; it results from antiparallel spin coupling and incomplete magnetic moment cancellation.

Ferrite A ceramic material that exhibits low loss at high frequencies, and which contains iron oxide mixed (with oxides or carbonates of one or more metals such as manganese, zinc, nickel or magnesium.

Ferroelectric material A dielectric material such as Rochelle salt and barium titanate with a domain structure containing dipoles (asymmetric distributions of electrical charge) which spontaneously align. Their domain structure makes them analogous to ferromagnetic materials. They exhibit hysteresis and usually the piezoelectric effect.

Ferroelectricity Spontaneous alignment of electric dipoles within a material under the influence of an electric field, resulting in a hysteresis loop when the direction of electric field is switched.

Ferromagnetism The property of a substance which is attracted to a magnet. Iron, cobalt, nickel, gadolinium, dysprosium and alloys containing these elements are ferromagnetic.

Ferroresonance The property of a transformer design in which the transformer contains two separate magnetic paths with limited coupling between them. The output contains a resonating tank circuit and draws power from the primary to replace power delivered to the load.

Ferroresonant transformer An AC regulator designed to operate at a given input frequency and which uses a resonant circuit to achieve regulation. Although the regulation is imprecise, the device itself is simple, very rugged and reliable. Typically operating at 50Hz or 60Hz, these "iron-core" (actually laminated silicon steel) transformers incorporate a capacitor in one of the secondaries that resonates with the inductance of the transformer. An excessively high input AC voltage which would otherwise result in excessive secondary voltage will cause the resonant circuit voltage to ring high enough to locally saturate the transformer core. Once saturated, the primary winding magnetic flux no longer links the secondary winding and the secondary voltage is limited.

Fertile material A material, which is not itself fissile (fissionable by thermal neutrons), that can be converted into a fissile material by irradiation in a reactor. There are two basic fertile materials: uranium-238 and thorium-232. When these fertile materials capture neutrons, they are converted into fissile plutonium-239 and uranium-233, respectively.

FET(Field effect transistor) A semiconductor device designed for fast, current switching applications.

FGD (Flue Gas Desulphurisation) FGD typically achieved by fitting scrubber units to plant to reduce its sulphur dioxide emissions.

Fick's first law The diffusion flux is proportional to the concentration gradient. This relationship is employed for steady-state diffusion situations.

Fick's second law The time rate of change of concentration is proportional to the second derivative of concentration. This relationship is employed in non-steady-state diffusion situations.

Field balancing equipment An assembly of measuring instruments for performing balancing operations on assembled machinery which is not mounted in a balancing machine.

Field effect transistor (FET) Transistor in which the resistance of the current path from source to drain is modulated by applying a transverse electric field between two electrodes.

Fifth wheel Load supporting plate mounted to the frame of a vehicle. Pivot mounted, it contains provision for accepting and holding the kingpin of a trailer, providing a flexible connection between the tractor and the trailer. Center of the fifth wheel should always be located ahead of the centerline of the rear axle.

File server A process that provides access to a file from remote devices. Also used to refer to the physical server itself, although the term server also implies other services than file provision in most client/server networks.

Filling solution A solution of defined composition to make contact between an internal element and a membrane or sample. The solution sealed inside a pH glass bulb is called an internal filling solution. This solution normally contains a buffered chloride solution to provide a stable potential and a designated zero potential point. The solution which surrounds the reference electrode internal and periodically requires replenishing is called the reference filling solution. It provides contact between the reference electrode internal and sample through a junction.

Film badge Photographic film used for measurement of ionizing radiation exposure for personnel monitoring purposes. The film badge may contain two or three films of differing sensitivities, and it may also contain a filter that shields part of the film from certain types of radiation.

Filter One or more discrete components positioned in a circuit to attenuate signal energy in a specified band of frequencies.

Filter block A hollow, vitrified clay masonry unit, sometimes salt-glazed, designed for trickling filter floors in sewage disposal plants.

Final charging voltage The voltage which a battery reaches at the end of a charging operation. In the case of constant voltage charging, this voltage is determined by the setting of the charging equipment.

Final drive ratio A fluid coupling consists of two fan-like impellers in a sealed, oil-filled housing. The input "fan" churns the oil, and the churning oil, in turn, twirls the output "fan." Such a coupling allows some speed difference between its input and output shafts. The automatic transmission's torque converter is based on the fluid coupling principle.

Fine line design Printed circuit design permitting two (rarely three) traces between adjacent DIP pins. It entails the use of a either dry film solder mask or liquid photoimageable solder mask (LPI), both of which are more accurate than wet solder mask.

Fine pitch Refers to chip packages with lead pitches below 0.050". The largest pitch in this class of parts is 0.8mm, or about 0.031". Lead pitches as small as 0.5mm (0.020") are used.

Finger A program that provides information about users on an Internet host (possibly may include a user's personal information, such as project affiliation and schedule).

Fire setting An ancient tunneling technique in which rock is heated with fire and then doused with cold water, causing the rock to fracture.

Firmware The combination of hardware device and computer instructions and/or computer data that resides as read-only software on the hardware device.

First article A sample part or assembly manufactured prior to the start of production for the purpose of ensuring that the manufacturer is capable of manufacturing a product which will meet the requirements. (Graf).

First ray This describes the first cuneiform and the first metatarsal which move about a common axis. This axis angulates from the posterior, medial and dorsal to anterior lateral and plantar. No exact angles have been established for this axis however, it is angulated approximately 45o form the frontal and sagittal planes. The axis angulates only slightly from the transverse plane. Movement at the first ray is primarily dorsiflexion and plantar flexion but some accompanying eversion and inversion also occurs respectively.

Fissile material Although sometimes used as a synonym for fissionable material, this term has acquired a more restricted meaning. Namely, any material fissionable by thermal (slow) neutrons. The three primary fissile materials are uranium-233, uranium-235, and plutonium-239.

Fission (fissioning) The splitting of a nucleus into at least two other nuclei and the release of a relatively large amount of energy. Two or three neutrons are usually released during this type of transformation.

Fission gases Those fission products that exist in the gaseous state. In nuclear power reactors, this includes primarily the noble gases, such as krypton and xenon.

Fission products The nuclei (fission fragments) formed by the fission of heavy elements, plus the nuclide formed by the fission fragments' radioactive decay.

Fissionable material Commonly used as a synonym for fissile material, the meaning of this term has been extended to include material that can be fissioned by fast neutrons, such as uranium-238.

Fixed structural positions These are single plane structural abnormalities which are usually congenital in origin. It would be impossible for the foot, or part of the foot to move to such a position about a normal axis of motion. There are many types and only the more common are listed here.

Fixed suspended solids (FSS) is the matter remaining from the suspended solids analysis which will not burn at 550°C. It represents the non-filterable inorganic residue in a sample.

Flag Any of various types of indicators used for identification of a condition or event; for example, a character that signals the termination of a transmission.

FLAGS Far North Liquids and Gas System. Name given to subsea pipeline running from the Brent field to the St Fergus terminal in NE Scotland.

Flash 1. To turn a vector photoplotter lamp on for a brief but precise duration and then off, during which time the relative positions of the lamp and film remain fixed. This exposes the film with the image of a small object (the size and shape of which is controlled by the transparent portion of an aperture).

2. n. A small image on film created in such wise or as directed by a command in a Gerber file.) The maximum size (x or y dimension)for a flash varies from one photoplotting shop to another, but is commonly ½ inch.

Flash point Flash Point Temperature is the lowest temperature at which a liquid releases sufficient vapour that can be ignited by an energy source.

Flex circuit Flexible circuit, or flexcircuit; a printed circuit made of thin, flexible material.

Flexible circuitry An array of conductors bonded to a thin, flexible dielectric. It has the unique property of being a three-dimensional circuit that can be shaped in multiplanar configurations, rigidized in specific areas, and molded to backer boards for specific applications. As an interconnect, the main advantages of flex over traditional cabling are greater reliability, size and weight reduction, elimination of mechanical connectors, elimination of wiring errors, increased impedance control and signal quality, circuit simplification, greater operating temperature range, and higher circuit density. In many applications, lower cost is another advantage of using flexible circuits. From Comparison of Printed Flexible Circuitry and Traditional Cabling, an article By JACK LEXIN - vice president for marketing and sales at L.E. Flex Circuits , Carlsbad, Calif. appearing in inter-Connection Technology , December 1992 .

FlexRay bus A fault tolerant, high speed data communications bus designed for complex automotive control applications.

Flip chip A mounting approach in which the chip (die) is inverted and connected directly to the substrate rather than using the more common wire bonding technique. Examples of this kind of flip-chip mounting are beam lead and solder bump .

Flip flop Binary device whose outputs change value only in response to an input pulse.

Float charge An arrangement in which the battery and the load are permanently connected in parallel across the DC charging source, so that the battery will supply power to the load if the charger fails. Compensates for the self-discharge of the battery.

Float life The expected lifetime in hours of a battery when used in a float charge application.

Float voltage The voltage required for retaining a charged battery in a fully charged condition. This is also known as float charging.

Floating output ungrounded output of a power supply where either output terminal may be referenced to another specified voltage.

Flocculant settling Settling in which particle concentrations are sufficiently high that particle agglomeration occurs. This results in a reduction in the number of particles and an increase in average particle mass. As agglomeration occurs higher settling velocities result.

Flooded lead acid cell In "flooded" batteries, the oxygen created at the positive electrode is released from the cell and vented into the atmosphere. Similarly, the hydrogen created at the negative electrode is also vented into the atmosphere. This can cause an explosive atmosphere in an unventilated battery room. Furthermore the venting of the gasses causes a net loss of water from the cell. This lost water needs to be periodically replaced. Flooded batteries must be vented to prevent excess pressure from the build up of these gasses.

Floppy disk A small, flexible disk carrying a magnetic medium in which digital data is stored for later retrieval and use.

Flow The progressive achievement of tasks along the value stream so that a product proceeds from design to launch, order to delivery, and raw materials into the hands of the customer with no stoppages, scrap or backflows.

Flow battery A battery in which the electrolyte flows or is pumped through the electrodes

Flow velocity/flow rate The distance traveled by a packet of fluid in a unit of time. The speed depends on where within the liquid the flow is measured. For example, the surface of a river will be flowing at the fastest speed, while at the very bottom of the river, it will not be flowing at all. Hence, a velocity gradient can be measured. In a closed pipe, liquid at the center of a pipe will be flowing the fastest, while at the edge of the pipe the velocity is zero. The average flow velocity can be calculated by dividing the flow rate by the cross sectional area.

Flowline Pipeline carrying reservoir fluid on the seabed from a well to a riser.

Flowmeter A device for monitoring and measuring the flow of a substance (typically fluid or gas).

Fluid shear stress In flowing fluid all the molecules of the fluid "rub" against one another as they travel down the tube. Fluid molecules also exert a "rubbing" force on the walls of the tube. This rib or shear force can be calculated for flow in a tube if you know the fluid velocity, roughness of the tube, fluid properties, and diameter of the tube.

Fluidization The suspension of particles by sufficient upward velocity of the fluid. During fluidization the gravity force is overcome by a combination of buoyancy and fluid friction.

Fluorescence Luminescence which persists less than a second after the exciting cause has been removed. If the luminescence persists significantly longer it is called phosphorescence.

Flushing When data is output to a text file it is usually buffered to make processing more efficient, flushing forces any items in the buffer to be actually written to the file.

Flux A term applied to the amount of some type of particle (neutrons, alpha radiation, etc.) or energy (photons, heat, etc.) crossing a unit area per unit time. The unit of flux is the number of particles, energy, etc., per square centimeter per second.

Flyback converter A power supply switching circuit which normally uses a single transistor. During the first half of the switching cycle the transistor is on and energy is stored in a transformer primary; during the second half of the switching cycle this energy is transferred to the transformer secondary and the load.

Flywheel battery A flywheel stores kinetic energy in a high speed (up to 100,000 rpm) rotating cylinder and is "charged" and "discharged" via an integral motor/ generator. High power availability but low capacity.

FMEA Failure Modes and Effects Analysis. Identifies failure modes - their causes and effects - the safeguards already incorporated and the potential additional measures that could be considered.

Foam A polymer that has been made porous (or spongelike) by the incorporation of gas bubbles.

FOB Free On Board. Incoterm used to describe responsibilities for carriage; risk and cost.

Foldback current limiting A power supply output protection circuit whereby the output current decreases with increasing overload, reaching a minimum at short circuit. This minimizes internal power dissipation under overload conditions. Foldback current limiting is normally used with linear regulators and is not necessary to protect switching regulators. However, many switchers incorporate foldback to protect the load form excessive current.

Follower A component in a cam system that is driven through a pattern of displacements as it rests against a moving contoured surface.

Foot flat During stance phase, after heel contact, the rest of the foot comes down onto the ground at foot flat, which generally occurs at around 80% of the gait cycle, just before toe off on the other limb. This is referred to as the load acceptance phase of gait.

Foot switches These are used to record the timing of gait. With a switch attached to the heel and one beneath the forefoot, it is possible to measure, the timing of heel contact, foot flat, heel off and toe off, and the duration of stance phase. Foot switches are usually connected through a trailing wire to a microcomputer or alternatively either a radio transmitter or a portable recording device may be used to collect data, which is transferred to software for analysis.

Footprint 1. The pattern and space on a board taken up by a component.

2. Decal.

Force This is a vector quantity, which means that it has both magnitude and direction. Force is measured in the newton (N) and the force of gravity acting on a mass of 1kg. is 9.81 N. F=ma

Force platform (Force Plate) This instrument measures the direction and magnitude of the ground reaction force beneath the foot.

Forced vibration Vibration of a system caused by an imposed force. Steady-state vibration is an unchanging condition of periodic or random motion.

Forefoot adductus This is a single plane fixed position where the angle formed by the longitudinal bisection of the rearfoot and the metatarsals is angled toward the midline of the body away from the longitudinal bisection of the rearfoot.

Forefoot angle The angle formed between the longitudinal bisection of the metatarsus and the longitudinal bisection of the rearfoot. This angle is normally an adductus angle. The forefoot angle is usually equal and opposite to the digital angle. Any alteration of the forefoot angle will proportionately alter the digital angle in the opposite direction. The forefoot angle is influenced by altered position or subluxation of the midtarsal and subtalar joints. The size of the angle is determined by the difference between the metatarsus and the lesser tarsus angles.

Forefoot equinus A fixation of the forefoot in a position it would assume if plantarflexed. The distal aspect of the foot is farther away from the tibia. This is usually observed on non weightbearing and disappears on weightbearing. The condition must not be confused with metatarsus equinus, a rare condition, where there is a plantar flexion at the midtarsal joints. This condition is also referred to as Shaeffer's Foot.

Forefoot plane The plantar plane of the forefoot is represented by a plane on the plantar surface of the 2nd, 3rd, and 4th metatarsal heads. It is usual for the first and fifth metatarsal heads to share this plane, but when this is not the case, the plane made by the 2nd.,3rd., and 4th., metatarsal heads are used to represent the forefoot plane.

Forefoot supinatus Describes a forefoot fixed in an inverted position about the longitudinal axis of the mid tarsal joint, when the rest of the foot is in the subtalar neutral position. Tonic spasm of the anterior tibial muscle simultaneously dorsiflexes the first ray and temporarily inverts the forefoot around the longitudinal axis of the midtarsal joint. Clinically where the condition is indistinguishable from forefoot varus, symptoms subside with forefoot wedging.

Forefoot valgus Describes a fixed structural abnormality in which the plantar aspect of the forefoot is everted, on the frontal plane, relative to the plantar aspect of the rearfoot, when the calcaneum is vertical and the mid tarsal joints are locked and fully pronated.

Forefoot varus This fixed plane deformity is seen when the forefoot plane is everted to the rearfoot i.e. the fifth metatarsal head is more dorsal than the first metatarsal head. This is a frontal plane deformity which can be compensated by abnormal rearfoot pronation.

Form The defined configuration of an item including the geometrically measured configuration, density, and weight or other visual parameters which uniquely characterize an item, component or assembly. For software, form denotes the language, language level and media.

Formal testing The process of conducting testing activities and reporting results in accordance with an approved test plan making provision for customer involvement.

Formaldehyde offgassing APA trademarked structural wood panels, such as plywood and oriented strand board, are made with phenol formaldehyde adhseives, which should not be confused with urea formaldehyde adhesvies. Formaldehyde-related problems have been associated with certain urea formaldehyde adhesives but not with the phenol formaldehyde adhesives. In fact, because formaldehyde levels associated with phenolic resin-bonded products are so low, these products are exempted from the Department of Housing and Urban Development testing and certification requirements.

Formation Electrochemical processing of a cell electrode(or plate) between manufacturing and first discharge, which transforms the active material into its useable form.

Formula quantity Strategic special nuclear material in any combination in a quantity of 5000 grams or more computed by the formula, grams = (grams contained U-235) + 2.5 (grams U-233 + grams plutonium). This class of material is sometimes referred to as a Category I quantity of material.

Forward bias The conducting bias for a p-n junction rectifier that assures electron flow to the n side of the junction.

Forward converter A power supply switching circuit that transfers energy to the transformer secondary when the switching transistor is on.

Four wheel On a vehicle equipped with Four-Wheel Anti-Lock Brakes, all four wheels are equipped with speed sensors. When these sensors determine that the wheels are decelerating so rapidly that lockup may occur, the Electro-Hydraulic Control Unit (EHCU) is activated. The EHCU then modulates the brake pressure in the appropriate brake lines by means of the solenoid-operated valves. This is intended to prevent wheel lockup and help the vehicle maintain directional stability during potentially hazardous braking situations

Four wheel drive (4WD) In a Four Wheel Drive system, a secondary transmission assembly, called a transfer case, is driven from the main transmission. The transfer case distributes power to both axles to drive all four wheels. It is the heart of the Four-Wheel Drive system. Four-Wheel Drive can be full-time, in which power is delivered to both axles at all times or part-time, where the driver selects two or four wheel drive. Four wheel drive is often combined with independent suspension systems and off-road type tires to enhance driveability on rough, off-road terrain, or on-road driveability in unfavorable driving conditions.

Free wheel diode A diode in a pulse-width modulated switching power supply that provides a conduction path for the counter electromotive force of an output choke.

Four wheel independent suspension A type of suspension in which all wheels are mounted to separate suspension members with no rigid axle connecting them. Therefore a disturbance affecting one wheel has no effect on the opposite wheel. Four wheel independent suspension reduces the un-sprung weight, improves ride and handling over rough surfaces and permits room for a larger trunk.

FR 2 A NEMA grade of Flame-Retardant industrial laminate having a substrate of paper and a resin binder of phenolic. It is suitable for printed circuit board laminate and cheaper than the woven glass fabrics such as FR-4.

FR 4 A NEMA grade of Flame-Retardent industrial laminate having a substrate of woven-glass fabric and resin binder of epoxy. FR-4is the most common dielectric material used in the construction of PCBs in the USA. Its dielectric constant is from 4.4 to 5.2 at below-microwave frequencies. As frequency climbs over 1 GHz, the dielectric constant of FR-4 gradually drops.

FR 6 Fire-Retardant glass-and-polyester substrate material for electronic circuits. Inexpensive; popular for automobile electronics.

A comparison of many types of PCB laminates can be found at the ILNorplex (Industrial Laminates Norplex) Laminate Selector chart .

Fractionation The separation of crude oil into its more valuable and usable components through distillation.

Fragmentation The breaking of a datagram into several smaller pieces, usually because the original datagram was too large for the network or software.

Frame Usually refers to the completed Ethernet packet, which includes the original data and all the TCP/IP layers' headers and trailers (including the Ethernet's).

Frame check sequence (FCS) A mathematical function used to verify the integrity of bits in a frame, similar to the Cyclic Redundancy Check (CRC).

Free body diagram This is used to represent a shapeless "lump" floating in space in complete equilibrium, no matter the number of forces acting on the body. The sum of the clockwise forces equals the number of the anticlockwise forces etc. This type of diagram is very useful to depict complex forces acting on a body, and provides a simple model to analyse the effects of turning forces acting on the body.

Free energy (G) A thermodynamic quantity that is a function of the enthalpy (H), the Kelvin temperature (T) and the entropy (S) of a system; G=H-TS. At equilibrium, the free energy is at a minimum. Under certain conditions the change in free energy for a process is equal to the maximum useful work.

Free radical A radical.

Freeware Software that is made available by the author at no cost to anyone who wants it (although the author retains rights to the software).

Freezing point The temperature at which the substance goes from the liquid phase to the solid phase.

Frenkel defect In an ionic solid, a cation-vacancy and cation-interstitial pair.

Frequency Number of times per second that a quantity representing a signal, such as a voltage, changes state. Also, the number of waves (cycles) per second that pass a given point in space.

Frequency modulated output A transducer output which is obtained in the form of a deviation from a center frequency, where the deviation is proportional to the applied stimulus.

Frequency of vibration The number of cycles occurring in a given unit of time. RPM - revolutions per minute. CPM- cycles per minute.

Frequency output An output in the form of frequency which varies as a function of the applied input.

Frequency response Two relations between sets of inputs and outputs. One relates frequencies to the output-input amplitude ratio; the other relates frequencies to the phase difference between the output and input.

Frequency, natural The frequency of free (not forced) oscillations of the sensing element of a fully assembled transducer.

FROG A depression in the bed surface of a brick. Sometimes called a panel.

Front wheel drive (FWD) A drive system where the engine and transaxle components apply the driving force to the front wheels rather than the rear wheels. Benefits of Front-Wheel drive include: Maximized passenger space. Enhanced cargo area. excellent drive traction; particularly on wet or slippery surfaces, since the drive is through the front wheels, which carry a heavier load.

FTA Fault Tree Analysis. A method of calculating the probability of an event from the probabilities or frequencies of its causal events.

Fuel cell An electrochemical generator in which the reactants are stored externally and may be supplied continuously to a cell.

Fuel cycle The series of steps involved in supplying fuel for nuclear power reactors. It can include mining, milling, isotopic enrichment, fabrication of fuel elements, use in a reactor, chemical reprocessing to recover the fissionable material remaining in the spent fuel, reenrichment of the fuel material, refabrication into new fuel elements, and waste disposal.

Fuel gauge An indication of the State of Charge (SOC) or how much charge is remaining in a battery. Also called a Gas Gauge.

Fuel injection, electronic A computer-controlled method of delivering fuel under pressure. The computer monitors signals from coolant temperatures, manifold vacuum, exhaust oxygen sensor, and engine cranking sensor. It "tells" the injectors to release and adjust the fuel to yield an air/fuel mixture assuring engine operation well matched with emission requirements, optimum fuel economy and overall vehicle performance.

Fuel pump A mechanical or electrical device that draws fuel from the fuel tank and delivers it to the carburetor or injectors.

Fuel reprocessing The processing of reactor fuel to separate the unused fissionable material from waste material.

Fuel rod A long, slender tube that holds fissionable material (fuel) for nuclear reactor use. Fuel rods are assembled into bundles called fuel elements or fuel assemblies, which are loaded individually into the reactor core.

Fuel temperature coefficient of reactivity The change in reactivity per degree change in the fuel temperature. The physical property of fuel pellet material (uranium-238) that causes the uranium to absorb more neutrons away from the fission process as fuel pellet temperature increases. This acts to stabilize power reactor operations. This coefficient is also known as the Doppler coefficient.

Fugitive emissions Emissions (air pollutants) released to the air other than those from stacks or vents.

They are often due to equipment leaks; evaporative processes and windblown disturbances.

Full bridge A Wheatstone bridge configuration utilizing four active elements or strain gages.

Full bridge converter A power switching circuit in which four power switching devices are connected in a bridge configuration to drive a transformer primary.

Function Three mode PID controller. A timeproportioning controller with integral and derivative functions. The integral function automatically adjusts the system temperature to the set point temperature to eliminate droop due to the time proportioning function.

Functional area A distinct group of system performance requirements which, together with all other such groupings forms the next lower-level breakdown of the system on the basis of function.

Functional requirements Functional Requirements capture the intended behavior of the system or product - what the system will do. This behavior may be expressed as functions, tasks, or services the system or product is required to perform. Therefore, functional requirements do not include performance characteristics, operating conditions, use cases, and specifications.

Furring Method of finishing the interior face of a concrete or masonry wall: it consist of wood strips which provide space for insulation, allow space for measures to prevent moisture transmittance, and provide a flat, level substrate for finishing.

Fuse Safety protective device that permanently opens an electric circuit when overloaded.

Fusion reaction A reaction in which at least one heavier, more stable nucleus is produced from two lighter, less stable nuclei. Reactions of this type are responsible for enormous release of energy, as in the energy of stars, for example.

Fuzzy logic A method of deriving precise answers from vague data.

G

G The force of acceleration due to gravity equal to 32.1739 ft/sec^2 or 386 in./sec^2.

GA General Arrangement. Common name given to a drawing showing the general arrangement of a piece of equipment.

Gage factor A measure of the ratio of the relative change of resistance to the relative change in length of a piezoresistive strain gage.

Gage length The distance between two points where the measurement of strain occurs.

Gage pressure Absolute pressure minus local atmospheric pressure.

Gage pressure transducer A transducer which measures pressure in relation to the ambient pressure.

Gain margin The gain of a control loop at the frequency for which there is 360 degrees of phase shift around the control loop.

Gain Ratio of an output signal to an input signal.

Galling 1. A condition whereby excessive friction between mating parts results in localized welding; subsequent spalling and a further roughening of the rubbing surfaces may follow;

2. A severe form of scuffing associated with gross damage to the surfaces or failure.

Galvanic cell 1. A cell in which chemical change is the source of electrical energy; it usually consists of two dissimilar conductors in contact with each other and with an electrolyte, or of two similar conductors in contact with each other and with dissimilar electrolytes;

2. A cell or system in which a spontaneous oxidation-reduction reaction occurs, the resulting flow of electrons being conducted in an external part of the circuit.

Galvanic corrosion Corrosion associated with the current of a galvanic cell consisting of two dissimilar conductors in an electrolyte or two similar conductors in dissimilar electrolytes; where the two dissimilar metals are in contact, the resulting reaction is referred to as couple action.

Galvanic couple A pair of dissimilar conductors, commonly metals, in electrical contact.

Galvanic current The electric current that flows between metals or conductive nonmetals in a galvanic couple.

Galvanized steel A specially zinc-coated steel used on many major painted panels and in key unpainted areas of a vehicle to help prevent rust and corrosion.

Gamma radiation High-energy, short wavelength, electromagnetic radiation emitted from the nucleus. Gamma radiation frequently accompanies alpha and beta emissions and always accompanies fission. Gamma rays are very penetrating and are best stopped or shielded by dense materials, such as lead or depleted uranium. Gamma rays are similar to x-rays.

Gamma ray Short-wavelength electromagnetic radiation, similar to x-rays but of nuclear origin, with a range of wavelength from about 10^{-14} 10^{-10} m.

Gantt chart Gantt Chart is a diagram used in project management, where the x axis is time and the y axis shows tasks to be performed to complete the project. Each task is displayed as a horizontal bar spanning the time period during which it is expected to take place. Arrows may be drawn from one task to another to indicate dependencies (when one task can't be begun until another is completed). The Gantt chart was developed by Charles Gantt in 1917.

Gap insurance Insurance that will cover the difference between the replacement cost paid by conventional insurance and what is owed on the lease in the case the car is totaled or stolen.

Gas A substance possessing perfect molecular mobility and the property of indefinite expansion, as opposed to a solid or liquid; any such fluid or mixture of fluids other than air. Normally, these formless substances completely fill the space, and take the shape of, their container.

Gas centrifuge A uranium enrichment process that uses a large number of rotating cylinders in a series. These series of centrifuge machines, called trains, are interconnected to form cascades. In this process, UF6 gas is placed in a drum or cylinder and rotated at high speed. This rotation creates a strong gravitational field so that the heavier gas molecules (containing U-238) move toward the outside of the cylinder and the lighter gas molecules (containing U-235) collect closer to the center. The stream that is slightly enriched in U-235 is withdrawn and fed into the next higher stage, while the slightly depleted stream is recycled back into the next lower stage. Significantly more U-235 enrichment can be obtained from a single unit gas centrifuge than from a single unit gaseous diffusion barrier.

Gas chromatography The separation and identification of individual chemical components from a sample. A typical machine costs over $250,000.

Gas cooled reactor A nuclear reactor in which a gas is the coolant.

Gas filled shock absorbers A nitrogen gas chamber is used to pressurize the shock absorber in place of the traditional air/oil combination. Gas filled shock absorbers provide more stable damping in a variety of conditions and thus improves ride and road contact.

Gas gauge An electrical circuit which indicates the amount of charge remaining in a battery.

Gas metal arc welding An arc welding process that produces coalescence of metals by heating them with an arc between a continuous filler metal electrode and the workpieces; shielding is obtained entirely from an externally supplied gas.

Gas stripping Gas transfer of an undesirable gas from a water stream to the atmosphere.

Gaseous diffusion plant A facility where uranium hexafluoride gas is filtered. Uranium-235 is separated from uranium-238, increasing the percentage of uranium-235 from 1 to about 3 percent. The process requires enormous amounts of electric power.

Gassing The generation of a gaseous product at one or both electrodes as a result of the electrochemical action. In Lead Acid batteries gassing produces hydrogen and oxygen.

Gate 1. A device or element that has the ability to block or pass a signal.

2. A device having one output channel and two or more input channels that performs a logic function.

3. A control electrode in a semiconductor device such as a triac, or FET.

Gateway In Internet terms, a gateway is a device that routes datagrams. More recently used to refer to any networking device that translates protocols of one type network into those of another network.

Gauge repeatability and reproducibility Gauge Repeatability and Reproducibility (GR&R) is the evaluation of a gauging instrument's accuracy by determining whether the measurements taken with it are repeatable and reproducible. Repeatability is the variation in measurement obtained with one measurement instrument when used several times by an appraiser while measuring the identical characteristic on the same part. Reproducibility is the variation in the averages of the measurements made by the different appraisers using the same measuring instrument when measuring the identical characteristic on the same part.

Gauss Measure of flux density in Maxwells per square centimeter of cross- sectional area. One gauss is iC-4 Tesla.

Gear ratio The number of revolutions a driving (pinion) gear requires to turn a driven (ring) gear through one complete revolution. For a pair of gears, the ratio is found by dividing the number of teeth on the driven gear by the number of teeth on the driving pinion gear.

Geiger mueller counter A radiation detection and measuring instrument. It consists of a gas-filled tube containing electrodes, between which there is an electrical voltage, but no current, flowing. When ionizing radiation passes through the tube, a short, intense pulse of current passes from the negative electrode to the positive electrode and is measured or counted. The number of pulses per second measures the intensity of the radiation field. It was named for Hans Geiger and W. Mueller, who invented it in the 1920s. It is sometimes called simply a Geiger counter or a G-M counter and is the most commonly used portable radiation instrument.

Gel cell A battery which uses gelled electrolyte, an aqueous electrolyte that has been fixed by the addition of a gelling agent.

Gene A given segment of the DNA molecule that contains the code for a specific protein.

General availability The point in the product life cycle when production has been ramped-up to sufficient volumes and when product issues have been resolved so that the product is made available to all interested customers.

Generation (net) The gross amount of electric energy produced less the electric energy consumed at a generating station for station use.

Generator A device that converts mechanical energy into electrical energy. It can produce either AC or DC electricity. Seldom used in automotive applications, it has been replaced by the alternator.

Genu recurvatum This describes a hyperextension of the knee. Hyperextension of the knee is resisted by both collateral ligaments; both cruciate ligaments; the oblique popliteal ligament; the posterior aspect of the joint capsule; the menisci; the hamstrings and gastrocnemius: and the configeration of the condyles. The greater the moment of force tending to produce hyperextension, the greater will be the tensile stress in the ligaments. Paralysis of either the hamstrings or quadriceps tends to produce recurvatum.

Geochemistry Hybrid science which applies chemical principles and techniques to geologic/geotechnical studies in order to understand how chemical elements are distributed and act inside the crust, mantle, and core of the earth. Over billions of years, chemical differentiation of the earth's crust has created vast rafts of silica-rich rocks, the continents, which float on iron- and magnesium-rich rocks of the ocean bottoms.

Geometric dimensioning and tolerancing Geometric Dimensioning and Tolerancing (GD&T) - ANSI-Y14.5 standard for showing the dimensioning and tolerancing on a drawing considering the functions or relationships of part features. GD&T depicts the geometric relationship of part features (instead of the Cartesian relationship), allowing the maximum tolerance which permits full function of the product.

Geotechnical engineering Geotechnical engineering uses the principles of mechanics to analyze and predict the behavior of earth materials. Expertise in this area is gained through a degree in civil engineering and is required for the design and construction of most, if not all, civil structures.

Gerber file Data file used to control a photoplotter . Named after Gerber Scientific Co., who made the original vector photoplotter .

Gibbs free energy The Gibbs free energy is the Helmholtz free energy plus the product of the system volume and the external pressure. Changes in the Gibbs free energy at a constant pressure thus include work done against external pressure as a system undergoes volumetric changes. This proves convenient for describing equilibria in gases and liquids at a constant pressure (e.g., at one atmosphere), but is of little use in describing machine-phase chemical processes. Changes in the Gibbs free energy caused by a change in the applied pressure (at constant volume) have no direct physical significance.

Glass An amorphous solid obtained when silica is mixed with other compounds, heated above its melting point, and then cooled rapidly.

Glass transition temperature (Tg) The temperature at which, upon cooling, a noncrystalline ceramic or polymer transforms from a supercooled liquid to a rigid glass.

Glob top A blob of non-conductive plastic, often black in color, which protects the chip and wire bonds on a packaged IC and also on a chip on board . This specialized plastic has a low coefficient of thermal expansion so that ambient temperature changes will not rip loose the wire bonds it is designed to protect. In high-volume chip on board production, these are deposited by automated machinery and are round. In prototype work, they are deposited by hand and can be custom-shaped; however, in designing for manufacturability, one assumes a prototype product will "take-off" and ultimately have high market demand, and so lays out chip on board to accommodate a round glob top with adequate tolerance for machine-driven "slop-over".

Global warming The long-term warming of the plant due to increases in greenhouse gases which trap reflected light preventing it from exiting to space.

Go/No Go gauges Go/No-Go Gauges are gauges that provides categorical data about whether one or more dimension of a workpiece is within specification limits.

Goniometer A measuring instrument used to calculate the angular relationship between the bones which make up anatomical joints.

Gopher An application that allows you to access publicly available information on Internet hosts that provide Gopher service.

Government open system interconnection profile (GOSIP) A government standard that uses the OSI reference model.

GPS Relative navigation A method of formation flying using the distant constellation of Global Positioning System satellites. Radio signals are received and timed from several GPS satellites to triangulate position. Each FASTRAC twin then shares its raw information with the other so that the location of the entire formation can be tracked at all times.

Graceful degradation Graceful Degradation is the quality of a product, system or design such that when something goes wrong, it happens a little at a time and with plenty of opportunity to take action to correct the problem or at least protect against its worst consequences.

Grain boundary The interface separating two adjoining grains having different crystallographic orientations.

Grain growth The increase in average grain size of a polycrystalline material: for most materials, an elevated temperature heat treatment is necessary.

Grain size The average grain diameter as determined from a random cross section.

Granite Granites are dense and hard rocks that resist wear and often have a speckled appearance. While geologically the term "granite" is really restricted to certain crystalline rocks of igneous origin made up of quartz, feldspar and mica, in the stone and masonry industry the term is used for almost all igneous rocks with an interlocking granular texture. For example, black granite refers to black, fine-grained igneous rocks such as basalt and diabase, which scientifically speaking are not granite. Granites have low porosity and permeability, which gives most of them a high resistance to weathering, although some may be susceptible to iron staining. Granite is often used in construction and lanscaping componenets that are in contact with the ground (steps, walkways, even foundations) in predominantly brick, limestone or sandstone buildings.

Graphical user interface Graphical User Interface - an interface to a computer that uses icons to represent desktop objects, such as documents and programs, that the user can access and manipulate with a pointing device, such as a mouse.

Gravity A general term for the phenomenon of attraction between things having mass. The attraction between our planet and a human-sized object causes the object to fall.

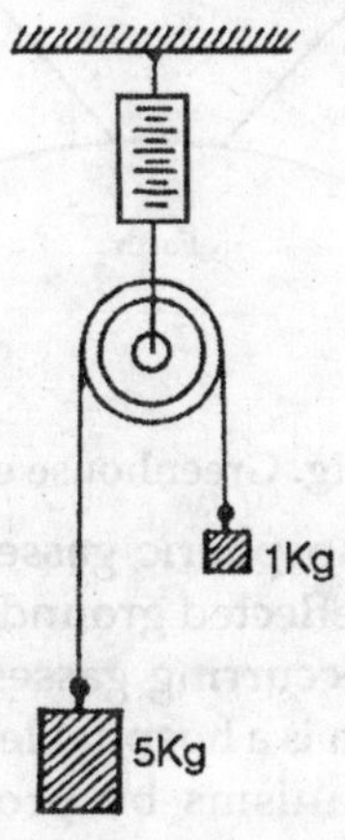

Fig. Gravity

Gravity dam A dam constructed so that its great weight resists the force of water pressure.

Gray (Gy) The international system (SI) unit of absorbed dose. One gray is equal to an absorbed dose of 1 Joule/kilogram (one gray equals 100 rads).

Green ceramic body A ceramic piece, formed as a particulate aggregate, that has been dried but not fired.

Greenhand Someone who is new to the offshore oil industry - more commonly used on drill rigs.

Sometimes referred to as green hat on production platforms.

Greenhouse effect The transmission of shortwave solar radiation by the atmosphere coupled with the selective adsorption of longer-wavelength terrestrial radiation, especially by water vapor, carbon dioxide, and methane, (CO_2, $H_2O(g)$, CO_4) etc, resulting in warming of the atmosphere.

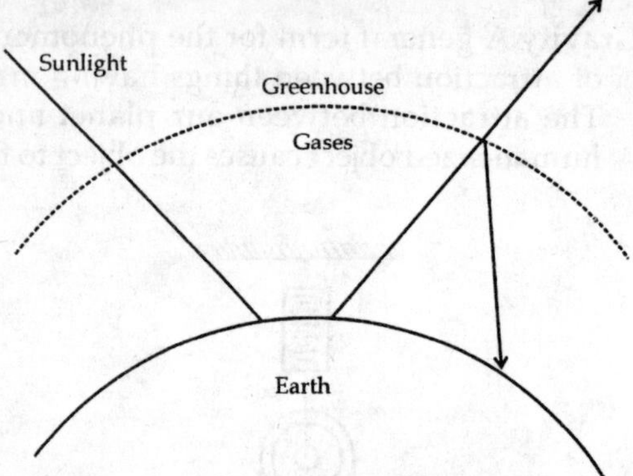

Fig. Greenhouse effect

These atmospheric gasses trap incoming solar and reflected ground radiation. These naturally occurring gasses make sure that planet earth is a hospitable environment for living organisms by providing average temperatures of 15°C (288K) rather than -21°C (255K) as refleced by the Stefan Boltzman equation. The 33K difference results solely out of these gases present in the atmosphere. The overall heat budget of planet earth at the current level is in dynamic equilibrium with an average of 20% of solar radiation absorbed within the atmosphere, an average 30% reflected by the clouds and ground surfaces, and the remaining 50% absorbed at ground level. Altogether, about 70% incoming radiation is re-emitted back into space (planetary thermal radiation). Antropogenic influences are the result of a buildup of CO_2 in the atmosphere (e.g. through the burning of fossil fuels) in which CO_2 traps solar heat beneath the atmospheric layers; it eventually could lead to increased global temperatures, and changed climatic patterns. The table below reveals the contribution of greenhouses gases accounting for the 33K increase in the earth's atmosphere; molecular gas contribution [%] contribution [K] the release of additional, manmade CO_2 will definitely have an increase to the naturally occurring low levels of this gas (350ppm).

Greenhouse gases Gases which trap solar radiation. Of the solar energy entering the earth's atmosphere a portion is reflected back and a portion penetrates onto the earth's surface. The portion reflected back from the earth's surface is at a different wavelength that when it entered. Carbon dioxide and other gases, which pass solar radiation, absorb this reflected radiation, increasing the earth's temperature. This is much like a greenhouse, hence the name.

Ground A conducting connection, whether intentional or accidental, by which an electric circuit or equipment is connected to earth, or to some conducting body that serves in place of earth. (National Electric Code)

Ground loop An unintentionally induced feedback loop or crosstalk caused by two or more circuits sharing a common electrical ground.

Ground reaction When the body makes contact with the ground, according to Newtons Third Law, there is an equal reaction to the action of body contact. In other words, the force of ground contact is met with an equal and opposite ground reaction force. No force is recorded at that instant. Walking on different surfaces however and after the initial contact will change the type of reaction e.g. tramping on a hard concrete surface would tend to increase ground reaction compared to walking on a soft grassy surface. Ground reactive forces are measured on a force platform and recorded in newtons (N).

Ground state The lowest-energy state of a system. The electronic ground state of a system cannot reduce its energy by an electronic transition, but may contain vibrational energy (kinetic and potential energy associated with the motions and positions of its atoms); extended systems at ordinary temperatures are always vibrationally excited, and so "ground state" is often taken to mean "electronic ground state."

Grounded Connected to or in contact with earth or connected to some extended conductive body which serves instead of the earth.

Grounded junction A form of construction of a thermocouple probe where the hot or measuring junction is in electrical contact with the sheath material so that the sheath and thermocouple will have the same electrical potential.

Groundwater Water which is contained in geologic strata. Also properly written as two words, ground water.

Group A set of linked atoms in a molecule; a defined substructure. Typically, a set that is usefully regarded as a unit in chemical reactions of interest.

Group technology Group Technology is a coding and classification system to identify similarities in part geometry, features, characteristics and processes. Its is used to aid in design retrieval, part standardization, manufacturing cell design, and production scheduling.

Group velocity In wave propagation, the speed of the waveform (e.g., of a peak) can be different from the speed of a group of waves (e.g., of a set of ripples in water). The latter is the group velocity, and is the speed of propagation of information and wave energy. The waveform speed is the phase velocity.

Groupthink An undesirable condition in which all members of a group (e.g. a project team) begin to think alike or pretend to think alike. No members are then willing to raise objections or concerns about a project even though they are legitimate and based on hard data.

Groupware Groupware is any type of software designed for groups and for communication, including email, video conferencing, workflow, chat, and collaborative editing systems. This technology may be used to communicate, cooperate, coordinate, solve problems, compete, or negotiate. While traditional technologies such as the telephone qualify as groupware, the term is ordinarily used to refer to a specific class of technologies relying on modern computer networks, such as email, newsgroups, videophones, or chat. Groupware technologies are typically categorized along two primary dimensions: a)whether users of the groupware are working together at the same time ("realtime" or "synchronous" groupware) or different times ("asynchronous" groupware), and b) whether users are working together in the same place ("collocated" or "face-to-face") or in different places ("non-collocated" or "distance").

Grout A thin mixture of cementitious material and aggregate to which sufficient water is added to produce pouring consistency without segregation of the constituents. 1. High-Lift Grouting: The technique of grouting masonry in lifts up to 12 ft.

GUI (Graphical User Interface) A computer interface based on graphical symbols rather than text. Windowing environments and Macintosh environments are GUIs.

Guinier preston (G-P) zone A small precipitation domain in a supersaturated metallic solid solution; it has no well-defined crystalline structure of its own and contains an abnormally high concentration of solute atoms; the formation of G-P zones constitutes the first stage of precipitation and is usually accompanied by a change in properties of the solid solution in which they occur.

Gunpowder Any of several low-explosive mixtures used as a blasting agent in mining and tunneling; the first such explosive was black powder, which consists of a mixture of potassium nitrate, sulfur, and charcoal.

Gyration The range of gyration is a measure of the distribution of mass in a body segment. This is the rotational equivalent to mass, and is determined by the size, shape and density of the segment. The range of gyration is often converted to the Moment of Inertia by the Parallel Axis Theorem.

H Variously, the symbol for Henry, Magnetic Field Strength, or Magnetomotive Force.

Hacking Originally referred to playing around with computer systems; now often used to indicate destructive computer activity.

Half bridge Two active elements or strain gages.

Half bridge converter A switching power supply design in which two power switching devices are used to drive the transformer primary.

Half cell reaction The electrochemical reaction between the electrode and the electrolyte.

Half duplex One way at a time data communication; both devices can transmit and receive data, but only one at a time.

Half life The time in which one half of the atoms of a particular radioactive substance disintegrate into another nuclear form. Measured half-lives vary from millionths of a second to billions of years. Also called physical or radiological half-life.

Half life, effective The time required for a radionuclide contained in a biological system, such as a human or an animal, to reduce its activity by one-half as a combined result of radioactive decay and biological elimination.

Half thickness Any given absorber that will reduce the intensity of an original beam of ionizing radiation to one-half of its initial value.

Half wave rectifier A circuit element, such as a diode, that rectifies only one-half the input ac wave to produce a pulsating dc output.

Hall effect When a conductor carrying an electric current is placed in an external magnetic field perpendicular to the current there is voltage drop across the conductor at right angles to the current which is proportional to the magnetic field. Used to measure magnetic field strength.

Halogen headlamp A sealed-beam headlamp with a small inner bulb filled with halogen which surrounds a tungsten filament. Halogen headlamps may increase luminous intensity at the road surface by 50 to 80 percent, as compared to the long-conventional sealed-beam headlamp systems. Many halogen headlamp systems incorporate high-beam and low-beam in one element, enhancing their serviceability.

Handshake An interface procedure that is based on status/data signals that assure orderly data transfer as opposed to asynchronous exchange.

Hardenability A measure of the depth to which a specific ferrous alloy may be hardened by the formation of martensite upon quenching from a temperature above the upper critical temperature.

Hardness The measure of a material's resistance to deformation by surface indentation or by abrasion. There are various scales in use to express hardness. The Mohs scale is qualitative and somewhat arbitrary and ranges from 1 on the soft end for talc to 10 for diamond. Quantitative scales are the Rockwell (HR), Brinell (indicated by HB), Knoop (HK) and Vickers (HV). Knoop and Vickers are referred to as microhardness testing methods on the basis of load and indenter size.

Hardware Articles made of material, such as aircraft, ships, tools, computers, vehicles, fittings, and their components (mechanical, electrical, electronic, hydraulic, pneumatic). Computer software and technical documentation are excluded.

Hardware address The low usually corresponding to the unique identifier of the network interface card (NIC). Ethernet addresses are 48 bits.

Harmonic oscillator A system in which a mass is subject to a linear restoring force, like an ideal spring. A harmonic oscillator vibrates at a fixed frequency, independent of amplitude.

Hash table A method used for implementing associative arrays, which allows the keys to be converted to numbers for internal storage purposes.

Hastelloy A widely used nickel-molybdenum-chromium alloy. Offers excellent resistance to wet chlorine, hypochlorite bleach, ferric chloride and nitric acid.

Hazard assessment Evaluating the effects of a stressor or determining a margin of safety for an organism by comparing the concentration that causes effects with an estimate of exposure to the organism.

Head loss The loss of pressure in a flow system measured using a length parameter (i.e., inches of water, inches of mercury).

Head pressure Pressure in terms of the height of fluid, P = yrg, where r = fluid density and y = the fluid column heights. Expression of a pressure in terms of the height of fluid, r = yrg, where r is fluid density and y = the fluid column height. g = the acceleration of gravity.

Head, reactor vessel The removable top section of a reactor pressure vessel. It is bolted in place during power operation and removed during refueling to permit access of fuel handling equipment to the core.

Heads up display The digital projection of pertinent instrument data onto the lower portion of the windshield on the driver's side. The driver does not have to take his eyes off the road to read his instrument panel.

Health physics The science concerned with the recognition, evaluation, and control of health and environmental hazards that may arise from the use and application of ionizing radiation.

Heap leach A method of extracting uranium from ore using a leaching solution. Small ore pieces are placed in a heap on an impervious material (plastic, clay, asphalt) with perforated pipes under the heap. Acidic solution is then sprayed over the ore, dissolving the uranium. The solution in the pipes is collected and transferred to an ion-exchange system for concentration of the uranium.

Heat As defined in thermodynamics, heat is the energy that flows between two systems as a result of temperature differences (a system contains neither heat nor work, but can produce heat or do work). Heat thus differs from thermal energy.

Heat capacity The ratio of the heat input to the temperature increase in a system.

Heat engine A device that converts heat energy into mechanical work. Energy is transferred from a source at high temperature, part of it is converted into work and the rest is transferred to sink at a lower temperature.

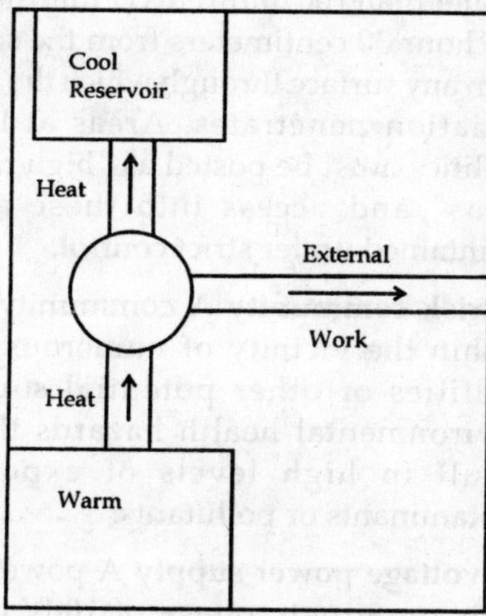

Fig. Heat engine

Heat exchanger Any device that transfers heat from one fluid (liquid or gas) to another fluid or to the environment.

Heat pump A device that absorbs heat from a cool environment and gives it off to a region of higher temperature.

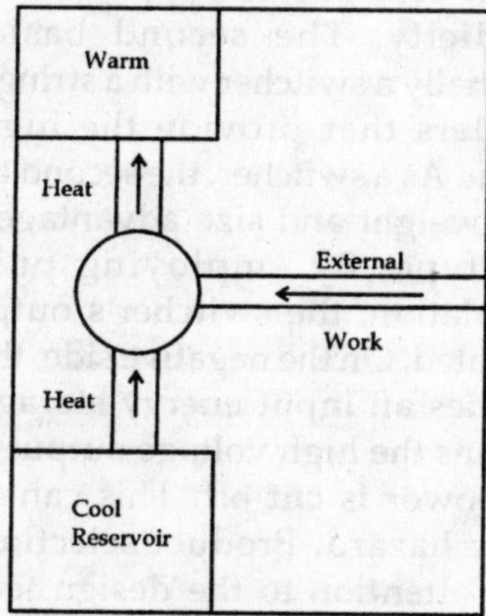

Fig. Heat pump

Heat sink 1. Thermodynamic. A body which can absorb thermal energy.

2. Practical. A finned piece of metal used to dissipate the heat of solid state components mounted on it.

Heat transfer The process of thermal energy flowing from a body of high energy to a body of low energy. Means of transfer are conduction; the two bodies contact. Convection; a form of conduction where the two bodies in contact are of different phases, i.e. solid and gas. Radiation: all bodies emit infrared radiation.

Heavy water (D_2O) Water containing significantly more than the natural proportions (one in 6,500) of heavy hydrogen (deuterium, D) atoms to ordinary hydrogen atoms. Heavy water is used as a moderator in some reactors because it slows down neutrons effectively and also has a low probability of absorption of neutrons.

Henry (H) Unit of inductance. One henry (H) is the inductance of a closed circuit in which an electromotive force of 1 volt is produced when the electric current in the circuit varies uniformly at the rate of 1 ampere per second.

Henry's law The amount of gas dissolved in a solution is directly proportional to the pressure of the gas above the solution.

Hertz (Hz) The SI unit of measurement for frequency, named in honor of Heinrich Hertz who discovered radio waves. One hertz equals one cycle per second.

Heterotrophic A group of organisms which obtain carbon for synthesis from other organic matter or proteins.

Hexadecimal Refers to a base sixteen number system using the characters 0 through 9 and A through F to represent the values. Machine language programs are often written in hexadecimal notation.

Hexagonal close packed (HCP) The unit cell is of hexagonal geometry and is generated by the stacking of close-packed planes of atoms.

HHV Heating Value is defined as the amount of energy released when a fuel is burned completely in a steady-flow process and the products are returned to the state of the reactants.

The heating value is dependent on the phase of water/steam in the combustion products. If H_2O is in liquid form the heating value is called HHV (Higher Heating Value).

Hibernation state A state in which the the status of the various functions of a circuit has been saved in memory and the circuit has been switched off save energy. When power is reapplied, data taken from the memory is used to restore the circuit to the status it had before switch off.

Hierarchical design Hierarchical Design is a design methodology where portions of large designs are divided into manageable sections or sub-blocks that may be created, represented symbolically, designed, and then connected together when completed. This methodology allows different parts of the design to be worked on in parallel.

High enriched uranium Uranium enriched to 20 percent or greater in the isotope uranium-235.

High level waste Radioactive materials at the end of a useful life cycle that should be properly disposed of, including:

1. The highly radioactive material resulting from the reprocessing of spent nuclear fuel, including liquid waste directly in reprocessing and any solid material derived from such liquid waste that contains fission products in concentrations;

2. Irradiated reactor fuel; and

3. Other highly radioactive material that the Commission, consistent with existing law, determines by rule require permanent isolation.

High-level waste (HLW) is primarily in the form of spent fuel discharged from commercial nuclear power reactors. It also includes HLW from activities and a small quantity of reprocessed commercial HLW .

High radiation area Any area with dose rates greater than 100 millirems (1 millisievert) in one hour 30 centimeters from the source or from any surface through which the ionizing radiation penetrates. Areas at licensee facilities must be posted as "high radiation areas" and access into these areas is maintained under strict control.

High risk community A community located within the vicinity of numerous sites of facilities or other potential sources of environmental health hazards that may result in high levels of exposure to contaminants or pollutants.

High voltage power supply A power supply with an output voltage of 100V or more. High Voltage power supplies often use air as the primary dielectric medium, avoiding the weight and serviceability problems associated with other dielectrics. Most high voltage supplies are of two types. The first type is simply a step-up power transformer with rectified and filtered (but otherwise unregulated) high voltage secondary. This is the most straightforward, efficient, and lowest cost power supply because of its simplicity. The second basic type is essentially a switcher with a string of voltage doublers that provide the high voltage output. As a switcher, the second type enjoys heat, weight and size advantages over the first type. By employing pulse width modulation, the switcher's output can be regulated. On the negative side, the switcher includes an input energy storage cap that sustains the high voltage output even if the AC power is cut off. This can result in a safety hazard. Product selection requires extra attention to the design features and construction techniques that contribute to safety and protection against failures. Pulse-width modulation and full-wave/ symmetrical cascade high voltage multipliers, with surge limiting resistors, contribute to high efficiency and reliability.

Highly accelerated life test Highly Accelerated Life Test (HALT) is a process developed to uncover design defects and weaknesses in electronic and mechanical assemblies using a vibration system combined with rapid high and low temperature changes. The purpose of HALT is to optimize product reliability by identifying the functional and destructive limits of a product. HALT addresses reliability issues at an early stage in product development.

Hindered (Zone) settling Settling in which particle concentrations are sufficient that particles interfere with the settling of other particles. Particles settle together as a body or structure with the water required to traverse the particle interstices.

Hip hiking A gait modification where the pelvis is lifted on the side of the swinging leg, by contraction of the spinal muscles and the lateral abdominal wall. Hip hiking is commonly used to slow walking with weak hamstrings, since the knee tends to extend prematurely, and thus to make the leg too long towards the end of swing phase. It is seldom employed for limb lengthening due to plantarflexion of the ankle

Hipot test (High Potential Test) A test performed by applying a high voltage for a specified time to two isolated points in a device to determine adequacy of insulating materials. For power supplies, it is often a test to determine if the breakdown voltage of a transformer or power supply exceeds the minimum requirement.

Hold Meter HOLD is an external input which is used to stop the A/D process and freeze the display. BCD HOLD is an external input used to freeze the BCD output while allowing the A/D process to continue operation.

Hole In a semiconductor, the term used to describe the absence of an electron; has the same electrical properties as an electron except that it carries a positive charge. (Graf).

Home page The document that your World Wide Web browser loads when it starts up. It should have links to other documents that you use frequently. Also, the main entry point to a site is sometimes called its home page (the default first page for that site).

Homogeneous and heterogeneous assays A homogeneous assay does not require a separation step to remove free antigen from bound antigen and relies upon the fact that the function of the label is modified upon binding, leading to a change in signal intensity. Because of high background signal a heterogeneous approach incorporating a separation step of bound and unbound makes the detection limit lower, approaching the values obtained by RIA. The homogeneous assay is less technically demanding.

Hooke's law Defines the basis for the measurement of mechanical stresses via the strain measurement. The gradient of Hooke's line is defined by the ratio of which is equivalent to the Modulus of Elasticity E (Young's Modulus).

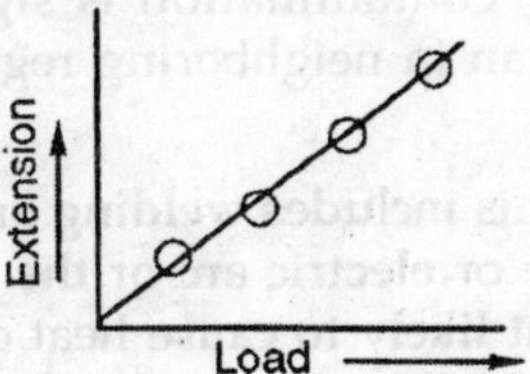

Fig. Hooke's Law

Horse power (Hp) The rate of doing work. 1 Hp = 746 Watts or 550 foot pounds per second.

Host address A unique number assigned to identify a host on the Internet (also called IP address or dot address). This address is usually represented as four numbers between 1 and 254 and separated by periods, for example, 192.58.107.230.

Host name A unique name for a host that corresponds to the host address.

Hot A colloquial term meaning highly radioactive.

Hot isostatic pressing 1. A process for simultaneously heating and forming a compact in which the powder is contained in a sealed flexible sheet metal or glass enclosure and subjected to equal pressure from all directions at a temperature high enough to permit plastic deformation and sintering;

2. A process that subjects a component (casting, powder forgins, etc.) to both elevated temperature and isostatic gas pressure in an autoclave; simultaneous application of heat and pressure virtually eliminates internal voids and microporosity through a combination of plastic deformation, creep, and diffusion.

Hot pressing A method used to densify a material, whereby heat and pressure are applied simultaneously, and the pressure is typically applied unidirectionally via rigid tooling.

Hot spot The region in a radiation/contamination area where the level of radiation/contamination is significantly greater than in neighboring regions in the area.

Hot work This includes welding or the use of any flame or electric arc or the use of any equipment likely to cause heat or flame or spark.

HP High Pressure. Often used with reference to a utility or vent line e.g. HP air supply.

Human factors Human Factors refers to the characteristics of human beings that are applicable to the design of systems and devices of all kinds. It furthers serious consideration of knowledge about the assignment of appropriate functions for humans and machines, whether people serve as operators, maintainers, or users in the system. And, it advocates systematic use of such knowledge to achieve compatibility in the design of interactive systems of people, machines, and environments to ensure their effectiveness, safety, and ease of performance.

Hurdle rate The minimum return on investment or internal rate of return percentage a new product must meet or exceed for it to be approved for investment with development.

HVAC Heating Ventilating and Air Conditioning.

Hybrid Hybrid circuit. Any circuit made by using a combination of the following component manufacturing technologies: monolithic IC , thin film , thick film and discrete component.

Hybrid electric vehicle (HEV) A vehicle which has two forms of motive power one of which is electric.

Hydrocarbon Any organic compound composed entirely of carbon and hydrogen. Two examples are methane gas and octane.

Hydrogen bond A hydrogen atom covalently bound to an electronegative atom (e.g., nitrogen, oxygen) has a significant positive charge and can form a weak bond to another electronegative atom; this is termed a hydrogen bond.

Hydrogen ion activity (aH+) Activity of the hydrogen ion in solution. Related to hydrogen ion concentration (CH^+) by the activity coefficient for hydrogen (H^+).

Hydrogen sulfide (H_2S) Gas emitted during organic decomposition. Also a byproduct of oil refining and burning. Smells like rotten eggs and, in heavy concentration, can kill or cause illness.

Hydrometer A device used for measuring the specific gravity of a fluid. In the case of lead acid batteries the specific gravity provides a measure of the state of charge of the cell.

Hydrophobic force Water molecules are linked by a network of hydrogen bonds. A nonpolar, nonwetting, surface (e.g., wax) cannot form hydrogen bonds. To form their full complement of hydrogen bonds, the nearby water molecules must form a more orderly (hence lower entropy) network. This both increases free energy and causes forces that tend to draw hydrophobic surfaces together across distances of several nanometers.

Hydrotest A pressure test of piping; pressure vessels or pressure-containing parts; usually performed by pressurising the internal volume with water.

Hygrometer An instrument for measuring humidity. Often confused with a hydrometer.

Hypereutectoid alloy For an alloy system displaying a eutectoid, an alloy for which the concentration of solute is greater than the eutectoid composition.

Hyperlinks The areas (words or graphics) in an HTML document that cause another document to be loaded when the user clicks them.

Hypermobility Any motion occurring in a joint in response to the reactive force of gravity at a time when that joint should be stable under such a load. This term is often misused to describe extra movement as seen in a contortionist, for example. Hypermobility is synonymous with instability.

Hypertext An online document that has words or graphics containing links to other documents. Usually, selecting the link area on-screen (with a mouse or keyboard command) activates these links.

Hypoeutectoid alloy For an alloy system displaying a eutectoid, an alloy for which the concentration of solute is less than the eutectoid composition.

Hypolimnion The lower layer of a lake.

Hysteresis A property of physical and chemical systems that do not instantly follow the forces applied to them, but react slowly, or do not return completely to their original state. In the case of magnetic systems, when an external magnetic field is applied to a magnetic material, the material becomes magnetised absorbing some of the external field. When the external field is removed the material remains magnetised to some extent, retaining some magnetic field.

Hysteresis loop A closed curve that shows, for each value of magnetizing force, two values of the magnetic flux density in a cyclically magnetized material: one when the magnetizing force is increasing, the other when it is decreasing.

Hysteresis loss Energy dissipated due to molecular friction as domains move through cycles of magnetization.

I Symbol for electric Current.

I/O Input/output information transfer between computer and peripherals such as keyboard or printer.

Icon A graphic functional symbol display. A graphic representation of a function or functions to be performed by the computer.

IEC The International Electrotechnical Commission (IEC) is the leading global organization that prepares and publishes international standards for all electrical, electronic and related technologies.

IGBT Insulated Gate Bipolar Transistor. It has the output switching and conduction characteristics of a bipolar power transistor but is voltage controlled like the MOSFET giving the high current carrying capability of the bipolar transistor but the ease of control of the MOSFET.

Immobilised electrolyte A construction technique used in lead-acid batteries. The electrolyte (the acid) is held in place against the plates instead of being a free-flowing liquid. The two most common techniques are Gel Cell and Absorbed Glass Mat.

Impedance (Z) Total resistance to flow of an alternating current as a result of resistance and reactance. Also the ratio of voltage to current in AC circuits, containing both resistance and reactance terms, usually expressed as ohms.

Impedance matching (Z Match) The connection across a source impedance of a matching impedance to allow optimum undistorted energy transfer.

Impedance testing Determination of the battery's internal impedance by measuring the voltage drop across a cell when it carries a sample alternating current.

In shoe device These instruments are used to measure the interface between the plantar surface of the foot and the superior surface of the foot orthosis. They usually consist of a ultra thin electronic sensors placed on paper thin templates and inserted into the subjects shoes. Collected data is conveyed to computer software for translation.

In situ In its original place; unmoved unexcavated; remaining at the site or in the subsurface.

In situ flushing Introduction of large volumes of water, at times supplemented with cleaning compounds, into soil, waste or ground water to flush hazardous contaminants from a site.

In situ leach A process using a leaching solution to extract uranium from underground ore bodies in place (in other words, in situ). The leaching agent, which contains an oxidant such as oxygen with sodium carbonate, is injected through wells into the ore body in a confined aquifer to dissolve the uranium. This solution is then pumped via other wells to the surface for processing.

In situ oxidation Technology that oxidizes contaminants dissolved in ground water, converting them into insoluble compounds.

In situ stripping Treatment system that removes or "strips" volatile organic compounds from contaminated ground or surface water by forcing an airstream through the water and causing the compounds to evaporate.

In situ vitrification Technology that treats soil in place at extremely high temperatures, or at more than 3,000 degrees F.

In vitro From the Latin for "in glass," isolated from the living organism and artificially maintained, as in a test tube.

In vivo From the Latin for "in one that is living," occurring within the living.

Inclusion Foreign particle present as an undesirable impurity in a material.

Incoloy A nickel-chromium alloy. Noted for good strength and excellent resistance to oxidation and carburisation in high-temperature atmospheres.

Inconel A nickel-chromium-iron alloy. Noted for having high temperature strength while maintaining excellent corrosion resistance.

Independent suspension A term used to refer to any type of suspension system that allows each of the two wheels of a given axle to move up and down independently of each other.

Index of refraction (n) The ratio of the velocity of light in a vacuum to the velocity in some medium.

Indirect costs Costs that are incurred for common or joint objectives that can not be readily identified to a single cost objective and readily treated as a direct cost. A cost that is allocated as opposed to being traced.

Individual plant examination (IPE) As requested by the NRC in Generic Letter 88-20, "Individual Plant Examination for Severe Accident Vulnerabilities" (November 23, 1988), a risk analysis that considers the unique aspects of a particular nuclear power plant, identifying the specific vulnerabilities to severe accident of that plant.

Individual plant examination for external events (IPEEE) While the "individual plant examination" takes into account events that could challenge the design from things that could go awry internally (in the sense that equipment might fail because components do not work as expected), the "individual plant examination for external events" considers challenges such as earthquakes, internal fires, and high winds.

Induced current Current that flows as a result of an Induced EMF (Electromotive Force).

Induced radioactivity Radioactivity that is created when stable substances are bombarded by ionizing radiation. For example, the stable isotope cobalt-59 becomes the radioactive isotope cobalt-60 under neutron bombardment.

Inductance (in Henry, H) That property of an electric circuit which tends to oppose change in current in the circuit. One henry (H) is the inductance of a closed circuit in which an electromotive force of 1 volt is produced when the electric current in the circuit varies uniformly at the rate of 1 ampere per second.

Inductive charging A charger in which the charging current is induced by an external induction coil into a secondary transformer winding housed within the battery together with rectifying and charge control circuits.

Inductive circuit Circuit in which an EMF is produced by a changing current.

Inductive reactance (XL) Opposition to a changing current as a result of inductance:

Inductor Energy storage circuit component consisting of a coil of wire and possibly a magnetic material.

Industrial design Industrial Design is the design that is done in companies and consultancies by people trained in industrial design, or in art and design schools in general. Industrial design focuses on the physical form and interactive properties as opposed to the functioning of the product or system.

Industrial engineering Industrial engineering identifies the most effective ways for an organization to use the basic factors of people, processes, technology, materials, information, and energy to make or process a product. Industrial engineers plan, design, implement, and manage integrated production and service delivery systems that ensure performance, reliability, and maintainability.

Inertia This is a descriptive term used to explain the resistance offered by a body to attempt to set it in motion, or to stop it if it is already moving. Inertia has no physical quantitative value.

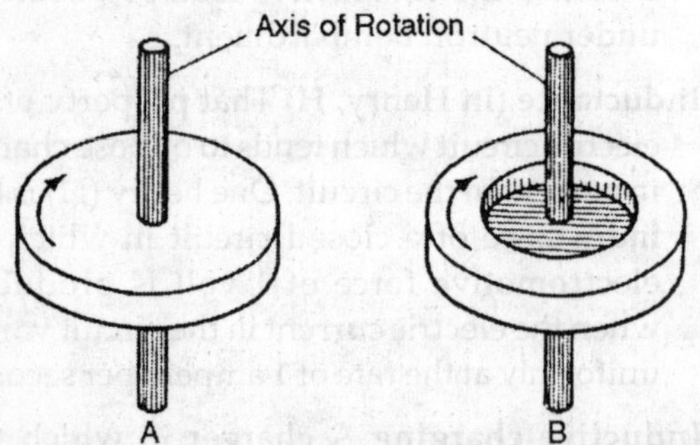

Fig. Moment of inertia

Infectious disease A disease caused by pathogenic organisms.

Inferior Allied to structures near the feet; as contrasted to superior, nearer the head.

Infiltraticn The movement of water from the surface of the land through the unsaturated zone and into the groundwater. This occurs during and immediately after precipitation events. It can also occur at the bottom of lakes and rivers.

Influent The fluid entering a system, process, tank, etc. An effluent from one process can be an influent to another process.

Infra red radiation The spectrum of the heat radiated by a warm body.

Infrared An area in the electromagnetic spectrum extending beyond red light from 760 nanometers to 1000 microns (106 nm). It is the form of radiation used for making non-contact temperature measurements.

Infrared reflow Technique in which long wavelength light (IR) serves as the heat source to reflow solder and form solder joints.

Inhibitor A substance added to the electrolyte to prevent or slow down an unwanted electrochemical process. Used to prevent corrosion of the electrodes or the formation of dendrites.

Initial sequence number (ISN) A number defined during the startup of a connection using TCP. Used to number datagrams.

Initial unbalance Initial unbalance is that unbalance of any kind that exists in the rotor before balancing.

Initiate In TCP/IP, to send a request for something (usually a connection).

Inorganic flux An aqueous flux solution of inorganic acids and halides.

Input AC The sinewave input voltage normally specified in volts RMS. An input waveform that is not sinusoidal, such as a square wave or distorted sinewave wave (from a UPS) can affect power supply operation and must be defined. Contact Professor Power for information on derating of standard catalog power supplies for voltage, frequency, and wave shape variations beyond those specified in the catalog.

Input DC The DC level normally specified in volts. Lower DC voltages can cause heating and limit output power. Contact Professor Power for information on derating at low input voltage.

Input impedance The resistance of a panel meter as seen from the source. In the case of a voltmeter, this resistance has to be taken into account when the source impedance is high; in the case of an ammeter, when the source impedance is low.

Input line filter A low-pass or band-reject filter at the input of a power supply which reduces line noise fed to the supply or fed onto the input power lines. This filter may be external to the power supply.

Input reflected ripple current (DC/DC converters) The AC ripple current measured at the input and generated by the switching action of the converter. Low ESR (Effective Series Resistance) input capacitors are used to reduce the input ripple voltage caused by the ripple current.

Input resistance (Impedance) The input resistance of a pH meter is the resistance between the glass electrode terminal and the reference electrode terminal. The potential of a pH-measuring electrode chain is always subject to a voltage division between the total electrode resistance and the input resistance.

Input voltage range The range of input voltage values for which a power supply or device operates within specified limits.

Inrush current The peak instantaneous input current drawn at turn on by a power supply.

Inrush current limiting The characteristic of a circuit that limits inrush current when a power supply is turned on.

Insert mouldings Plastic parts containing metal inserts used to simplify product assembly and reduce costs. Inserts made from metal or other materials are placed in the mould prior to the injection of plastic. The plastic flows around the inserts and fixes their position.

Insertion point (in lithography context) Adaptation of a new lithography technique is referred to as the insertion point of that technique.

Instrinsically safe Term applied to equipment and wiring which is incapable of releasing sufficient electrical or thermal energy under normal or abnormal conditions to cause ignition of a specific hazardous atmospheric mixture in its most easily ignited concentration.

Insulation Non-conductive materials used to separate electric circuits.

Insulation resistance The resistance measured between two insulated points on a transducer when a specific dc voltage is applied at room temperature.

Insulator (electric) A nonmetallic material that has a filled valence band at 0 K and a relatively wide energy band gap; consequently the room-temperature electrical conductivity is very low.

Integral A form of temperature control.

Integrated circuit A microcircuit that consists of interconnected elements inseparably associated and formed in-situ on or within a single substrate, usually silicon, to perform an electronic circuit function.

Integrated logistics support (ILS) A disciplined, unified and iterative approach to management and technical activities necessary to: · integrate support considerations into system and equipment design; · develop support requirements that are related consistently to design, readiness objectives and to each other; · acquire required support; and · provide required support during the operational phase at minimum cost.

Integrated magnetics A magnetic component in which separate circuit elements share a common core segment.

Integrated plant evaluation An evaluation that considers the plant as a whole rather than system by system.

Integrated service digital network (ISDN) A set of standards for integrating multiple services (voice, data, video, and so on).

Intelligent battery Battery containing circuitry enabling some communication between the battery and the application or with the charger.

Intelligent charger Charger which is able to react to inputs from an intelligent battery to control or optimise the charging process.

Intelligent energy manager IEM A system for reducing the demands that power hungry applications place on the battery.

Interchangeability error A measurement error that can occur if two or more probes are used to make the same measurement. It is caused by a slight variation in characteristics of different probes.

Interdiffusion Diffusion of atoms of one metal into another metal.

Interface The functional and physical characteristics required to exist at a common boundary. In development, a relationship among two or more entities in which the entities share, provide, or exchange data. An interface is not a software unit, or other system component; it is a relationship among them.

Interface control Interface control comprises the delineation of the procedures and documentation, both administrative and technical, contractually necessary for identification of functional and physical characteristics between two or more configuration items which are provided by different contractors/acquiring agencies, and the resolution of the problems thereto.

Interface design compatibility Intra-system and intersystem design compatibility of engineering interfaces shall be delineated as interface requirements in appropriate specifications. Interface control requirements and drawings related to: · the major system elements of a prime contractor's contractual responsibility; · other equipment, computer software, facilities, and procedural data furnished by the acquirer;

Intergranular fracture Fracture of polycrystalline materials by crack propagation along grain boundaries.

Interior gateway protocol (IGP) A protocol used by gateways in an autonomous system to transfer routing information.

Interleaved transformer windings The segmentation of portions of the same winding by another to improve or control inductive coupling.

Intermediate solid solution A solid solution or phase having a composition range that does not extend to either of the pure components of the system.

Intermolecular Describes an interaction (e.g., a chemical reaction) between different molecules.

Internal energy The sum of the kinetic and potential energies (including electromagnetic field energies) of the particles that make up a system.

Internal impedance Resistance to the flow of AC current within a cell. It takes into account the capacitive effect of the plates forming the electrodes.

Internal reference electrode (Element) The reference electrode placed internally in a glass electrode.

Internal resistance (Apparent) Quotient of the difference of voltage across battery terminals to the corresponding difference of current. It should be observed that the internal resistance is not constant but varies with state of charge, temperature and the testing method.

International organization for standardization (ISO) An international body composed of individual countries' standards groups that focuses on international standards.

Internet A collection of networks connected together that span the world that uses the NFSNET as its backbone. The Internet is the specific term for a more general internetwork or collection of networks. The term used to describe all the worldwide interconnected TCP/IP networks.

Internet activities board (IAB) The Internet group that coordinates the development of the TCP/IP protocol suite.

Internet address A 32 digit in the format 4X3

Internet control message protocol (ICMP) A control and error message protocol that works in conjunction with the Internet Protocol (IP).

Internet explorer A Microsoft Windows 95 Web browser.

Internet protocol (IP) The part of TCP/IP that handles routing.

Internet research task force (IRTF) A part of the IAB that concentrates on research and development of the TCP/IP protocol suite.

Internic The NSFNET manager sites on the Internet that provide information about the Internet.

Interpreter A system program that converts and executes each instruction of a high-level language program into machine code as it runs, before going onto the next instruction.

Interrupt To stop a process in such a way that it can be resumed.

Interstitial diffusion A diffusion mechanism that causes atomic motion from interstitial site to interstitial site.

Interstitial site Octahedral and tetrahedral open spaces within a close-packed arrangment of atoms or ions in which a cation can fit.

Interstitial solid solution A solid solution wherein relatively small solute atoms occupy interstitial positions between the solvent or host atoms.

Intramolecular Describes an interaction (e.g., a chemical reaction) within a single molecule. Intramolecular interactions between widely separated parts of a molecule resemble intermolecular interactions in most respects.

Intrinsic Characterizes pure undoped semiconductor; electrical conductivity depends only on temperature and the band gap energy.

Intrinsically safe An instrument which will not produce any spark or thermal effects under normal or abnormal conditions that will ignite a specified gas mixture.

Inversion This describes motion in the frontal plane, where the plantar surface of the foot, or part of the foot, is tilted to face the midline of the body i.e. the plantar surface of the foot is directed toward the medial sagittal plane. The axis of motion lies on the sagittal and transverse planes. A fixed inverted position is referred to as a varus deformity.

Inverted A foot, or part of a foot, is in an inverted position when the foot or part is tilted in the frontal plane, so the plantar surface of the foot, or part of the foot faces toward the midline of the body and away from the transverse plane, or away from some other specified reference point.

Inverter An electrical circuit which generates a sine-wave output (regulated and without breaks) using the DC current supplied by the rectifier-charger or the battery. The primary elements of the inverter are the DC/AC converter, a regulation system and an output filter.

Investment casting This process is based on surrounding (investing) an expendable pattern, typically wax, with a ceramic mold and then removing (by melting or vaporizing) the pattern prior to pouring molten alloy into the mold; also known as lost wax and precision casting.

Iodine spiking factor The magnitude of a rapid, short-term increase in the appearance rate of radioiodine in the reactor coolant system. This increase is generally caused by a reactor transient that results in a rapid drop in reactor coolant system pressure relative to the fuel rod internal pressure.

Ion 1. An atom that has too many or too few electrons, causing it to have an electrical charge, and therefore, be chemically active.

2. An electron that is not associated (in orbit) with a nucleus.

Ion exchange A common method for concentrating uranium from a solution. The uranium solution is passed through a resin bed where the uranium-carbonate complex ions are transferred to the resin by exchange with a negative ion like chloride. After build-up of the uranium complex on the resin, the uranium is eluted with a salt solution and the uranium is precipitated in another process.

Ionic bonding A coulombic interatomic bond that exists between two adjacent and oppositely charged ions; one of the primary types of atomic bonding in ceramics.

Ionic mobility Defined similarly to the mobility of nonelectrolytic particles, viz., as the speed that the ion obtains in a given solvent when influenced by unit power.

Ionic strength The weight concentration of ions in solution, computed by multiplying the concentration of each ion in solution (C) by the corresponding square of the charge on the ion (Z) summing this product for all ions in solution and dividing by 2:ionic strength - 1/2 _ Z2 C.

Ionization The process of adding one or more electrons to, or removing one or more electrons from, atoms or molecules, thereby creating ions. High temperatures, electrical discharges, or nuclear radiations can cause ionization.

Ionization chamber An instrument that detects and measures ionizing radiation by measuring the electrical current that flows when radiation ionizes gas in a chamber, making the gas a conductor of electricity.

Ionizing radiation Any radiation capable of displacing electrons from atoms or molecules, thereby producing ions. Some examples are alpha, beta, gamma, x-rays, neutrons, and ultraviolet light. High doses of ionizing radiation may produce severe skin or tissue damage.

Ionophore A macro-organic molecule capable of specifically solubilizing an inorganic ion of suitable size in organic mediums.

IP address A 32 digit in the format 4X3

IP datagram The basic unit of information passed through a TCP/IP network. The datagram header contains source and destination IP addresses.

IPTS-48 International Practical Temperature Scale of 1948. Fixed points in thermometry as specified by the Ninth General Conference of Weights and Measures which was held in 1948.

IR drop The voltage drop across a battery due to its internal impedance.

Iron A chemical element (Fe); one of the cheapest and most used metals.

Iron core A general class of cores that contain iron. Sometimes used to refer to cores made only from steel laminations.

Irradiation Exposure to radiation.

Irreversible reaction A reaction in which the reactant(s) proceed to product(s), but the products react at an appreciable rate to reform reactant(s).

ISC Integrated Service Contract. A contract likely to include design and project services; maintenance; upgrades as well as reliability and integrity management.

More extensive than an Alliance Contract.

ISE (Ion selective electrode) Ions in solution are quantified by measuring the change in voltage (i.e. potentiometric) resulting from the distribution of ions (by ion exchange controlled by the ion exchange current io) between a sensing membrane (the ion selective membrane) and the solution. This potential is measured at zero current with respect to a reference electrode which is also in contact with the solution. The potential measured is proportional to the logarithm of the analyte concentration. The oldest and best known ISE is the pH sensor based on a glass membrane. More recently, polymeric membranes have been formed incorporating ionophores rendering the membrane specific to certain ions only.

ISFET Ion Sensitive Field Effect Transistor: a logical extension of ISE's. They can be conceptualized by imagining that the lead from an ion-selective electrode, attached via a cable to a FET in the high impedance input stage of a voltmeter, is made shorter until no lead exists and the selective membrane is attached directly to the FET..For an ISFET, the property measured is the lateral conductivity between two opposing doped regions (the source and drain) surrounding the active area. The underlying change is a change in flat-band voltage.

ISO 10303 An ISO technical standard for product data representation and exchange commonly referred to as STEP or the Standard for the Exchange of Product Model Data.

ISO9000 The International Standards Organization quality assurance system is gaining popularity around the world. ISO9000 is the overall standard. ISO9001 is the most stringent standard, which requires a complete quality system including engineering and design, resource planning, documentation, installation and service. ISO9002 contains quality requirements for product production and installation. ISO9003 pertains only to inspection and testing. ISO9004 gives detailed directions for implementing each of the ISO9000 standards.

ISO 9001 A quality assurance standard issued by the International Organization for Standards.

Standard is titled; Quality systems - Model for quality assurance in design/ development, production, installation and servicing.

ISO A network of national standards institutes from 148 countries working in partnership with international organizations, governments, industry, business and consumer representatives.

ISO reference model The seven functions within each layer.

ISO/TS 16949:2002 An ISO technical standard titled "Quality management system - particular standards for the application of ISO 9001:2000 for automotive production and relevant service part organizations". This standard replaces QS-9000 and harmonizes requirements for automotive manufacturers internationally.

Isobaric Describes barometric pressure

ISODE (ISO Development Environment) An attempt to develop software that enables OSI protocols to run on TCP/IP.

Isoelectronic Two molecules are described as isoelectronic if they have the same number of valence electrons in similar orbitals, although they may differ in their distribution of nuclear charges (e.g., H-CN and H-N+C-. KineticPertaining to the rates of chemical reactions. A fast reaction is said to have fast kinetics; if the balance of products in a reaction is controlled by reaction rates rather than by thermodynamic equilibria, the reaction is said to be kinetically controlled.

Isolation The electrical separation between input and output of a power supply by means of the power transformer. The isolation resistance (normally in megaohms) and the isolation capacitance (normally in picofarads) are generally specified and are a function of materials and spacings employed throughout the power supply.

Isolation transformer A transformer with a one-to-one turns ratio.

Isolation voltage The maximum ac or dc specified voltage that may be continuously applied between isolated circuits.

Isomerism The phenomenon whereby two or more polymer molecules or mer units have the same composition but different structural arrangements and properties.

Isomers Two or more different compounds with the same chemical formula but different structure and characteristics.

Isomorphous Having the same structure. In the phase diagram sense, isomorphicity means having the same crystal structure or complete solid solubility for all compositions.

Isopotential point A potential which is not affected by temperature changes. It is the pH value at which dE/dt for a given electrode pair is zero. Normally, for a glass electrode and SCE reference, this potential is obtained approximately when immersed in pH 7 buffer.

Isotactic A type of polymer chain configuration where all side groups are positioned on the same side of the chain molecule.

Isothermal At a constant temperature. In an isothermal process heat is, if necessary, supplied or removed from the system at just the right rate to maintain constant temperature.

Isotope Any two or more forms of an element having identical or very closely related chemical properties and the same atomic number but different atomic weights or mass numbers.

Isotope separation The process of separating isotopes from one another, or changing their relative abundances, as by gaseous diffusion or electromagnetic separation. Isotope separation is a step in the isotopic enrichment process.

Item A non-specific term used to denote any product, including systems, subsystems, assemblies, subassemblies, units, sets, accessories, computer programs, computer software or parts.

Izod test A type of impact test in which a V-notched specimen, mounted vertically, is subjected to a sudden blow delivered by the weight at the end of a pendulum arm; the energy required to break off the free end is a measure of the impact strength or toughness of the material.

J

J Symbol for Joule.

Jacket Steel structure placed on the seabed with a deck supporting drilling and/or production facilities.

JAM An Ethernet term for communicating with all devices on a network on which a collision has occurred.

Jitter A term used with 10BaseT (twisted of phase with one another.

Joint A device connecting two or more adjacent parts of a structure; a roller joint allows adjacent parts to move controllably past one another; a rigid joint prevents adjacent parts from moving or rotating past one another.

Josephson effect The flow of electric current through nonconductive material when placed between two superconductors. Used to detect very weak magnetic fields.

Joule This is a SI unit of measurement used to record quantity of work done. When a force moves an object over a distance, the work done is calculated as the product of the force and the distance. For example, if a force of two newtons moves an object three meters, the work done is 2 (n) x 3 (meters) = 6 newton/meters or Joules. This unit can also be used to measure energy, which is the capacity to work.

Joule heating The I2R loss or heating effect of a current I flowing through a resistance **R**.

Journal A journal is that part of a rotor that is in contact with or supported by a bearing in which it revolves.

Junction The point in a thermocouple where the two dissimilar metals are joined.

Junction field effect transistor (JFET) A low power semiconductor having a conductive channel whose resistance is controlled by the reverse voltage on the gate channel junction.

Just in time A system for producing and delivering the right items at the right time in the right amounts. The key elements of just in time are flow, pull, standard work and takt time.

JV Joint Venture. A collaboration between two or more companies in a contract.

K When referring to memory capacity, two to the tenth power (1024 in decimal notation).

Kaizen A Japanese word that means "continuous improvement." It refers to small, incremental improvement of an activity to create more value with less waste. A kaizen event is a highly focused, action-oriented workshop that typically involves a team of five to 15 individuals. It usually lasts three to five days. The goal of a kaizen event is to concentrate on improving one specific process.

Kalman filter A mathematical technique for deriving accurate information from inaccurate data.

Kanban A Japanese word that means "card" or "visible record." It refers to a small card attached to boxes of parts that regulates pull by signaling upstream production and delivery.

Kelvin (K) 1. Unit of temperature in the International System of Units (SI) equal to the fraction 1/273.16 of the thermodynamic temperature of the triple point of water. The Kelvin temperature scale uses Celsius degrees with the scale shifted so 0 K is at absolute zero. Add 273.16 to any Celsius value to obtain the corresponding value in Kelvins.

2. A technique using 4 terminals to isolate current carrying leads from voltage measuring leads.

Kelvin bridge An electrical circuit for measuring very low impedances such as battery internal impedance, contact resistance and resistance of circuit elements such as wires and cables. Also known as the Kelvin Connection for voltage sensing.

KEMA Notified Body based in the Netherlands.

Kerberos An authentication scheme developed at MIT used to prevent unauthorized monitoring of logins and passwords.

Kerogen A fossilized organic material present in oil shale and some other sedimentary rocks.

Ketones Organic compounds with two hydrocarbon groups bonded to a carbonyl group.

KEY In an associative array, a series of unique keys are associated with values.

Keyed connectors Plug and socket pairs with a unique mechanical profile which can only be mated with eachother in a particular orientation and which do not allow mating with connectors of a different design.

Killed steel Steel treated with a strong deoxidizing agent, such as aluminum or silicon, in order to reduce the oxygen content to a level so no reaction occurs between carbon and oxygen during solidification.

Kiln A furnace in which ceramics are fired.

Kilo A Greek prefix meaning "thousand" in the nomenclature of the metric system. This prefix multiplies a unit by 1000.

Kilobyte (kB) 210 (= 1024, or about one thousand) bytes of information.

Kilohertz (kHz) One thousand cycles per second.

Kilovolt The unit of electrical potential equal to 1000 volts.

Kilowatt hour (kwh) 1000 watthours. Kilovolt amperes (kva): 1000 volt amps.

Kinematics This describes motion but without reference to the forces involved. An example of a kinetic instrument is a camera, which can be used to observe the motion of the trunk and the limbs during walking, but which gives no information on the forces.

Kinesology The science of movements considered especially as therapeutic or hygiene agencies.

Kinetic energy Energy associated with mass in motion, i.e., 1/2 rV2 where r is the density of the moving mass and V is its velocity.

Kinetic This is the study of force, moments, mass and acceleration but without the detailed knowledge of the position and orientation of the objects involved. A forceplate is used to measure force under the foot during gait.

Kinetic molecular theory A model that assumes that an ideal gas is composed of tiny particles (molecules) in constant motion.

Kistler force plate A force plate is used to measure ground reaction force as the subject walks across it. The force plate consists of a top and bottom plate, supported by four legs. Each leg is a sensitive instrument which can measure force and the completed data is formed by their output. The instrument measures both medial and lateral as well as vertical force. For the purpose of biomechanics, the force is measured on a x, y and z axes, where x is the direction of walking; y is taken as vertical; and z axis is taken to be the right-handed orthogonal axis system. All force plates achieve the same objectives the Kistler use piezzo-electric while AMTI use strain gauges.

Kitting A process in which assemblers are supplied with kits—a box of parts, fittings and tools—for each task they perform. This eliminates time-consuming trips from one parts bin, tool crib or supply center to another to get the necessary material.

Knoop hardness test An indentation hardness test using calibrated machines to force a rhombic-based pyramidal diamond indenter having specified edge angles, under specified conditions, into the surface of the test material and to measure the long diagonal after load removal.

Knowledge based engineering Knowledge-Based Engineering is a set of design automation tools that capture design knowledge and rules to automate the design process.

Knowledge management The overall management process to capture, organize, manage and disseminate knowledge in an organization to improve enterprise effectiveness by avoiding mistakes and avoiding the time to relearn needed knowledge. Since product development is very knowledge intensive, knowledge management offers tremendous leverage and opportunity for improvement.

KVA Kilovolt amperes (1000-volt amps).

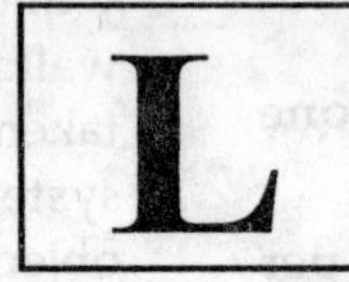

L Variously, the symbol for Inductance and length, and the abbreviation for liter.

Label or marker A problem endemic in immunoassays is the absence of a chemical signal created by the antibody-antigen binding, in contrast with an enzyme-substrate binding reaction which produces a chemical reaction product. As a result of this absence, the use of a label or marker is usually required to detect the bound antibody-antigen complex. Several markers have been established for use in immunoassays. Examples of such markers are:

- Particles (e.g. latex, gold particles, erythrocytes)

- Metal and dye sols (e.g. Au, Palanil®; Luminous Red G)

- Chemiluminescent and bioluminescent compounds (e.g. Luciferase/luciferins, Luminol and derivatives, Acridinium esters, Peroxidase)

- Electrochemical active species (ions, redox species, ionophores)

- Fluorophores (e.g. Dansyl chloride DANS, Rare earth metal chelates, Umbelliferones)

- Chromophores

- Enzymes (e.g. Alkaline phosphatase, â-D-Galactosidase, Peroxidase), substrates, cofactors

- Liposomes

- Iodine-125, tritium, 14C, 75Se, 57Co.

Lag 1. A time delay between the output of a signal and the response of the instrument to which the signal is sent.

2. A time relationship between two waveforms where a fixed reference point on one wave occurs after the same point of the reference wave.

Laminar flow Streamlined flow of a fluid where viscous forces are more significant than inertial forces, generally below a Reynolds number of 2000.

LAN (Local Area Network) A collection of devices connected to enable communications between themselves on a single physical medium. A network of computers that is limited to a (usually) small physical area, like a building.

Landfills: 1. Sanitary landfills are disposal sites for non-hazardous solid wastes spread in layers, compacted to the smallest practical volume and covered by material applied at the end of each operating day.

2. Secure chemical landfills are disposal sites for hazardous waste, selected and designed

to minimize the chance of release of hazardous substances into the environment.

Lapping A surface finishing operation used to achieve a fine polish and close tolerances.

Large scale integration (LSI) The combining of about 1,000 to 10,000 circuits on a single chip. Typical examples of LSI circuits are memory chips and microprocessor.

Laser Acronym for light amplification by stimulated emission of radiation; a source of concentrated coherent light generated by stimulating electronic or molecular transitions to lower energy levels.

Laser photoplotter (also "laser plotter") A photoplotter which simulates a vector photoplotter by using software to create a raster image of the individual objects in a CAD database, then plotting the image as a series of lines of dots at very fine resolution. A laser photoplotter is capable of more accurate and consistent plots than a vector photoplotter.

Laser soldering Method of soldering in which the heat required to reflow a solder interconnection is provided by a laser (YAG or CO_2); the solder joints are heated sequentially and cooled rapidly.

Laser trimming A method for adjusting the value of thin- or thick-film resistors by using a computer-controlled laser system.

Latent heat Expressed in BTU per pound. The amount of heat needed (absorbed) to convert a pound of boiling water to a pound of steam.

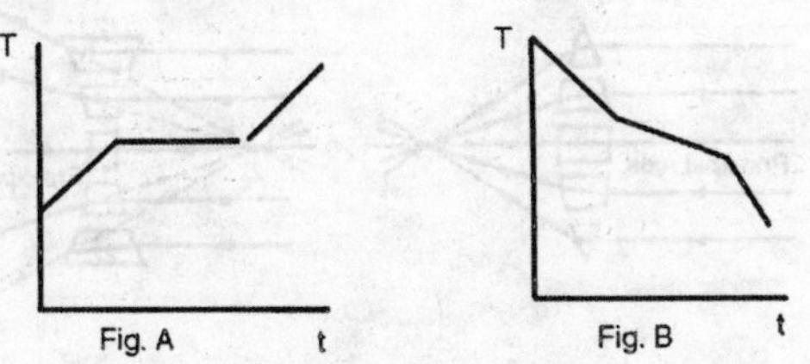

Fig. Latent Heat

Lateral At, belonging to, or pertaining to the side; situated on either side of the cardinal sagittal plane.

Lateral movement Horizontal movement of a structure or part of it, earth, etc.

Lateral support Vertical or horizontal means of bracing walls by columns, pilasters, cross walls, beams, floors, roofs, etc.

Lattice The space arrangement of atoms in a crystal.

Lattice parameter The length of any side in a crystal structure's unit cell.

Layer winding The method of winding a transformer or choke whereby conductors are layered one top of another, commonly separated by an insulating layer.

L-C Filter A low pass filter that consists of an inductance (L) and a capacitance (C).

LCD Liquid Crystal Display — display device employing light source and electrically alterable optically active thin film.

LCL Lower Control Limit is the lower limit used within statistical process control that define the constraints of common cause variations. When a parameter value falls below the lower control limit, it flags the occurrence of special causes contributing to variation. There is a similar Upper Control Limit and the mean to show the central tendency.

LDO (Low Drop Out) Regulator An LDO is a type of linear regulator. Dropout voltage is the minimum input to output voltage differential required for the regulator to sustain an output voltage within 100mV of its nominal value.

Leachate A liquid generated in landfills. It is the result of water seeping into and through the wastes. As the water contacts the waste materials it dissolves part of the organic and inorganic matter contained in the landfill. If this leachate is allowed to exit the bottom of the landfill, it will carry contaminants to the groundwater and/or adjoining surface water.

Leaching The act of dissolving the soluble portion of a solid mixture by some solvent. An example is the dissolving of inorganic or organic contaminants from refuse in a landfill by infiltrating rain water.

Lead 1. Soft and heavy metallic element used since Roman times.

2. Short, insulated electrical conductor.

3. The section of masonry wall built up as a guide for laying the balance of the wall. A line is attached to leads as a guide for constructing a wall between them.

4. Hanging (usually vertical) guide beam that aligns the hammer and the pile.

5. The distance a screw thread advances in one complete turn

Lead acid cell Secondary cell which uses lead peroxide and sponge lead for plates, and sulfuric acid water for electrolyte.

Lead resistance DC resistance in the leads of a circuit element or device.

Lead time The total time a customer must wait to receive a product after placing an order.

Leakage The loss of all or parts of a useful agent, as of the electric current that flows through an insulator or the magnetic flux that passes outside useful flux circuits.

Leakage current 1. The ac or dc current flowing from input to output and/or chassis of an isolated device at a specified voltage.

2. The reverse current in semiconductor junctions.

Leakage flux Magnetic lines of force that go beyond their intended path and do not serve their intended purpose.

Leakage inductance Self inductance in a transformer caused by leakage flux.

Leakage rate The maximum rate at which a fluid is permitted or determined to leak through a seal. The type of fluid, the differential Limits of Error: A tolerance band for the thermal electric response of thermocouple wire expressed in degrees or percentage defined by ANSI specification MC-96.1 (1975).

Lean manfacturing Lean Manufacturing is a operations philosophy that aims to synchronize production with demand, thereby minimizing inventory and cycle time. Lean Manufacturing is supported during product development with approaches such as robust design, mistake-proofing and standardization.

Learning bridge A network bridge device that has the function of a bridge and the capability to monitor the network in order to determine which nodes are connected to it, and adjust routing data accordingly.

Lease A contract granting the use of a car for a specified period of time in return for a set fee. Leases may be classified as open ended or closed ended.

Leased line A dedicated communication line between two points. Usually used by organizations to connect computers over a dedicated telephone circuit.

LEL Lower Explosive Limit. Below the LEL a mixture of substance and air lacks sufficient fuel (substance) to burn.

Lens A transparent object with two refracting surfaces. Usually the surfaces are flat or spherical (spherical lenses). Sometimes, to improve image quality. Lenses are deliberately made with surfaces which depart slightly from spherical.

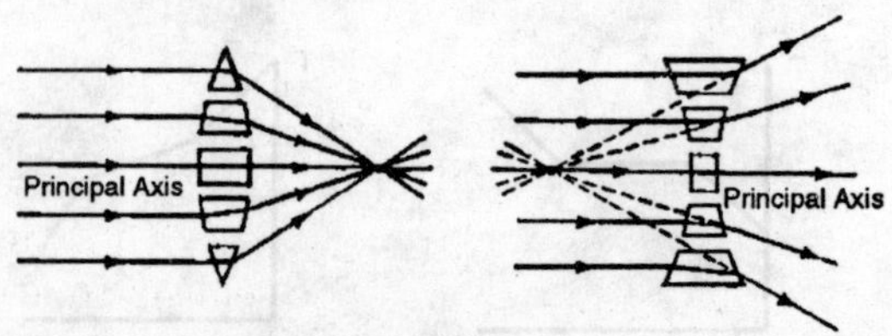

Fig. Lenses

LER Local Equipment Room.

Lesser tarsus angle The angle between the longitudinal bisection of the lesser tarsus and the longitudinal bisection of the rearfoot. This angle is normally an abducted angle.

Lethal dose (LD) The dose of radiation expected to cause death to 50 percent of an exposed population within 30 days (LD 50/30). Typically, the LD 50/30 is in the range from 400 to 450 rem (4 to 5 sieverts) received over a very short period.

Lever rule In a phase diagram, the relative proportions of the conjugate phases, at a stated value of temperature and pressure, or both, is such that a state of mechanical balance would obtain, if the corresponding weight of each phase were placed upon its composition point upon the tie-element (tie-line, tie-triangle, etc.) and the fulcrum were located at the gross composition point of the mixture.

LHV Heating Value is defined as the amount of energy released when a fuel is burned completely in a steady-flow process and the products are returned to the state of the reactants.

The heating value is dependent on the phase of water/steam in the combustion products. When H_2O is in vapor form the heating value is called LHV (Lower Heating Value).

Licensing basis The collection of documents or technical criteria that provides the basis upon which the NRC issues a license to possess radioactive materials, conduct operations involving emission of radiation, use special nuclear materials, or dispose of radioactive waste.

Life (lifetime) The length of time the sensor can be used before its performance changes.

Life cycle All the phases through which an item passes from the time it is initially developed until the time it is either deemed unserviceable in its current state and use or disposed of as being excess to all known materiel requirements.

Life cycle cost Life Cycle Cost is the total cost to the customer of acquiring, operating, and disposing of a product/ system over its full life. These costs include development, acquisition, installation, training, operation, suppoıt, and disposal.

Life test A test in which a device is subjected to actual or accelerated use to obtain an estimate of life expectancy.

Lifetime energy throughput The total amount of energy in Watthours which can be taken out of a rechargeable battery over all the cycles in its lifetime before its capacity reduces to 80% of its initial capacity when new.

LIFO (Last-In First-Out) A queue where the last item placed in the queue is the first item processed when the queue is processed.

Ligand In protein chemistry, a small molecule that is (or can be) bound by a larger molecule is termed a ligand. In organometallic chemistry, a moiety bonded to a central metal atom is also termed a ligand; the latter definition is more common in general chemistry.

Light emitting diode (LED) A semiconductor device that radiates in the visible spectrum when energized by an electric current. Color is determined by the electroluminescent characteristics of the materials used in fabricating the devices, and by the addition of various dopants. For example, copper-doped zinc sulphide emits light in the 555 nanometer (green) range, the area of peak sensitivity of the human eye.

Light metal One of the low-density metals, such as aluminum, magnesium, titanium, beryllium, or their alloys.

Light water Ordinary water as distinguished from heavy water.

Light water reactor A term used to describe reactors using ordinary water as coolant, including boiling water reactors (BWRs) and pressurized water reactors (PWRs), the most common types used in the United States.

Limb rotation During locomotion the entire limb as a unit undergoes internal and external rotation commensurate with the transfer of weight and mechanical motions that pass through it. The limb generally internally rotates for the first 25% of gait then begins to externally rotate until toe off.

Lime 1. Various classes of calcium oxide alone or in combinations with magnesium oxide, containing impurities such as silica, iron, aluminum oxide, etc. This lime is industrially produced by obtained by calcining forms of calcium carbonate (as shells or limestone), with the final product a white caustic lumpy powder which is used as an industrial alkali, construction materials (such as plasticizers for mortar or plaster), etc. Also known as quicklime or unslaked lime.

2. Quicklime to which sufficient water has been added as to convert the oxides to hydroxides, primarily to calcium hydroxide, a white crystalline strong alkali, Ca(OH)2, that is used especially to make mortar and plaster and to soften water

Lime putty The plastic material resulting after the lime hydration process stops and the semi-fluid mass cools off. In t his form the hydrated lime is ready to be added to masonry mortar mixes (as a plasticizer) or can be used by itself as either base- or finish-coat plaster, especially in historical restorations.

Lime, hydrated Quicklime to which sufficient water has been added as to convert the oxides to hydroxides. This reaction, called slaking, develops sufficient heat to bring the entire semifluid mass to a boil.

Limestone Limestone's are sedimentary rocks consisting mainly of calcite (calcium carbonate) and/or dolomite (calcium magnesium carbonate), They are usually formed from fragments of shell, coral and other marine organisms, Some finely grained and compact limestones - for example, oolite - are created from chemical precipitates. Sometimes called greystone, limestones were popular for building because they combined relatively easy workability with good weather resistance, Sulfur oxides in today's acid rain converts limestone to friable gypsum, however

Limit of detection The smallest measurable input. This differs from resolution, which defines the smallest measurable change in input. For a temperature measurement, this would provide an indication of the lowest temperature a sensor could generate an output in response to.

Limited rights Rights to use, duplicate, or disclose technical data, in whole or in part, by or for the acquirer, with the express limitation that such data shall not without written permission of the party asserting limited rights, be: released or disclosed.

LCC Life Cycle Cost.

Limited slip differential A differential in a rear-drive vehicle fitted with a mechanism that limits the speed and torque differences between its two outputs. Limited slip ensures that some torque is always distributed to both wheels, even when one is on a very slipper surface.

Limiting condition for operation The section of Technical Specifications that identifies the lowest functional capability or performance level of equipment required for safe operation of the facility.

Limiting safety system settings Settings for automatic protective devices related to those variables having significant safety functions. Where a limiting safety system setting is specified for a variable on which a safety limit has been placed, the setting will ensure that automatic protective action will correct the abnormal situation before a safety limit is exceeded.

Limits of error A tolerance band for the thermal electric response of thermocouple wire expressed in degrees or percentage defined by ANSI specification MC-96.1 (1975).

Limnology The study of freshwater ecosystems.

LIN Bus Local Interconnect Network An automotive industry standard for on-board vehicle communications. It is a single wire, serial communications bus which is used for networking intelligent sensors and actuators.

Line 1. Medium for transmission of electricity between circuits or devices.

2. The voltage across a power transmission line.

Line conditioner A circuit or device designed to improve the quality of an ac line.

Line frequency regulation The percentage change in output for a specified change in the line frequency at specified load values, with all other factors constant.

Line power supply 1. A power supply in which the ac line voltage is rectified and filtered without using a line frequency isolation transformer.

2. A power supply switched into service upon line loss to provide power to the load without significant interruption.

Line regulation The change in value of DC output voltage resulting from a change in AC input voltage over a specified range, or from low line to high line or from high line to low line. Normally specified as the + or

Line transient A perturbation outside the specified operating range of an input or supply voltage.

Linear Aside from its geometric meaning, linear describes systems in which an output is directly proportional to an input. In particular, a linear elastic system is one in which the internal displacements are (at equilibrium) directly proportional to applied forces.

Linear charger Charger which uses a series regulator. The simplest and cheapest type but less efficient than a Switch mode charger.

Linear heat generation rate The heat generation rate per unit length of fuel rod, commonly expressed in kilowatts per foot (kw/ft) of fuel rod.

Linear power supply An electronic power supply employing linear regulation techniques. "Linears" can provide fast transient response, very precise output regulation, very low output PARD (noise), and excellent isolation from the source voltage. They are also larger (about 10X) heavier (about 10X) and generate more heat (about 5X) compared to an equivalent switcher. Linear supplies typically utilize a 50Hz - 60Hz transformer to provide an isolated lower unregulated secondary voltage that is then regulated with a "linear pass device". The pass device is usually a transistor controlled with a local feedback loop.

Linear regulation A regulation technique wherein the control device, such as a transistor, is placed in series or parallel with the load. The output voltage is regulated by varying the effective resistance of the control device to dissipate the power resulting from the difference between input and output voltage time the current.

Linear regulator A power transistor or device connected in series with the load of a constant voltage power supply in such a way that the feedback to the series regulator changes its voltage drop as required to maintain a constant dc output.

Linearity 1. The ideal property wherein the change in the value of one quantity is directly proportional to the change in the value of another quantity, the result of which, when plotted on graph, forms a straight line.

2. Commonly used in reference to Linearity Error.

Linearity error The deviation of the output quantity from a specified reference line.

Link A generic term referring to a connection between two end points.

Lintel Beam placed over an opening in a wall.

Lipids Water-insoluble substances than can be extracted from cells by nonpolar organic solvents.

Liquid Matter in a fluid state but relatively incompressible. An ideal liquid offers no permanent resistance to a shear stress but is incompressible. It has then a constant volume and incompletely fills any container of less than this volume. A real liquid is appreciably compressible, and the liquid state of a substance might be defined as the denser, and less compressible, phase of the two-phase fluid system which can exist in equilibrium at temperatures below the critical temperature.

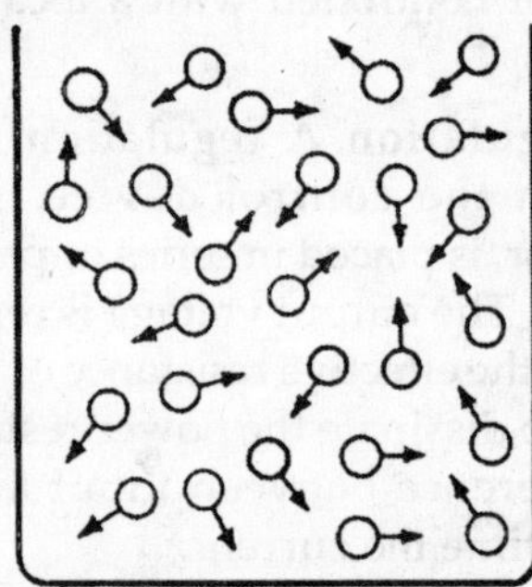

Molecular Models of Liquids

Fig. Liquid

Liquid junction potential The potential difference existing between a liquid-liquid boundary. The sign and size of this potential depends on the composition of the liquids and the type of junction used.

Liquid photoimageable solder mask (LPI) A mask sprayed on using photographic imaging techniques to control deposition. It is the most accurate method of mask application and results in a thinner mask than dry film solder mask. It is often preferred for dense SMT.

Liquidus The line on the phase equilibrium diagram above which only liquids are stable and below which some solid is present; the lowest temperature at which a metal or alloy is completely liquid.

LIST A list is a series of values separated by commas; lists are often enclosed in parentheses to avoid ambiguity and these parentheses are often necessary.

Lithium ion cell A secondary lithium cell in which both the negative and positive electrodes are lithium insertion (intercalation) compounds. Also known as rocking chair, shuttlecock or swing cell.

Lithium polymer cell A lithium ion cell with a solid polymer electrolyte.

LNG Liquefied Natural Gas. Natural gas (mainly methane) refrigerated to reach liquid phase suitable for transportation in specialised vessels or by pipeline.

Load Capacitance, resistance, inductance or any combination thereof, which, when connected across a circuit determines current flow and power used.

Load cell A device that converts a load into an electrical signal (usually measured in mV per V excitation). Mainly used in weighing applications.

Load cell systems Instruments used to measure pressure under foot. A forceplate sensitised by many hundreds of load cells which measure vertical force beneath a particular area of the foot. Dividing the force by the area of the cell gives the mean pressure beneath the foot in that area. Many different types of load cells have been used, including resistive and capacitative strain gauges, conductive rubber, piezoelectric materials and a photoelastic optical system, and a number of different methods have been used to display the output of data. The Musgrave and Sheffield Pedobarograph are in common use.

Load impedance The complex resistance to the flow of current posed by a load that exhibits both the reactive and resistive characteristics.

Load regulation 1. Static: The change in output voltage as the load is changed from specified minimum to maximum and maximum to minimum, with all other factors held constant.

2. Dynamic: The change in output voltage expressed as a percent for a given step change in load current. Initial and final current values and the rates of change must be specified. The rate of change shall be expressed as current/unit of time, e.g., 20 amperes A/u second. The dynamic regulation is expressed as a +/- percent for a worst case peak-to-peak deviation for dc supplies, and worst case rms deviation for ac supplies.

Lobaset An Ethernet term meaning a maximum transfer rate of 10 Megabits per second that uses baseband signaling and twisted pair cabling.

Local control Control over the stabilized output signal by means located within or on the power supply. May or may not be calibrated.

Local sensing Using the power supply output voltage terminals as the error-sensing points to provide feedback to the voltage regulator.

Local variables Local variables can only be accessed in the current block and in subroutines called from that block.

Lock up torque converter A torque converter that contains a special clutch that forms a solid connection between the engine output shaft and the transmission input shaft when a certain, pre-set speed is attained. This reduces transmission friction losses and increases efficiency.

Logarithmic scale A method of displaying data (in powers of ten) to yield maximum range while keeping resolution at the low end of the scale.

LOGIC Leading Oil & Gas Industry Competitiveness. An industry funded organisation that works with companies throughout the O&G industry to stimulate collaboration and improve competitiveness.

Logic ground Common return or reference point for logic signals.

Logic high A voltage representing a logic value of one 1. in positive logic.

Logic Inhibit/Enable A referenced or isolated logic signal that turns a power supply output off or on.

Logic low A voltage representing a logic value of zero (0) in positive logic.

Logical Conveys an abstract concept in a simpler manner, such as using a logical machine name instead of its physical address.

Logical link control (LLC) The upper part of the data link sublayer protocol that is responsible for governing the exchange of data between two end points.

Logical operators This term is used to mean Boolean operators, that is, those dealing with True/False values.

London dispersion force An attractive force caused by quantum-mechanical electron correlation. For example, a neutral spherical molecule (such as a single argon atom) has no charge and produces no external electric field, yet a pair of molecules has a distribution of electron configurations weighted toward those with lesser electron-electron repulsions; this creates a small net attraction.

Long term stability The output voltage change of a power supply, in percent, due to time only, with all other factors held constant. Long-term stability is a function of component aging.

Longitudinal bisections For the convenience of anatomists and clinicians, the relationship between parts of the foot can be described by a series of theoretical lines drawn through the foot.

Longitudinal bisection of the digits This is a line bisecting the shaft of the proximal phalanx of the second toe. The proximal phalanx is bisected proximally at the base and distally at the neck. The longitudinal bisection of all the digits are normally parallel to the longitudinal bisection of the rearfoot and to each other.

Longitudinal bisection of the lesser tarsus This is a line perpendicular to the line transecting the lesser tarsus. The transection line is determined by bisecting the distance from the lateral proximal aspect of the fifth metatarsal, excluding the styloid process, to midpoint between the medial proximal aspect of the base of the first metatarsal and the anterior medial aspect of the effective articular surface of the head of the talus with the navicular. A perpendicular to this line is the longitudinal axis to the lesser tarsus.

Longitudinal bisection of the metatarsus This is the line bisecting the shaft of the second metatarsal. The second metatarsal is bisected proximally at the base and distally at the neck.

Longitudinal bisection of the tarsus This line connects the longitudinal bisection of the posterior surface of the calcaneum, to the dorsal anterior aspect of the calcaneum.

Loop In a pressurized water reactor, the coolant flow path through piping from the reactor pressure vessel to the steam generator, to the reactor coolant pump, and back to the reactor pressure vessel. Large PWRs may have as many as four separate loops.

Loop resistance The total resistance of a thermocouple circuit caused by the resistance of the thermocouple wire. Usually used in reference to analog pyrometers which have typical loop resistance requirements of 10 ohms.

LOPA Layers of Protection Analysis. A semi-quantitative risk analysis technique.

LOPA results are intended to be conservative (overestimating the risk) and usually adequate to understand the required safety integrity level.

Lordosis This is the forward curvature of the spine, causing the normal concavity of the lower lumber region of the back

Loss of coolant accident (LOCA) Those postulated accidents that result in a loss of reactor coolant at a rate in excess of the capability of the reactor makeup system from breaks in the reactor coolant pressure boundary, up to and including a break equivalent in size to the double-ended rupture of the largest pipe of the reactor coolant system.

Low level waste A general term for a wide range of wastes having low levels of radioactivity. Industries; hospitals and medical, educational, or research institutions; private or government laboratories; and nuclear fuel cycle facilities (e.g., nuclear power reactors and fuel fabrication plants) that use radioactive materials generate low-level wastes as part of their normal operations. These wastes are generated in many physical and chemical forms and levels of contamination.

Low-level radioactive wastes containing source, special nuclear, or byproduct material are acceptable for disposal in a land disposal facility. For the purposes of this definition, low-level waste has the same meaning as in the Low-Level Radioactive Waste Policy Act, that is, radioactive waste not classified as high-level radioactive waste, transuranic waste, spent nuclear fuel, or byproduct material as defined in section 11e.(2) of the Atomic Energy Act (uranium or thorium tailings and waste).

Low population zone (LPZ) An area of low population density often required around a nuclear installation before it's built. The number and density of residents is of concern in emergency planning so that certain protective measures (such as notification and instructions to residents) can be accomplished in a timely manner.

Lowest replaceable item The lowest level component at which maintenance will be performed or discrete configuration control enforced.

LPG Liquefied Petroleum Gas. Butane and propane mixture separated from well fluid stream. LPG can be transported under pressure in refrigerated vessels.

LPI Stands for Liquid PhotoImageable. Refers to liquid photoimageable solder mask.

LSA Scale Low Specific Activity scale found adhering to pipe and equipment internals is mainly due to radium-226 produced from the decay of naturally occurring uranium-238.

LSD (Least Significant Digit) The rightmost active (non-dummy) digit of the display.

LS-TTL Compatible For digital input circuits, a logic 1 is obtained for inputs of 2.0 to 5.5 V which can source 20 μA, and a logic 0 is obtained for inputs of 0 to 0.8 V which can sink 400 μA. For digital output signals, a logic 1 is represented by 2.4 to 5.5 V with a current source capability of at least 400 μA; and a logic 0 is represented by 0 to 0.6 V with a current sink capability of at least 16 MA. "LS" stands for low-power Schottky.

LS-TTL Unit load A load with LS-TTL voltage levels, which will draw 20 μA for a logic 1 and -400 μA for a logic 0.

LTI Lost Time Incident. Accident resulting in personnel not being able to work as a result of their injury is called an LTI.

Luminescence Defined as the mission of light from a substance in an electronically excited state. Depending on whether the excited state is singlet or triplet, the emission is called fluorescence (less than one second decay) or phosphorescence (longer than 1 second decay). Depending on the source, molecules get the needed extra energy from different types of luminescence are distinguished: radioluminescence, photoluminescence (in the same category are fluorescence and phosphorescence), chemiluminescence and bioluminescence, electrochemiluminescence, sonochemi-luminescence and thermoluminescence.

M Mega; one million. When referring to memory capacity, two to the twentieth power (1,048,576 in decimal notation). M Symbol tar Mutual Inductance.

Machine language Instructions that are written in binary form that a computer can execute directly. Also called object code and object language.

Machine-phase chemistry The chemistry of systems in which all potentially reactive moieties follow controlled trajectories (e.g., guided by molecular machines working in vacuum).

MacPherson strut A suspension system that consists of a combination coil spring and shock absorber in one compact unit at each wheel. With this "independent" suspension design, road shocks at one wheel are not transferred to the opposite wheel. MacPherson struts use fewer parts, meaning a reduction on weight and fewer elements that could wear out.

Macromolecule A huge molecule made up of thousands of atoms.

Magnetic amplifier Transformer type device employing a control circuit to vary the magnetic core saturation, thus varying the output voltage of the amplifier.

Magnetic field strength (designated by H) (A/m) Magnetic field produced by a current, independent of the presence of magnetic material. The units of H are ampere-turns per meter, or just amperes per meter.

Magnetic flux Total number of lines of magnetic force.

Magnetic flux density The number of lines of flux per cross-sectional area of a magnetic circuit expressed in Gauss or Tesla. Sometimes referred to as Magnetic Induction.

Magnetic flux density or magnetic induction (designated by B) The magnetic field produced in a substance by an external magnetic field. The units of B are tesla (T). One tesla is the magnetic flux density given by a magnetic flux of 1 weber per square meter. One weber is a magnetic flux that, linking a circuit of 1 turn, would produce in it an electromotive force of 1 volt if it were reduced to zero at a uniform rate in 1 second. Both B and H are field vectors. One henry (H) is the inductance of a closed circuit in which an electromotive force of 1 volt is produced when the electric current in the circuit varies uniformly at the rate of 1 ampere per second. The magnetic field strength and flux density are related according to: $B = \mu H$, where μ is the permeability.

Magnetic induction The use of a magnetic field to generate currents or voltages in a conductor sometimes referred to as Magnetic Flux Density.

Magnetic resonance imaging (MRI) A method of looking inside the human body without using surgery, harmful dyes or x-rays based on Nuclear Magnetic Resonance (NMR).

Magnetic susceptibility (÷m) The proportionality constant between the magnetization M and the magnetic field strength H. The magnetic susceptibility is unitless.

Magnetization (M) The total magnetic moment per unit volume of material. Also, a measure of the contribution to the magnetic flux by some material within an H field. The magnitude of M is proportional to the applied field as: M = ÷m × H, with ÷m the magnetic susceptibility.

Magnetizing current The no-load current in the transformer primary winding that is required to magnetize the core. Sometimes called exciting or excitation current.

Magnetohydrodynamic generator MHD The production of electricity by passing a conducting fluid or plasma through a magnetic field.

Magnetostrictive material A material that changes dimension in the presence of a magnetic field or generates a magnetic field when mechanically deformed.

Maintainability A characteristic of design and installation which inherently provides for an item to be retained in, or restored to a specified condition within a given period of time, when the maintenance is performed in accordance with prescribed procedures and resources. Synonymous with Serviceability and Supportability.

Major flooding Flood conditions resulting in extensive inundation and property damage. Typically characterized by the evacuation of people and livestock and the closure of both primary and secondary roads.

Major review A formal demonstration and confirmation across the system that supports or constitutes a program milestone event.

Majordomo Software that automates the management of electronic mailing lists.

Malleable cast iron White cast iron that has been heat treated to convert the cementite into graphite clusters; a relatively ductile cast iron.

Malleolar torsion When the knee is extended and the subtalar joint in neutral, the angular relationship between the transverse bisection of the malleoli and the frontal plane of the knee describes malleolar torsion. External Malleolar Torsion of 13-18o is present in normal adults. While this relationship is primarily structural, it is also influenced by the position of the subtalar and ankle joints. Closed chain pronation of the subtalar joint decreases external malleolar torsion, whereas Closed chain supination increases external malleolar torsion.

Man a unix Command that provides information about the UNIX command entered in the parameter command. (The man command is short for manual entry.)

Management information base (MIB) A database used by SNMP containing configuration and statistical information about devices on a network.

Mandrel (Balancing Arbor) An accurately machined shaft on which work is mounted for balancing.

Manhattan algorithm An algorithm to determine a cross street for an avenue address in Midtown Manhattan New York City, or for the length of a trip from one address in Manhattan to another. If you know the building addresses for where you are and where you want to go in Manhattan, you can call a cab company and find out what it will cost you. An algorithm is used to get the answer, because in Manhattan the street and avenue numbers do not necessarily correspond intuitively to the building numbers.

Manhattan length The length of the two sides of a right triangle as a distance between two points, as opposed to the hypotenuse.. (Derived from the Manhattan algorithm for determining the length of a taxicab trip following streets and avenues on the island of Manhattan, NY.) Routing of traces in orthagonal patterns in a PCB design, or in a semiconductor chip, follows the same pattern as streets and avenues in a city. The minimum distance between two component leads, or two nodes on a chip, when routing on 90 degrees is the Manhattan length.

Advanced PCB auto-routers permit specification of maximum length of classes of nets as a percentage of Manhattan length. For example, one could specify clocks as 120% and random nets as 160% of Manhattan length. (This percentage, expressed as a ratio, becomes the "Manhattan coefficient", ie. a Manhattan coefficient of 1.2 means the routed length is 120% of the Manhatten length.) Specifying such limits on the auto-router prevents long and circuitous routes.

Manual reset (Adjustment) The adjustment on a proportioning controller which shifts the proportioning band in relationship to the set point to eliminate droop or offset errors.

Manufacturability The characteristic of a product's design that facilitates the fabrication of the product's components and their assembly into the overall product.

Manufacturing engineering Manufacturing engineering includes all aspects of manufacturing operations, including the behavior and properties of materials and materials processes; the design of products, equipment, and tooling necessary for their manufacture; management of manufacturing enterprises; and, the design and operation of manufacturing systems

Marble Geologically, marble refers to certain crystalline rocks composed primarily of calcite or dolomite-metamorphosed limestone. In the stone and construction industries on the other hand, the term "marble" refers to any rock that can take a high polish. This even includes a few limestones and granites. The purest from is white statuary marble, best exemplified by product mined in the Carrara area of Italy. Commonly found marbles contain various impurities which produce a whole spectrum of patterns and colors. Because is softer than most other construction stones, marble can be worked more easily and is used mainly for ornamentation. The downfall is that marbles are often more expensive. They also tend to be sensitive to exposure to atmospheric pollutants and acids from any sorce, so many are not suitable for exterior use even when protective transparent coatings are used.

Margining The ability to temporarily shift output voltage by a specified amount (often +/- 5%) for system testing.

Mariner's compass It is a type of compass. It is used by mariners for navigational purposes. It consists of a pivoted compass card in a gimbalmounted, non-ferrous metal bowl.

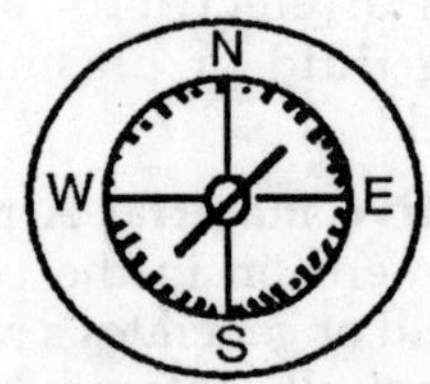

Fig. Mariner's compass

Martensite A metastable iron phase supersaturated in carbon that is the product of a diffusionless (athermal) transformation from austenite.

Mask Pattern on glass, like a photographic negative, for producing integrated-circuit elements on semiconductor wafer.

Mason To the French maçon (Latin matio or machio), "a builder of walls" or "a stone-cutter" masonry Brick, stone, concrete, hollow-tile, concrete-block, gypsum block, or other building units, bonded together with mortar (usually).

Masonry A building material such as stone, clay, brick, or concrete.

Masonry cement A mill-mixed cementitious material to which sand and water must be added.

Masonry unit Natural or manufactured building units of burned clay, concrete, stone, glass, gypsum, etc.Hollow Masonry Unit: One whose net cross-sectional area in any plane parallel to the bearing surface is less than 75 percent of the gross.Modular Masonry Unit: One whose nominal dimensions are based on the 4 in. module.Solid Masonry Unit: One whose net cross-sectional area in every plane parallel to the bearing surface is 75 percent or more of the gross.

Mass The amount of matter contained within a body. The term is often confused with weight, but mass and weight are two separate entities. Weight is a measure of the force exerted by gravity on an object.

Mass balance An organized accounting of all inputs and outputs to an arbitrary but defined system. Stated in other terms, the rate of mass accumulation within a system is equal to the rate of mass input less the rate of mass output plus the rate of mass generation within the system.

Mass customization Mass Customization is the evolutionary step beyond mass production in manufacturing whereby products can be customized to the needs of individual customers while achieving most of the economies of mass production. It is based on principles such as product line rationalization, standardization of components, modular design, postponement of customization to late in the production cycle, multi-function assemblies, use of software to customize product operation, etc.

Mass energy equation The equation developed by Albert Einstein, which is usually given as $E = mc^2$, showing that, when the energy of a body changes by an amount E (no matter what form the energy takes), the mass (m) of the body will change by an amount equal to E/c^2. The factor c squared, the speed of light in a vacuum (3 x 10 to the eighth power), may be regarded as the conversion factor relating units of mass and energy. The equation predicted the possibility of releasing enormous amounts of energy by the conversion of mass to energy. It is also called the Einstein equation.

Mass flow rate Volumetric flowrate times density, i.e. pounds per hour or kilograms per minute.

Mass operator Operator added to the Lagrangian of a quantized field theory in order to cancel out certain divergences. Combines with the term representing the mechanical mass to give the observed mass.

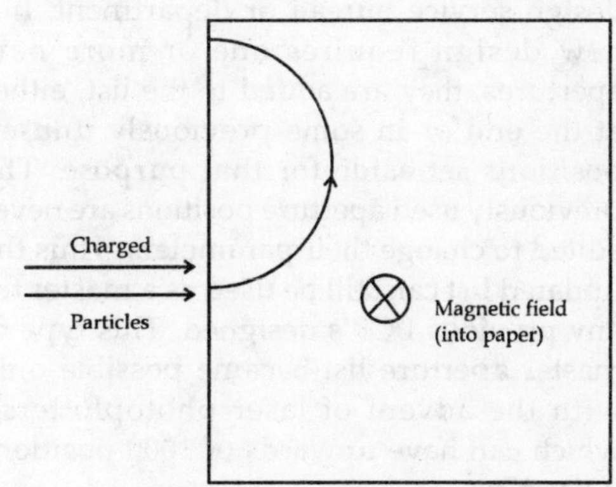

Fig. Mass spectrometer

Mass number The number of nucleons (neutrons and protons) in the nucleus of an atom. Also known as the atomic weight.

Mass spectrometer A device which produces a mass spectrum of a sample to find out its composition by ionizing the sample and separating ions of differing masses and recording their relative abundance by measuring intensities of ion flux. Mass spectroscopy allows detection of compounds by separating ions by their unique mass. A typical machine costs around $250,000.

Mass storage A device like a disk or magtape that can store large amounts of data readily accessible to the central processing unit.

Mass transfer (Diffusional) boundary layer The mass transfer boundary layer lies between an interface of some kind (e.g., a gas-liquid interface or the fluid-biofilm interface) and the region of the fluid where concentrations can be assumed to be relatively constant (i.e., well-mixed). In considerations of mass transfer, the diffusional boundary layer is where the action is.

Mass transfer and momentum boundary layers are not necessarily the same, but are often closely related.

Master aperture list 1. An aperture list which is used for every PCB designed by a PCB design service bureau or department. If a new design requires one or more new apertures, they are added to the list, either at the end or in some previously unused positions set aside for that purpose. The previously used aperture positions are never edited to change their parameters. Thus the updated list can still be used as a master for any previous PCB's designed. This type of master aperture list became possible only with the advent of laser photoplotters , which can have upwards of 1000 positions if need be.

2. Any aperture list which is used with two or more PCB's would be called the master aperture list for that set of PCB's.

Master/Slave operation Interconnection of two or more regulated supplies in which one (the master) controls the others (the slaves).

Match A string that does fit a specified pattern.

Materiel A generic term covering systems, equipment, stores, supplies and spares, including related documentation, manuals, computer hardware and software.

Matrix The continuous phase in a composite or two-phase alloy microstructure in which a second phase is dispersed.

Maximum contaminant level (MCL) The maximum allowable concentration of a given constituent in potable water.

Maximum dependable capacity (gross) In a nuclear power reactor, dependable main-unit gross generating capacity, winter or summer, whichever is smaller. The dependable capacity varies because the unit efficiency varies during the year due to temperature variations in cooling water. It is the gross electrical output as measured at the output terminals of the turbine generator during the most restrictive seasonal conditions (usually summer).

Maximum elongation The strain value where a deviation of more than ±5% occurs with respect to the mean characteristic (diagram of resistance change vs strain).

Maximum excitation The maximum value of excitation voltage or current that can be applied to the transducer at room conditions without causing damage or performance degradation beyond specified tolerances.

Maximum load 1.The highest allowable output rating specified for any or all outputs of a power supply under specified conditions including duty cycle, period and amplitude.

2. The highest specified output power rating of a supply specified under worst case conditions.

Maximum operating temperature The maximum temperature at which an instrument or sensor can be safely operated.

Maximum transmission unit (MTU) The largest datagram that can be handled by a specific network. The MTU can change over different networks, even if the transport is the same (such as Ethernet).

Maxwell's equations (JC Maxwell; 1864) Four elegant equations which describe classical electromagnetism in all its splendor. They are:

Gauss' law The electric flux through a closed surface is proportional to the algebraic sum of electric charges contained within that closed surface; in differential form, div E = rho, where rho is the charge density.

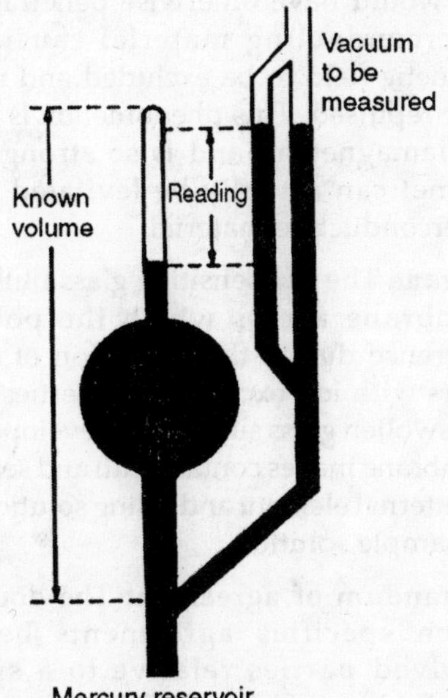

Fig. Mcleod gauge

Gauss' law for magnetic fields The magnetic flux through a closed surface is zero; no magnetic charges exist. In differential form, div B = 0.

Faraday's law The line integral of the electric field around a closed curve is proportional to the instantaneous time rate of change of the magnetic flux through a surface bounded by that closed curve; in differential form,

curl E = -dB/dt, where d/dt here represents partial differentation.

Ampere's law, modified form The line integral of the magnetic field around a closed curve is proportional to the sum of two terms: first, the algebraic sum of electric currents flowing through that closed curve; and second, the instantaneous time rate of change of the electric flux through a surface bounded by that closed curve; in differential form,

curl H = J + dD/dt, where d/dt here represents partial differentiation. In addition to describing electromagnetism, his equations also predict that waves can propagate through the electromagnetic field, and would always propagate at the same speed — these are electromagnetic waves; the speed can be found by computing $(em)^{-1/2}$, which is c, the speed of light in vacuum.

Mean Numerical average of data values.

Mean temperature The average of the maximum and minimum temperature of a processequilibrium.

Mean time between failure (MTBF) The average length of time between system failures, exclusive of infant mortality and rated end-of-life. An established method of calculating MTBF is described in Mil Handbook 217.

Mean time to repair (MTTR) The average time required to repair a product.

Measurand A physical quantity, condition, or property that is to be measured.

Measuring junction The thermocouple junction referred to as the hot junction that is used to measure an unknown temperature.

Mechanical Pertaining to the positions and motions of atoms, as defined by the positions of their nuclei. A purely mechanical device can be described in terms of atomic positions and motions without reference to electronic properties, save through their effect on the potential energy function.

Mechanical charging Charging by replacing one or more of the active chemicals in the cell.

Mechanical engineering Mechanical engineering applies the principles of mechanics and energy to the design of machines and devices. Perhaps the broadest of all engineering disciplines, mechanical engineering is generally combined into three broad areas- energy, structures and motion in mechanical systems, and manufacturing.

Mechanical hysteresis The difference of the indication with increasing and decreasing strain loading, at identical strain values of the specimen.

Mechanosynthesis Chemical synthesis controlled by mechanical systems operating with atomic-scale precision, enabling direct positional selection of reaction sites; synthetic applications of mechanochemistry. Suitable mechanical systems include AFM mechanisms, molecular manipulators, and molecular mill systems. Processes that fall outside the intended scope of this definition include reactions guided by the incorporation of reactive moieties into a shared covalent framework (i.e., conventional intramolecular reactions), or by the binding of reagents to enzymes or enzymelike catalysts.

Medium effect (f m) For solvents other than water the medium effect is the activity coefficient related to the standard state in water at zero concentration. It reflects differences in the electrostatic and chemical interactions of the ions with the molecules of various solvents. Solvation is the most significant interaction.

Mega A prefix that multiplies a basic unit by 1,000,000 (10 to the sixth power).

Megabyte (MB) 220 (1,048,576, or about one million) bytes of information.

Megacurie One million curies.

Megahertz (Mhz) One million cycles per second.

Megawatt (MW) One million watts.

Megawatt hour (MWh) One million watt-hours.

Meissner effect When a superconducting material is cooled below its critical temperature it will exclude or repel a magnetic field. A magnet moving by a conductor induces currents in the conductor. This is the principle upon which the electric generator operates. But, in a superconductor the induced currents exactly mirror the field that would have otherwise penetrated the superconducting material causing the magnetic field to be excluded and magnet to be repulsed. This phenomenon is known as diamagnetism and is so strong that a magnet can actually be levitated over a superconductive material.

Membrane The pH-sensitive glass bulb is the membrane across which the potential difference due to the formation of double layers with ion-exchange properties on the two swollen glass surfaces is developed. The membrane makes contact with and separates the internal element and filling solution from the sample solution.

Memorandum of agreement The document which specifies agreements between involved parties relative to a specific interface. The MOA delineates responsibilities of each organization and identifies formats and data elements required for a successful interface.

Memory effect Reversible, progressive capacity loss in nickel based batteries found in NiCad and to a lesser extent in NiMH batteries. It is caused by a change in crystalline formation from the desirable small size to a large size which occurs when the cell is recharged before it is fully discharged.

MEMS Stood originally for Micro-ElectroMechanical System — microscopic mechanical elements, fabricated on silicon chips by techniques similar to those used in integrated circuit manufacture, for use as sensors, actuators, and other devices. Today almost any miniaturized device (based on Si technology or traditional precision engineering, chemical or mechanical) is referred to as a MEMS device.

Metabolism The processes which sustain an organism, including energy production, synthesis of proteins for repair and replication.

Metal An opaque lustrous elemental chemical substance that is a good conductor of heat and electricity and, when polished, a good reflector of light; most elemental metals are malleable, ductile, and are generally denser than the other elemental substances; metals are structurally distinguished from nonmetals by their atomic bonding and electron availability; the electron band structure of metals is characterized by a partially filled valence band; the "free electrons" lost from the outer shells of metallic atoms are available to carry an electric current; the defining property of a metal is that it is an element with a positive thermal coefficient of resistivity, meaning the electrical resistivity of a metal continuously increases as temperature increases.

Metallic bond A primary interatomic bond involving the nondirectional sharing of nonlocalized valence electrons ("sea of electrons") which are mutually shared by all the atoms in the metallic solid.

Metallurgy The science and technology of metals and alloys.

Metastable A classical system is metastable if it is above its minimum-energy state, but requires an energy input before it can reach a lower-energy state; accordingly, a metastable system can act like a stable system, provided that energy inputs (e.g., thermal fluctuations) remain below some threshold. Systems with strong metastability are commonly described as stable. Quantum mechanical effects can permit metastable states to reach lower energies by tunneling, without an energy input; an associated, broader definition of metastable embraces all systems that have a long lifetime (by some standard) in a state above the minimum-energy state.

Metatarsus This consists of the five metatarsals.

Metatarsus angle The angle formed between the longitudinal bisection of the metatarsus and the longitudinal bisection of the lesser metatarsus. This angle is normally an adducted angle.

Metatarsus primus elevatus This is present in the first ray when the range of dorsiflexion exceeds the range of plantarflexion.

Meteorology The study of the atmosphere and weather of the lower atmosphere, below 100 km.

Method of correction A procedure whereby the mass distribution of a rotor is adjusted to reduce unbalance, or vibration due to unbalance, to an acceptable value. Corrections are usually made by adding material to, or removing it from, the rotor.

Metric Measures used to indicate progress or achievement.

Metric ton Approximately 2200 pounds in the English system of measurements.

MIC Monolithic Integrated Circuit.

Mica A transparent mineral used as window material in high-temperature ovens.

Micro A prefix that divides a unit into one million parts (0.000001).

Micro ball grid array A fine pitch ball grid array. Fine pitch for BGAs is anything less than 1.27 mm (50 mil) (some say 1.00 mm (39 mil)). SMTnet terms and definitions SMT in Line terms and definitions.

Micro BGA Micro Ball Grid Array.

Microamp One millionth of an ampere, 10-6 amps, μA.

Microbial transport The movement of individual cells or aggregates of cells through a system. Often, this involves displacement of freely suspended microbes from the bulk aqueous phase to a surface and vice versa, allowing dissemination of progeny throughout the system.

Microbially influenced corrosion (MIC) The result of the presence and/or activities of surface-associated organisms that promotes deterioration of a metal or alloy, generally taking the form of localized attack on the surface.

Microcomputer A computer which is physically small. It can fit on top of or under a desk; based on LSI circuitry, computers of this type are now available with much of the power currently associated with minicomputer systems.

Microcurie One millionth of a curie. That amount of radioactive material that disintegrates (decays) at the rate of 37 thousand atoms per second.

Microcycles Rapid, shallow charge and discharge cycles which occur in automotive battery applications, particularly those which involve regenerative braking.

Microdischarge plasma thruster An experimental ultra-miniature rocket motor developed at the University of Texas Plasma Research Lab. Helium gas is electrified and turned into a superhot gas known as plasma, then expelled out of a tiny hole to produce a rocket. The design is very small and efficient, producing a small amount of thrust for a long duration of time. The force produced, however, is less than a gentle breeze!.

Micron One millionth of a meter (0.000001), and another term for micrometer (10^{-6}).

Microphone Device that produces voltage or current in response to a sound wave.

Microprocessor Chip containing the logical elements for performing calculations and carrying out stored instructions.

Microscopy The investigation of microstructural elements using some type of microscope, e.g. scanning electron microscopy (SEM), light optical microscopy (LOM).

Microstructure In material engineering the structural features of a material such as grain boundaries, grain size and structure, subject to observation under a microscope, selective etching etc. In MEMS microstructure unfortunately is also used to designate a micromachined feature.

Microvoid coalescence (MVC) Occurs due to the nucleation of microvoids, followed by their growth and eventual coalescence; initiation is caused by particle cracking or interfacial failure between an inclusion or precipitate particle and the surrounding matrix.

Mid stance This term is used to describe the period of time between foot flat and heel off, or to define a particular event occurring during that period of time. It is in the latter sense, to mean the time at which the swing phase leg passes the stance phase leg, and the two feet are side by side.

Mid swing Mid swing on one side corresponds to mid stance on the other, it is the time when the swinging leg passes the stance phase leg, and the two feet are side by side. Swing phase occupies about 40% of the gait cycle.

Mil One thousandth of an inch (0.0254 mm).

Milestone A scheduled event for which some individual is accountable and that is used to measure progress. An important event representing the completion of a major work task or group of work tasks. Milestones are usually scheduled and can be used to measure progress. Reviews are often conducted upon the completion of a milestone.

Mill tailings Naturally radioactive residue from the processing of uranium ore into yellowcake in a mill. Although the milling process recovers about 93 percent of the uranium, the residues, or tailings, contain several naturally-occurring radioactive elements, including uranium, thorium, radium, polonium, and radon.

Miller indices A set of 3 integers (4 for hexagonal) that designate crystallographic planes, as determined from reciprocals of fractional axial intercepts.

Milliroentgen (mR) One thousandth of a roentgen (R). 1mR = 10^{-3} R = 0.001 R.

Millivolt Unit of electromotive force. It is the difference in potential required to make a current of 1 millampere flow through a resistance of 1 ohm; one thousandth of a volt, symbol mV.

MILNET (Military Network) A network that was originally part of ARPANET, now designated for exclusive military use in installations that require reliable network services.

MIME (Multi-Purpose Internet Mail Extensions) An extension to Internet mail that allows for the inclusion of non-textual data such as video and audio in e-mail.

Mineral insulated thermocouple A type of thermocouple cable which has an outer metal sheath and mineral (magnesium oxide) insulation inside separating a pair of thermocouple wires from themselves and from the outer sheath. This cable is usually drawn down to compact the mineral insulation and is available in diameters from .375 to .010 inches. It is ideally suited for high-temperature and severe-duty applications.

Minimization A comprehensive program to minimize or eliminate wastes, usually applied to wastes at their point of origin.

Minimum load 1. The lowest specified current to be drawn on a constant voltage power supply for the voltage to be in a specified range.

2. For a constant current supply, the maximum value of load resistance.

Minimum operating temperature The lowest ambient temperature at which the power supply will continuously operate safely and within specifications.

Minimum starting temperature The lowest ambient temperature at which a power supply will turn on and operate safely.

Mining engineering Mining engineering comprises all aspects of discovering, removing, and processing minerals from the earth. Mining engineers design the mine layout, supervise its construction, devise systems to transport minerals to processing plants, and develop plans to return the area to its natural state.

Mismatch Incorrect matching of load and source impedance.

Misreaction A chemical reaction that fails by yielding an unwanted product.

Mistake proofing Mistake-Proofing - improving product designs, tooling designs, or processes to prevent mistakes from being made or to quickly and easily detect or mitigate the effect of a mistake. Mistake proofing involves six principles: elimination, replacement, prevention, facilitation, detection, and mitigation

Mixed dislocation A dislocation that has both edge and screw components.

Mixed liquor suspended solids (MLSS) The total suspended solids concentration in the activated sludge tank.

Mixed liquor volatile suspended solids (MLVSS) The volatile pended solids concentration in the activated sludge tank.

Mixed oxide (MOX) fuel A mixture of uranium oxide and plutonium oxide used to fuel a reactor. Mixed oxide fuel is often called "MOX." Conventional nuclear fuel is made of pure uranium oxide.

MM2 A molecular mechanics program developed by Norman Allinger and coworkers; the "MM2 model" is the molecular potential energy function described by the equations, rules, and parameters embodied in that program.

Mobility (electron, and hole) The proportionality constant between the carrier drift velocity and applied electric field.

MODEM (Modulator Demodulator) An electronic device that allows digital computer data to be transmitted via analog phone lines.

Modem eliminator A device that functions as two modems to provide service for data terminal equipment (DTE) and data communications equipment (DCE).

Moderator A person who examines all submissions to a newsgroup or mailing list and allows only those that meet certain criteria to be posted. Usually, the moderator makes sure that the topic is pertinent to the group and that the submissions aren't flames.

Moderator temperature coefficient of reactivity As the moderator (water) increases in temperature, it becomes less dense and slows down fewer neutrons, which results in a negative change of reactivity. This negative temperature coefficient acts to stabilize atomic power reactor operations.

Module A self-contained part of a hardware item designed as a single replaceable unit, with a specific integral electronic function. It should require no installation other than mechanical mounting and completion of electrical connectors.

Modulus of elasticity (E) The ratio of stress to strain when deformation is totally elastic. Also the Young's modulus.

Modulus of rupture Breaking strength in a nonductile solid as measured by bending.

Moiety A portion of a molecular structure having some property of interest.

Moisture Essentially water, quantitatively (and sometimes qualitatively) determined by definite prescribed methods which may vary according to the nature of the material. For example, ASTM D 4263 (Standard test method for indication of moisture in concrete by the plastic sheet method) is a test method used to indicate the presence of capillary moisture in concrete.

Moisture barrier A special building paper fastened over the sheathing on the inside face of the cavity. It is lapped and taped to provide a third level of protection against moisture penetration. moisture contentTerm which in engineering indicates the percentage moisture content (which equals the weight of moisture divided by the weight of dry material multiplied by 100). The moisture content of a concrete aggregate, soil or mineral sample consists of two portions, namely, the free or surface moisture which can be removed by exposure to air, and the inherent moisture which is entrapped in the the material, and is removed by heating at a specific temperature (in the case of most coal for example, this temperature is 200 degrees F or 93.3 degrees C).

Molding (plastics) Shaping a plastic material by forcing it, under pressure at a high temperature, into a mold cavity.

Mole A number of instances of something (typically a molecular species) equaling ~6.022×10^{23}. Mole ordinarily means gram-mole; a kilogram-mole is ~6.022×10^{26}.

Molecule A set of atoms linked by covalent bonds. A macroscopic piece of diamond is technically a single molecule. (Sets of atoms linked by bonds of other kinds are sometimes also termed molecules.)

Molecular machine A mechanical device that performs a useful function using components of nanometer scale and defined molecular structure; includes both artificial nanomachines and naturally occurring devices found in biological systems.

Molecular manipulator A programmable device able to position molecular tools with high precision, for example, to direct a sequence of mechanosynthetic steps; a molecular assembler.

Molecular mechanics models Many of the properties of molecular systems are determined by the molecular potential energy function. Molecular mechanics models approximate this function as a sum of 2-atom, 3-atom, and 4-atom terms, each determined by the geometries and bonds of the component atoms. The 2-atom and 3-atom terms describing bonded interactions roughly correspond to linear springs.

Molecular mill A mechanochemical processing system characterized by limited motions and repetitive operations without programmable flexibility.

Moment This is the torque or the turning effect on the body. Moments are more than just force, but are the product of force and the perpendicular distance between them and the fulcrum or pivot point. The formula for calculating the moment of force is:

$M = F \times D$

Where M is the moment of force (Newton - meters,N.m); F is the force (Newtons, N); and D is the distance (Meters, m) from the fulcrum point.

Moment arm This is the perpendicular distance between the force and the fulcrum point.

Momentum The momentum of an object is calculated by multiplying its velocity by its mass.

Momentum (hydrodynamic) boundary layer The momentum boundary layer lies between an interface of some kind (e.g., a water/pipe wall interface or the fluid/biofilm interface) and the region of the fluid where the velocity is relatively constant or uniform when averaged over time. In considerations of shear stress and forces on biofilms, the momentum boundary layer (sometimes called the (laminar boundary layer) is where the action is. Mass transfer and momentum boundary layers are not necessarily the same, but are often closely related.

Monitoring Periodic or continuous surveillance or testing to determine the level of compliance with statutory requirements and/or pollutant levels in various media or in humans, plants and animals.

Monitoring of radiation Periodic or continuous determination of the amount of ionizing radiation or radioactive contamination present in a region, as a safety measure, for the purpose of health or environmental protection. Monitoring is done for air, surface and ground water, soil and sediment, equipment surfaces, and personnel (for example, bioassay or alpha scans).

Monoclonal antibodies Produced by injecting animals to elicit a response from lymphocytes to produce antibodies. Lymphocytes which produce antibodies with strong binding capability can be isolated and used to produce only one kind of antibody (monoclonal) on a permanent basis once the lymphocytes are immortalized. This is accomplished by fusing them (combining them genetically) with cancer cells which have the distinction of living indefinitely in a culture. Monoclonal antibodies can be produced repeatedly and collected for use in immunodetection.

Monolithic 1. Existing as one large, undifferentiated whole.

2. (Of an integrated circuit or its elements) built upon or formed within a single slice of silicon substrate.

Monolithic dome A dome composed of a series of arches, joined together with a series of horizontal rings called parallels.

Monolithic integrated circuit 1. An integrated circuit formed upon or within a semiconductor substrate with at least one of the circuit elements formed within the substrate.

2. A complete electronic circuit fabricated as an inseparable assembly of circuit elements

in a single small structure. It cannot be divided without permanently destroying its intended electronic function. (Graf)

Monomer A small molecule that may become chemically bonded to other monomers to form a polymer. From Greek mono "one" and meros "part".

Monovalent ion An ion with a single positive or negative charge (H^+, $C1^-$).

Monument Any design, scheduling or production technology with scale requirements necessitating that designs, orders and products be brought to the machine to wait in queue for processing. The opposite of a right-sized machine.

Moore's law After Gordon Moore:"The number of transistors per computer chip will double roughly every two years".

Morphology The microstructure of the solid phases of materials. The grain shapes and structure of crystals of the chemical components of a battery.

Mortar A plastic mixture of cementitious materials, fine aggregate and water.Fat Mortar: Mortar containing a high percentage of cementitious components. It is a sticky mortar which adheres to a trowel.High-Bond Mortar: Mortar which develops higher bond strengths with masonry units than normally developed with conventional mortar.

Mosaic A graphical interface to the World Wide Web (WWW).

MOSFET A Field Effect Transistor made using Metal Oxide Semiconductor technology. Controlled by voltage rather than current like a bipolar transistor. MOSFET's have a significantly higher switching speed than bipolar power transistors. Suitable for high power circuits, they generate almost no loss (little heat generation), enabling fast response, excellent linearity, and high efficiency. The positive temperature coefficient inhibits thermal runaway. (Degrades to an SFET - Smoke and Fire Emitting Transistor if subject to excessive voltages).

MOTD (message of the day) A message posted on some computer systems to let people know about problems or new developments.

Motherboard The pc board of a computer that contains the bus lines and edge connectors to accommodate other boards in the system. In a microcomputer, the motherboard contains the microprocessor and connectors for expansion boards.

Motions in the foot The motions of the foot are:

Abduction and Adduction which occur on the transverse plane;

Dorsiflexion and Plantar flexion which occur on the sagittal plane; and

Inversion and Eversion which occur on the frontal plane;

Mounting error The error resultant from installing the transducer, both electrical and mechanical.

Movable bridge A bridge in which the deck moves to clear a navigation channel; a swing bridge has a deck that rotates around a center point; a drawbridge has a deck that can be raised and lowered; a bascule bridge deck is raised with counterweights like a drawbridge; and the deck of a lift bridge is raised vertically like a massive elevator.

MSDS(Material safety data sheet) Information provided by battery or cell manufacturers about any hazardous materials used in their products.

MSRP Manufacturer's Suggested Price. MSRPs do not include applicable destination charges, state and local taxes, license fees, optional equipment or special items or services.

MTBF (Mean Time Between Failure) The average time between failures for a repairable product for a defined unit of measure (e.g., operating hours, cycles, miles, etc.).

MTF Multi-layer Thin Film .

Mueller bridge A high-accuracy bridge configuration used to measure three-wire RTDthermometers.

Mullite A substrate compound of alumina and silica ($3Al_2O_3 2SiO_2$).

Multi port fuel injection (MFI) Multi-Port Fuel Injection uses individual fuel injectors to spray fuel into each intake port, bypassing the intake manifold.

Multidisciplinary teamwork The timely and cooperative application of all appropriate disciplines in an open-communication, shared-information environment to effect people functioning as a team to achieve optimum solutions (people, product, and process) satisfy customer needs, objectives, and requirements.

Multihomed host A device attached to two or more networks.

Multimeter A portable test instrument which can be used to measure voltage, current, and resistance.

Multiple output power supply A power supply with two or more outputs.

Multiplexer A multiplexer is a device which enables several communications links or signals to share a single communications channel. At the receiving end of the link a demultiplexer separates the signals again. Various coding schemes are possible which enable the signals to be transmitted simultaneously or sequentially.

Mutual inductance (M) A measure of the amount of inductive coupling between two coils.

MUX Device for combining several signals or data streams into a single flow.

Mylar A registered trade mark of El Dupont De Nemours & Co. for polyethylene terephathalate. The polyester film is a common insulating material used in transformers, capacitors, etc.

N/C (No Connection) A connector point for which there is no internal connection.

Name resolution The process of mapping aliases to an address. The Domain Name System (DNS) is one system that does this.

Nano From the latin word meaning "dwarf". One billionth or 10^{-9}. One micron = 1000 nanometers. One nanometer is about the diameter of 3 to 6 atoms (depending on the element).

Nanobattery Very small battery built using nano technology. Of microscopic size 1 micron diameter they deliver 3.5 volts. The electrodes are ceramic or carbon particles and the electrolyte is a solid polymer impregnated in an aluminium oxide membrane.

Nanocurie One billionth 10^{-9} of a curie.

Nanomachine An artificial eutactic mechanical device that relies on nanometer-scale components.

Nanomechanical Pertaining to nanomachines.

Nanosatellite A Nanosatellite is a category of satellites corresponding to those with a mass from 1 to 10 kg.

Nanoscale On a scale of nanometers, from atomic dimensions to ~100 nm.

Nanosystem A eutactic set of nanoscale components working together to serve a set of purposes; complex nanosystems can be of macroscopic size.

Nanotechnology In recent general usage, any technology related to features of nanometer scale: thin films, fine particles, chemical synthesis, advanced microlithography, and so forth. As introduced by the author, a technology based on the ability to build structures to complex, atomic specifications by means of mechanosynthesis; this can be termed molecular nanotechnology.

Naphta An oil distillate. Naphta is an intermediate product between gasoline and kerosene.

It is known as a light product because of the low molecular weight of the hydrocarbons making it up.

Natural circulation The circulation of the coolant in the reactor coolant system without the use of the reactor coolant pumps. The circulation is due to the natural convection resulting from the different densities of relative cold and heated portions of the system.

Natural gas Petroleum in gaseous form consisting of light hydrocarbons often found in association with oil.

Navaid NAVigation AID. Beacon; wave off light etc installed on a platform or rig to warn ship and aircraft of installation.

Natural uranium Uranium as found in nature. It contains 0.7 percent uranium-235, 99.3 percent uranium-238, and a trace of uranium-234 by weight. In terms of the amount of radioactivity, it contains approximately 2.2 percent uranium-235, 48.6 percent uranium-238, and 49.2 percent uranium-234.

NB Nominal Bore. Term used when specifying a pipe size.

NBS National Bureau of Standards.

NC drill Numeric Control drill machine. A machine used to drill the holes in a printed board at exact locations, which are specified in a data file.

Necking Reduction of the cross-sectional area of a material in a localized area caused by uniaxial tension.

Negative 1. A reverse-image contact copy of a positive, useful for checking revisions of a PCB. If the negative of the current version is superimposed over a positive of an earlier version, all areas will be solid black except where changes have been made.

2. (Of a PCB image) Representing copper (or other material) as clear areas and absence of material as black areas. Typical of power and ground planes and solder mask.

Negative delta V (NDV) The NDV is the drop in the battery voltage which occurs when NiCad or NiMH cells reaches their fully charge state. Used to detect the end of the charging cycle in Nicads.

Negative electrode The electrode which has a negative potential. The anode.

Negative temperature coefficient A decreasing function with increasing temperature. The function may be resistance, capacitance, voltage, etc.

NEMA National Electrical Manufacturers Association. Leading US trade association which sets standards relating to electrically powered instrumentation.

NEMA Size case An older US case standard for panel meters, which requires a panel cutout of 3.93 x 1.69 inches.

NEMA-4 A standard from the National Electrical Manufacturers Association, which defines enclosures intended for indoor or outdoor use primarily to provide a degree of protection against windblown dust and rain, splashing water, and hose-directed water.

NEMA-7 A standard from the National Electrical Manufacturers Association, which defines explosion-proof enclosures for use in locations classified as Class I, Groups A, B, C or D, as specified in the National Electrical Code.

Nernst equation A mathematical description of electrode behavior: E is the total potential, in millivolts, developed between the sensing and reference electrodes; Ex varies with the choice of electrodes, temperature, and pressure: 2.3RT/nF is the Nernst factor (R and F are constants, n is the charge on the ion, including sign, T is the temperature in degrees Kelvin), and ai is the activity of the ion to which the electrode is responding.

Nernst factor (S, Slope); The term 2.3RT/nF is the Nernst equation, which is equal (at T = 25°C) to 59.16 mV when n = 1 and 29.58 mV when n - 2, and which includes the sign of the charge on the ion in the term n. The Nerst factor varies with temperature.

Net A collection of terminals all of which are, or must be, connected to each other electrically. Also known as a signal.

Net present value Net Present Value is a financial analysis technique that discounts a series of cash inflows (revenue) and outflows (investments and expenses) to determine the suitability of an investment. NPV is an evaluation technique used to screen new product development projects as part of portfolio management.

Net summer capability The steady hourly output that generating equipment is expected to supply to system load exclusive of auxiliary power, as demonstrated by tests at the time of summer peak demand.

Netiquette Network etiquette conventions used in written communications, usually referring to UseNet newsgroup postings but also applicable to e-mail.

Netlist List of names of symbols or parts and their connection points which are logically connected in each net of a circuit. A netlist can be "captured" (extracted electronically on a computer) from a properly prepared CAE schematic.

Network A group of computers that are connected to each other by communications lines to share information and resources.

Network file system (NFS) A protocol developed by Sun MicroSystems that enables clients to mount remote directories onto their own local filesystem.

Network information center (NIC) The Internet administration facility that controls the naming of networks accessible over the Internet.

Network information service (NIS) A set of protocols developed by Sun Microsystems used to provide directory services for network information.

Network interface card (NIC) A generic term for a networking interface board used to connect a device to the network. The NIC is where the physical connection to the network occurs.

Neumann's law The symmetry of the physical properties of a crystal must include the symmetry of the point group of the crystal.

Neutral The ac return somewhere connected to ground, but which should not be used for ground because it is a current-carrying path.

Neutral calcaneal stance position This is a reference position where the subject's feet are placed in their normal angle and base of gait, and with the subtalar in a neutral position. When observed from behind the concavity on the lateral side of the foot is parallel to the concavity on the lateral surface of the leg; and the lateral surface of the foot describes a straight line in the area of the calcaneal cuboid joint.

Neural network A powerful data modeling tool that is able to capture and represent complex input/output relationships. It is used as a basis for self learning systems.

Neutral position This refers to the position of a joint where the bones that make up the joint are placed in the optimal position for maximal movement.

Neutron An uncharged elementary particle, with a mass slightly greater than that of the proton, found in the nucleus of every atom heavier than hydrogen.

Neutron capture The reaction that occurs when a nucleus captures a neutron. The probability that a given material will capture a neutron is proportional to its neutron capture cross section and depends on the energy of the neutrons and the nature of the material.

Neutron chain reaction A process in which some of the neutrons released in one fission event cause other fissions to occur. There are three types of chain reactions: 1. Nonsustaining—An average of less than one fission is produced by the neutrons released by each previous fission (reactor subcriticality);

2. Sustaining—An average of exactly one fission is produced by the neutrons released by each previous fission (reactor criticality); and

3. Multiplying—An average of more than one fission is produced by the neutrons released by previous fission (reactor supercriticality).

Neutron flux A measure of the intensity of neutron radiation in neutrons/cm^2-sec. It is the number of neutrons passing through 1 square centimeter of a given target in 1 second. Expressed as nv, where n = the number of neutrons per cubic centimeter and v = their velocity in centimeters per second.

Neutron generation The release, thermalization, and absorption of fission neutrons by a fissile material and the fission of that material producing a second generation of neutrons. In a typical nuclear power reactor system, there are about 40,000 generations of neutrons every second.

Neutron leakage Neutrons that escape from the vicinity of the fissionable material in a reactor core. Neutrons that leak out of the fuel region are no longer available to cause fission and must be absorbed by shielding placed around the reactor pressure vessel for that purpose.

Neutron source Any material that emits neutrons, such as a mixture of radium and beryllium, that can be inserted into a reactor to ensure a neutron flux large enough to be distinguished from background to register on neutron detection equipment.

Neutron, thermal A neutron that has (by collision with other particles) reached an energy state equal to that of its surroundings, typically on the order of 0.025 eV (electron volts).

Newton (N) This is the Systeme Internationale unit for measuring force. The force of gravity applied to a mass of 1kg is equal to 9.81 N; one newton is the force exerted by a mass of about 102g. Force is a vector quantity and has both magnitude and direction e.g. the force of gravity is 10 N, downwards.

Newtonian physics The complete science of mechanical engineering is based on the three laws of force proposed by Sir Issac Newton. Neglecting the strange behaviour of atomic and sub atomic particles, all physical systems must obey all three laws of newton's Laws, at all times.

Newton's first law A body will continue in a state of rest or uniform motion in a straight line unless it is acted upon by an external force.

Newton's second law An external force will cause a body to accelerate in the direction of the force. The acceleration (a) is equal to the size of the force (F) divided by the mass (m) of the object, as in the equation:

$a = F/m$, or as it is more commonly used

$F = m \times a$.

Nicrosil/Nisil A nickel chrome/nickel silicone thermal alloy used to measure high temperatures. Inconsistencies in thermoelectric voltages exist in these alloys with respect to the wire gage.

Nitrification The biological oxidation of ammonia and ammonium sequentially to nitrite and then nitrate. It occurs naturally in surface waters, and can be engineered in wastewater treatment systems. The purpose of nitrification in wastewater treatment systems is a reduction in the oxygen demand resulting from the ammonia.

Nitroglycerin An explosive compound made from a mixture of glycerol and concentrated nitric and sulfuric acids, and an important ingredient of most forms of dynamite.

NMOS An acronym for n-channel metal-oxide-semiconductor, as in NMOS transistor and NMOS logic.

NMR (Normal-Mode Rejection) The ability of a panel meter to filter out noise superimposed on the signal and applied across the SIG HI to SIG LO input terminals. Normally expressed in dB at 50/60 Hz.

No load voltage Terminal voltage of battery or supply when no current is flowing in external circuit.

Noble gas A gaseous chemical element that does not readily enter into chemical combination with other elements. An inert gas. Examples are helium, argon, krypton, xenon, and radon.

Noble metal A metal with high resistance to chemical reaction, especially oxidation and solution by organic acids; sometimes called a precious metal.

Node A pin or lead which will have at least one wire connected to it. In a netlist, a node is described by a component reference desginator together with a pin number.

Noise The aperiodic random component on the power source output which is unrelated to source and switching frequency. Unless specified otherwise, noise is expressed in peak-to-peak units over a specified bandwidth.

Nominal capacity Used to indicate the average capacity of a battery. It is the average capacity when batteries are discharged at 0.2C within one hour of being charged for 16 hours at 0.1C and 20± 5°C. (or discharge at 0.05C for automotive batteries - SAE) Definition depends on the conditions.

Nominal dimension A dimension greater than a specified masonry dimension by the thickness of a mortar joint, but not more than 1/2 in.

Nominal group technique Nominal Group Technique, similar to brainstorming, is used by teams to generate ideas on a particular subject. Team members are asked to silently write down as many ideas as possible. Each member is then asked in turn to share one idea which is recorded. After all ideas are recorded, they are discussed and prioritized by the group.

Nominal value The stated or objective value of a quantity or component, which may not be the actual value measured.

Nominal voltage Used to indicate the voltage of a battery. Since most discharge curves are neither linear nor flat, a typical value is generally taken which is close to the voltage during actual use.

Non conformance The failure of a unit or product to conform to specified requirements.

Non developmental item Non-developmental item is a broad generic term that covers material available from a wide variety of sources with little or no developmental effort by the purchaser. NDIs include: · Items obtained from a domestic or commercial marketplace; · Items already developed and in use by users; · Items already developed to other standards and in agreement with the purchaser.

Non stochastic effect The health effects of radiation, the severity of which vary with the dose and for which a threshold is believed to exist. Radiation-induced cataract formation is an example of a non-stochastic effect (also called a deterministic effect).

Non vital plant systems Systems at a nuclear facility that may or may not be necessary for the operation of the facility (i.e., power production) but that would have little or no effect on public health and safety should they fail. These systems are not safety related.

Non-combustible material Any material which will neither ignite nor actively support combustion in air at a temperature of 1200 F when exposed to fire.

Noncrystalline The solid state wherein there is no long-range atomic order; sometimes the terms amorphous, glassy, and vitreous are used synonymously.

Nondestructive testing (NDT) A procedure for determining the quality or characteristics of a material, part, or assembly without permanently altering it or its properties; examples include ultrasonic and radiographic inspection.

Nonpoint source pollution (NPSP) Any pollution from a source which cannot be attributed to a particular discharge point, e.g. from agricultural crops, city streets, construction sites, etc.

Nonpower reactor Reactors used for research, training, and test purposes, and for the production of radioisotopes for medical and industrial uses.

Nonsteady state diffusion The diffusion condition for which there is some net accumulation or depletion of diffusing species; the diffusion flux is dependent on time.

NOR NOT-OR — logic gate whose output is the negation of that of the OR gate.

NORM Naturally Occurring Radioactive Material.

Normal (axial) Stress The force per unit area on a given plane within a body a = F/A.

Normal foot The term normal represents a set of circumstances whereby the foot will function in a manner which will not create adverse physical or emotional response in the individual. This definition applies when the lower extremity is used in an average manner and in an average environment, as dictated by the needs of the society at the moment.

Normal hydrogen electrode A reversible hydrogen electrode (Pt) in contact with hydrogen gas at 1 atmosphere partial pressure and immersed in a solution containing hydrogen ions at unit activity.

Normal mode rejection ratio The ability of an instrument to reject interference usually of line frequency (50-60 Hz) across its input terminals.

Normalcy The anatomical criteria for normal function are:

- the distal one third of the leg is vertical

- the subtalar joint rests in its neutral position i.e. it is neither pronated or supinated around its triplane axis.

- the bisection of the posterior surface of the calcaneum is considered to be approximately 3-4 degrees inverted to the vertical, when adopting a neutral position. This allows the plantar tubercles to make ground contact.

- the mid tarsal joint is in its maximum position of pronation. In this position the forefoot is considered to be locked against the rearfoot, and the plantar plane of the forefoot parallels the plantar plane of the rearfoot. Both planes are considered to be parallel to the transverse plane.

- metatarsals 2, 3, and 4 are in a totally dorsiflexed position. These metatarsals are considered to function in the sagittal plane and require to plantar flex during propulsion. metatarsals 1 and 5 are maintained in such a position that the plantar surface of their heads are on the same transverse plane as metatarsals 2, 3, and 4. The first and fifth rays function in all three body planes, but primarily displace sagittal plane motion.

In addition to the above, there must be adequate motion in the joints; correct direction of motion and normal neuromuscular function.

Normalizing For ferrous alloys, austenitizing above the upper critical temperature, then cooling in air; the objective of this heat treatment is to enhance toughness by refining the grain size.

NOT Logic gate whose output is binary 1 when its input is 0, and whose output is a 0 when its input is a 1.

Notified body Independent Testing Laboratory recognised to perform tests; audit quality systems and issue reports and certificates of conformity.

NOx Nitrogen Oxides. The term used to describe the sum of nitric oxide (NO); nitric dioxide (NO_2) and other oxides of nitrogen which play a major role in the formation of ozone.

Nozzle As used in power water reactors and boiling water reactors, the interface (inlet and outlet) between reactor plant components (pressure vessel, coolant pumps, steam generators, etc.) and their associated piping systems.

NPSH Net Positive Suction Head. Minimum suction pressure required by a pump to prevent cavitation.

NPT Normal Pressure and Temperature. NTP is used in many thermodynamic calculations and tabulations and is defined as 20 degrees Celsius and 1 atmosphere of pressure.

NPV Net Present Value. The cost of a product or system calculated in the present-day currency.

Found by subtracting a project's initial investment from the present value of the cash inflows discounted at a rate equal to the firm's cost of capital.

NRE Non-Recurring Engineering costs. A one time charge for design and implementation of custom battery packs or other products.

Nuclear energy The energy liberated by a nuclear reaction (fission or fusion) or by radioactive decay.

Nuclear engineering Nuclear engineering involves the design, construction, and operation of nuclear power plants for power generation, propulsion of nuclear submarines, and space power systems. Nuclear engineers are involved in the handling of nuclear fuels and the safe disposal of radioactive wastes.

Nuclear fission Occurs when the atomic nucleus splits into two or more smaller nuclei plus some by-products. These by-products include free neutrons and photons (usually gamma rays). Fission releases substantial amounts of energy (the nuclear binding energy). The neutrons released by the fission process may collide with other nuclei causing them in turn to undergo fission initiating to a chain reaction.

Nuclear force A powerful short-ranged attractive force that holds together the particles inside an atomic nucleus.

Nuclear fusion A process in which two nuclei join together to form a larger nucleus and releasing energy. It takes considerable energy to overcome the repulsion between the two positively charged nuclei to force them to fuse. The fusion of lighter nuclei, which creates a heavier nucleus and a free neutron, will generally release even more energy than it took to force them together. It is an exothermic process which could produce self-sustaining reactions.

Nuclear magnetic resonance (NMR) The interaction of atomic nuclei placed in an external magnetic field with an applied electromagnetic field oscillating at a particular frequency . Magnetic conditions within the material are measured by monitoring the radiation absorbed and emitted by the atomic nuclei. Used in MRI scanners and as a spectroscopy technique to obtain physical, chemical, and electronic properties of molecules.

Nuclear power plant An electrical generating facility using a nuclear reactor as its heat source to provide steam to a turbine generator.

Nuclear steam supply system The reactor and the reactor coolant pumps (and steam generators for a pressurized water reactor) and associated piping in a nuclear power plant used to generate the steam needed to drive the turbine generator unit.

Nuclear waste A particular type of radioactive waste that is produced as part of the nuclear fuel cycle (i.e., those activities needed to produce nuclear fission, or splitting of the atom). These include extraction of uranium from ore, concentration of uranium, processing into nuclear fuel, and disposal of byproducts. Radioactive waste is a broader term that includes all waste that contains radioactivity. Residues from water treatment, contaminated equipment from oil drilling, and tailings from the processing of metals such as vanadium and copper also contain radioactivity but are not "nuclear waste" because they are produced outside of the nuclear fuel cycle. NRC generally regulates only those wastes produced in the nuclear fuel cycle (uranium mill tailings, depleted uranium, spent fuel rods, etc.).

Nucleation The initial stage in a phase transformation, evidenced by the formation of small particles (nuclei) of the new phase, which are capable of growing.

Nucleon Common name for a constituent particle of the atomic nucleus. At present, applied to protons and neutrons, but may include any other particles found to exist in the nucleus.

Nucleotide A monomer of the nucleic acids composed of a five-carbon sugar, a nitrogen-containing base, and phosphoric acid.

Nucleus The small, central, positively charged region of an atom. Except for the nucleus of ordinary hydrogen, which has only a proton, all atomic nuclei contain both protons and neutrons. The number of protons determines the total positive charge or atomic number. This number is the same for all the atomic nuclei of a given chemical element. The total number of neutrons and protons is called the mass number.

Nuclide A general term referring to all known isotopes, both stable (279) and unstable (about 2,700), of the chemical elements.

Nutrient loading rate This is simply the mass rate at which substrate or nutrient is delivered to a system. If the inlet concentration of substrate in Sin (mass/ volume) and the fluid flow rate is F (volume/time), then the substrate loading is FSin (mass/time). An average measure of the rate of nutrient or substrate provision in a system is expressed as mass substrate per biofilm area per time.

O Symbol for oxygen.

O&G Oil and Gas.

O&M Operating and Maintenance. Usually in reference to a contract or budget e.g. an O&M contract.

O.D. Outside diameter.

OCA Offshore Contractors Association.

Octahedral position The void space among close-packed, hard sphere atoms or ions for which there are six nearest neighbors; an octahedron (double pyramid) is circumscribed by the lines constructed from centers of adjacent spheres.

Octal Pertaining to a base 8 number system.

Octane number A measurment of a fuel's resistance to spontaneous ignition.

The higher the octane number the greater fuel's resistance to spontaneous ignition.

Octane rating A unit of measurement on a scale intended to indicate the tendency of a fuel to detonate or knock based on the percentage of isooctane in the fuel. The higher the rating, the higher the percentage of isooctane and therefore the greater the resistance to detonation offered by the fuel.

OEM (Original equipment manufacturer) A company with the prime responsibility for conceiving, designing, manufacturing and distributing a particular product line.

Offset The difference in temperature between the set point and the actual process temperature. Also, referred to as droop.

Offset current The direct current that appears as an error at either input terminal of a dc amplifier when the input current source is disconnected.

Offset voltage The dc voltage that remains between the input terminals of a dc amplifier when the output voltage is zero.

Ofhc Oxygen-free high-conductivity copper. The industrial designation of the pure copper used in a Type T thermocouple.

OGP The International Association of Oil & Gas producers. Formed in 1974 to develop effective communications between the upstream industry and the complex network of international regulators.

Ohm (Ù) Unit of resistance. One ohm is the electrical resistance between two points of a conductor when a constant potential difference of 1 volt, applied to these points, produces a current of 1 ampere in the conductor.

Ohmeter An instrument used to measure electrical resistance.

Ohmic loss The voltage drop across the cell during passage of current due to the internal resistance of the cell. Also known as IR loss or IR drop.

Ohm's law The fundamental mathematical relationship between current (I), voltage (E) and resistance (~ discovered by George Simon Ohm. The passage of one Ampere through one Ohm produces one Volt.

One piece flow The opposite of batch production. Instead of building many products and then holding them in queue for the next step in the process, products go through each step in the process one at a time, without interruption. It improves quality and lowers costs.

OPEC Organisation of the Petroleum Exporting Countries - a permanent inter-governmental organization currently made up of 11 oil producing and exporting countries.

Its aim is to co-ordinate and unify the petroleum policies of the Member Countries and to determine the best means for safeguarding their individual and collective interests.

Open chain This describes a series of links free to move on each other but also capable of combining to support the body. The chain is said to be open when the distal link is free to move and closed when it is fixed. In open chain motion the calcaneum supinates and pronates about the subtalar joint.

Open circuit The lack of electrical contact in any part of the measuring circuit. An open circuit is usually characterized by rapid large jumps in displayed potential, followed by an off-scale reading.

Open circuit voltage (OCV) The difference in potential between the terminals of a cell or voltage when the circuit is open (no-load condition).

Open end lease A lease in which the lessor is not responsible for the value of a vehicle when the lease is done. Under this arrangement the lessee is responsible for the difference between the current actual market value at the end of the lease and the residual value. Also known as a "finance" lease.

Open shortest path first (OSPF) The basic Internet routing protocol for sending data over multiple paths. It uses the network's topology for routing decisions.

Open software foundation (OSF) A consortium of hardware and software vendors collaborating to produce technologies for device independent operation.

Open systems interconnection (OSI) A family of ISO data communications.

Operable A system, subsystem, train, component, or device is operable or has operability when it is capable of performing its specified functions and when all necessary attendant instrumentation, controls, electrical power, cooling or seal water, lubrication, or other auxiliary equipment that are required for the system, subsystem, train, component, or device to perform its functions are also capable of performing their related support functions.

Operand Argument to an operator (often an expression itself that must be evaluated first).

Operating system A collection of programs that controls the overall operation of a computer and performs such tasks as assigning places in memory to programs and data, processing interrupts, scheduling jobs and controlling the overall input/output of the system.

Operating temperature range The range of ambient baseplate or case temperatures through which a power supply is specified to operate safely and to perform within specified limits.

Operating voltage Voltage between the two terminals of the battery with a load connected.

Operational amplifier (Op amp) A high gain DC amplifier with a voltage gain of 100 to 100,000 or more and a very high (ideally infinite) input impedance and very low (ideally zero) output impedance. Op-amps are the basic building block of linear integrated circuits used for analogue circuit applications. They have positive and negative inputs which allow circuits which use feedback to achieve a wide range of functions.

Operational mode In a nuclear power reactor, an operational mode corresponds to any one inclusive combination of core reactivity condition, power level, and average reactor coolant temperature.

Operational pH The determination of sample pH by relating to pH measurements in a primary standard solution. This relationship assumes that electrode errors such as sensitivity and changes in asymmetry potential can be disregarded or compensated for, provided the liquid junction potential remains constant between standard and sample.

Operational power supply A power supply with a high open loop gain regulator which acts like an operational amplifier and can be programmed with passive components.

Operator Term used to describe a company appointed by venture stake holders to take primary responsibility for day-to-day operations for a specific plant or activity.

Optical isolation Two networks which are connected only through an LED transmitter and photoelectric receiver with no electrical continuity between the two networks.

Opto coupler A package that contains a light emitter and a photoreceptor used to transmit signals between electrically isolated circuits.

Opto isolator Also called opto-coupler. An isolation device using optical techniques to isolate the electrical connections between a transmitter and a receiver. Used to pass signals between high voltage and low voltage circuits and to replace switches and relays. Having no electrical connection they also help to cut down on ground loops.

Optoelectronic scanners A kinematic system of analysis which employs colour filter reflective markers. The system has three scanners arranged in a fixed orientation. Both scanning rate and the accuracy of the system are high, and its, output consists of "real time" 3 -D co-ordinates.

OR Logic gate whose output is a binary 1 if any of its inputs is a 1; zero otherwise.

Orbital In the approximation that each electron in a molecule has a distinct, independent wave function, the spatial distribution of an electron wave function corresponds to a molecular orbital. These, in turn, can be approximated as sums of contributions from the orbitals characteristic of the isolated atoms. An electron added to a molecule—or, similarly, one excited to a higher-energy state within a molecule—would occupy a state with a different wave function from the rest; an unoccupied state of this kind corresponds to an unoccupied molecular orbital. Orbital-symmetry effects on reaction rates arise when a reaction requires overlap between two lobes of the orbitals on each of two reagents: if the algebraic signs of the wave functions in the facing lobes do not match, bond formation between those orbitals is prohibited.

Ordering Positioning of host and substitution ions in an ordered, repetitious pattern rather than in a random arrangement.

Ore A natural mineral mined and treated for extraction of its components.

Organic compound Any compound containing carbon except for the carbonates (carbon dioxide, the carbonates and bicarbonates), the cyanides, and cyanates.

Organic nitrogen Nitrogen contained as amines in organic compounds such as amino acids and proteins.

Organization A unit within a company or other entity (e.g., Government agency or branch of service) within which many projects are managed as a whole. All projects within an organization share a common top-level manager and common policies.

Orientation Arrangements in space of the axes of a crystal lattice with respect to a coordinate system.

Orthosis (pl. Orthoses) This refers to the straightening of a deformity. Foot Orthotics (FO) can be classified by the material they are made from and can be devided into Function Foot Orthoses (FFO) and Accommodation Foot Orthoses (AFO). FFO's control joint movement by a series of levers known as wedges, they help reinstate normal gait patterns by deviating ground reactive forces through neutral positions of joints. AFO provide isobaric and isotatic function which dampen down peak forces and reduce dynamic friction on the foot surface.

Oscillations An oscillation is one complete period of vibratory or periodic motion, for example, the whole succession of states that takes place before the motion begins to repeat itself.

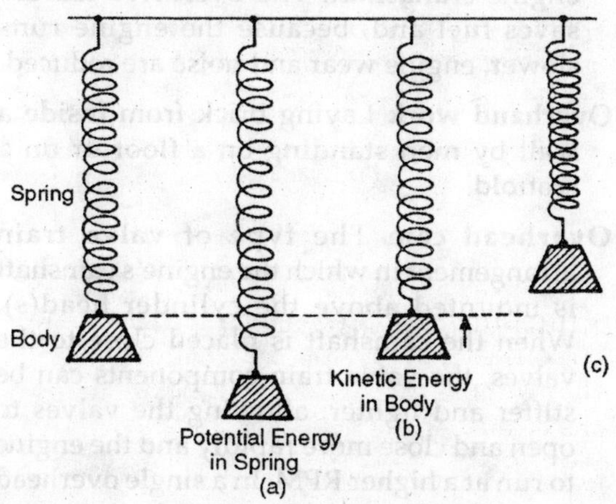

Fig. Oscillation

For example, one oscillation of a pendulum bob is a complete excursion from where it started back to its original position with the same velocity (magnitude and direction). The time of one oscillation is called one period and the number of oscillations per second (the reciprocal of the period) is called the frequency. The definition cannot be applied strictly to nonperiodic motions. In such motions the period is usually taken as the time between successive zeros of the displacement.

Oscillator Circuit that produces an alternating voltage (current) when supplied by a steady (DC) energy source.

OSHA (Occupational Safety and Health Administration) The department of the US government with the responsibility to ensure safety and healthful work environments.

Osmosis The diffusion of a solvent through a semi permeable membrane from a region of low solute concentration to a region of high solute concentration. The semi permeable membrane is permeable to the solvent, but not to the solute, resulting in a chemical potential difference across the membrane which drives the diffusion. The solvent flows from the side of the membrane where the solution is weakest to the side where it is strongest to equalise the concentration on both sides.

Output current limiting A protective feature that keeps the output current of a power supply within predetermined limits during overload to prevent damage to the supply or the load.

Output filter One or more discrete components used to attenuate output ripple and noise.

Output filter capacitor The capacitor(s) across the output terminals of a power supply.

Output impedance The impedance that a power supply appears to present to its output terminals.

Output LC Filter The low pass filter in the secondary of a switching power supply that smooths the rectified output to its average value. Also called an averaging filter.

Output noise The RMS, peak-to-peak (as specified) ac component of a transducer's dc output in the absence of a measurand variation.

Output range The specified range over which the output voltage or current can be adjusted.

Output rectifier A diode(s) used to convert ac to dc in the secondary of the transformer.

Output ripple and noise Output ripple is the periodic AC component imposed on the output voltage of a converter. Output noise refers to unwanted variations in the converter output that are unrelated to the switching frequency. Ripple and noise are typically specified together as a peak-to-peak value over a specified bandwidth. Ripple and noise is sometimes referred to as Periodic And Random Deviation (PARD). Care must be taken when measuring output noise and ripple not to induce errors in the test instrumentation. Attach a twisted wire pair (about 1 foot in length) with three twists per/inch between the converter outputs and an appropriate load. Connect a 33 uF electrolytic capacitor across the load. Using an oscilloscope with a minimum bandwidth of 20 MHz and a probe with the ground clip disconnected, measure the ripple at the connection of the load and twisted pair wires. This method eliminates the 'common mode noise" that interferes with measurements made directly at the converter output pins.

Output voltage The voltage measured at the output terminals of a power supply.

Output voltage accuracy the maximum allowable deviation of the DC output from its ideal or nominal value. Output voltage accuracy is sometimes called "output voltage tolerance" or "output voltage setpoint" and is commonly given as a percentage of output voltage.

Over charge Continuous charging of the battery after it reaches full charge. Generally overcharging will have a harmful influence on the performance of the battery which could lead to unsafe conditions. It should therefore be avoided.

Over current Exceeding the manufacturer's recommended maximum discharge current for a cell or battery.

Over discharge Discharging a battery below the end voltage or cut-off voltage specified for the battery.

Overvoltage 1. The potential difference between the equilibrium of an electrode and that of the electrode under an imposed polarization current.

2. A voltage that exceeds specified limits.

Overcurrent device A device capable of automatically opening an electric circuit, both under predetermined overload and short-circuit conditions, either by fusing of metal or by electro-mechanical means.

Overdrive A transmission in which the highest gear ratio is less than a one-to-one ratio. This means the drive shaft turns faster than the engine crankshaft. The overdrive feature saves fuel and, because the engine runs slower, engine wear and noise are reduced.

Overhand work Laying brick from inside a wall by men standing on a floor or on a scaffold.

Overhead cam The type of valve train arrangement in which the engine's camshaft is mounted above the cylinder head(s). When the camshaft is placed close to the valves, the valve train components can be stiffer and lighter, allowing the valves to open and close more rapidly and the engine to run at a higher RPM. In a single overhead cam (SOHC) layout, one camshaft actuates all of the valves in a cylinder head. In a dual overhead camshaft (DOHC) layout, one camshaft actuates the intake valves, and one camshaft operates the exhaust valves.

Overhead valve engine (OHV) An engine with both intake and exhaust valves placed directly over the piston. In this design, the camshaft is located in the block, and the valves are actuated by pushrods and rocker arms.

Overlap Orbitals lack sharply defined surfaces, declining in amplitude exponentially in their surface regions. When two orbitals are brought together, regions of substantial amplitude overlap. The resulting system can be described as two new orbitals, one formed by joining the two original orbitals without introducing a node in the wave function, and the other formed with a node between them. The nodeless joining reduces the energy of the electrons relative to the separate orbitals, resulting in a bonding interaction; joining with a node raises the energy, producing an antibonding interaction. If both new orbitals are occupied, antibonding forces dominate, resulting in overlap repulsion. Molecular mechanics models give an approximate description of overlap (and other) forces for a certain range of atoms and geometries.

Overlap repulsion A repulsive force resulting from the nonbonding overlap of two atoms.

Overload protection Power supply protection based on power supply input power. Many low cost switching power supplies have no method to control or limit output current. The easiest and least expensive protection method is sensing input power. If input power becomes excessive, you assume that one power supply output must have an excessive load. Some overload circuits permanently shut down the power supply; others will automatically recover when the fault is removed.

Overmoulding An injection moulding technique used to encapsulate and protect components or small sub-assemblies, usually by moulding a soft, flexible plastic over the components which must be able to withstand the temperatures and pressures of the moulding process. Used for cable connectors, gaskets, and for incorporating small components into cables. Two shot moulds may be used to provide soft plastic grips over a hard plastic shell.

Overshoot A transient change in output voltage in excess of specified output regulation limits, which can occur when a power supply is turned on or off, or when there is a step change in line or load.

Overvoltage protection (OVP) A power supply feature which shuts down the supply usually by firing a crowbar or clamping the output when an output voltage exceeds a preset level.

Oxidation A reaction in which there is an increase in valence resulting from a loss of electrons; often associated with the corrosion of metals, where the corroded metal forms an oxide; elevated temperatures increase the rate of oxidation.

Oxidative phosphorylation The synthesis of the energy storage compound adenosine triphosphate (ATP) from adenosine diphosphate (ADP) using a chemical substrate and molecular oxygen.

Oxygen consumption The measurement of oxygen uptake, this measurement is used routinely to calculate the metabolic cost of different activities.

P

Package 1. Decal or printed wiring board component.

2. A type of PCB component which contains a chip and acts to make a convenient mechanism for protecting the chip while on the shelf and after attachment to a PCB. With its leads soldered to a printed circuit board, a package serves as the electrical conduction interface between the chip and the board. An example is a DIP .

Pad A conductive area on a printed circuit board used for connection to a component lead or terminal area, or as a test point.

Paint Mixture of a pigment which is dispersed into a liquid (called a vehicle) designed for application to a substrate when spread in one or more thin coats. These layers dry out to a solid film, adherent to the substrate. Paint is designed to protect and/or decorate the surface it is applied to.

Panel Material (most commonly an glass/epoxy-copper laminate known as core) sized for fabrication of printed circuit boards. Panels come in many, many sizes, the most common being 12" by 18" and 18" by 24". Subtract 1/2" to 1" margins from the panel size to arrive at the space available for printed circuitry.

Panel wall Exterior, non-loadbearing wall wholly supported at each story. Panel walls are required to be self-supporting between stories. They must resist lateral forces such as wind pressures and must transfer these forces to adjacent structural members.

Panelize 1. To lay up more than one (usually identical)printed circuits on a pans. Individual printed circuits on a panel need a margin between them of 0.3". Some board houses permit less separation.

2. Lay up multiple printed circuits (called modules) into a sub-panel so that the sub-panel can be assembled as a unit. The modules can then be separated after assembly into discrete printed circuits.

PAPA Prepare to Abandon Platform Alarm.

Parallax An optical illusion which occurs in analog meters and causes reading errors. It occurs when the viewing eye is not in the same plane, perpendicular to the meter face, as the indicating needle.

Parallax error Apparent displacement of an object caused by a change in the position of the observer.

Parallel axis theorem The moment of inertia of any object about an axis through its center of mass is the minimum moment of inertia for an axis in that direction in space.

Parallel connection Connection together of the cell terminals of the same polarity of two or more cells to form a battery of higher capacity.

Parallel transmission Sending all data bits simultaneously. Commonly used for communications between computers and printer devices.

Paralleling Connecting the output of two or more supplies together, either for the purpose of delivering additional current or creating redundancy. The reliability of a power system is critical to overall system performance. Indeed, the requirements for power systems are generally higher than those for the system overall. A cost effective way to provide very high levels of reliability is to connect a number of independent power supply units in parallel, such that if one power supply fails, the remaining power supplies will continue delivering sufficient current to supply the maximum system load without any interruption. This architecture is often referred to as an (n+x) redundant power supply. If any one of several paralleled supplies can fail without causing the entire system to fail, the power system is said to be (n+1) redundant. If any two of several paralleled supplies can fail without causing the entire system to fail, the power system is said to be (n+2) redundan. By allowing the live removal or insertion of power supply boards to and from the system power bus, a "hot pluggable" redundant power supply system provides a practical way to achieve the zero down time required in critical applications such as mainframe computer systems and telecommunications systems. In practice, (n+x) redundancy can be difficult to achieve. Connecting power supplies in parallel and ensuring that the failure of one parallel supply unit does not affect the remainder presents particular difficulties, ruling out paralleling schemes such as master-slave configurations entirely.

Parallelogram of forces To simplify force systems, it is useful to use graphic models to resolve forces acting on a body. Two forces are drawn as a scale diagram, with the correct angle between them, the length of the line representing the magnitude of the force. In the parallelogram of forces, the two forces are used to draw a parallelogram, the diagonal of which shows the magnitude and direction of the resultant force.

Paramagnetism The property of a substance which is attracted to a magnet. It is similar to ferromagnetism except that the attraction is weaker. When a paramagnetic material is placed in a strong magnetic field, it becomes a magnet as long as the strong magnetic field is present. But when the strong magnetic field is removed the magnetic effect is lost. Below the substance's Curie temperature a paramagnetic material becomes ferromagnetic. Paramagnetism is exhibited by materials containing transition elements, rare earth elements and actinide elements. Liquid oxygen and aluminum are also examples of paramagnetic materials.

Parameter Restrictive factor. In engineering, a parameter may be a constant value in an equation or one of a set of measurable factors (such as humidity, temperature, color, or acidity) that define a system and determine or limits its behavior.

Paraplegia This is complete paralysis of the legs, usually due to disease or injury of the spinal column.

Pareto analysis/ diagram An analysis/ diagramming technique using frequency of occurrence to identify and display results generated by each identified cause. This analysis is commonly used to decide where to apply initial effort for maximum effect.

Pareto principle The Pareto principle suggests that 20% of a set of independent variables is responsible for 80% of the result. In quantitative terms, for example, 80% of the problems come from 20% of the causes (machines, raw materials, operators etc.). Therefore, effort aimed at the right 20% can solve 80% of the problems.

Parget 1. Plaster, whitewash, or roughcast for coating a wall.

2. Plasterwork especially in raised ornamental figures on walls

Parging The process of applying a coat of cement mortar to masonry.

Parity A technique for testing transmitting data. Typically, a binary digit is added to the data to make the sum of all the digits of the binary data either always even (even parity) or always odd (odd parity).

Parking apron A paved or unpaved airfield surface used for fixed wing aircraft parking. The area includes parking lanes, taxi lanes, exits, and entrances. Aircraft move under their own power to the parking spaces, where they may be parked and secured with tie-downs. Parking designed to distribute aircraft for the purpose of increased survivability (dispersed hardstands) are included in this category code. For inventory purposes, only the prepared surface is included.

Part One piece, or two or more pieces joined together which are not normally subjected to disassembly without destruction or impairment of designed use (examples: gear, screws, transistors, capacitors, integrated circuits.

Part or identifying number (PIN) Part or Identifying Number is an alpha-numeric designator which identifies parts, items, or bulk materials that are covered by a specification.

Partition Interior dividing wall, one story or less in height.

Partition function A function determined by the probability distribution (over phase space in the classical treatment; over quantum states in the quantum treatment) describing a thermally equilibrated system; many thermodynamic quantities can be expressed in terms of the partition function and its derivatives.

Parts per million (ppm) Parts (molecules) of a substance contained in a million parts of another substance (e.g., water).

Pass The motion of a spray gun or roller in one direction only.

Passivate To make a surface such as steel inert, usually by chemical or electrochemical means.

Passivation layer A resistive layer that forms on the electrodes in some cells after prolonged storage impeding the chemical reaction. This barrier must be removed to enable proper operation of the cell. Applying charge/discharge cycles often helps in preparing the battery for use.

Passive component A device which does not add energy to the signal it passes. Examples: resistor, capacitor, inductor. (Contrast with active component .

Paste The semi-fluid product of a dispersion process.

a: in coatings it is usually high in viscosity and may require dilution prior to application; a concentrated pigment dispersion used for shading;

b: lime putty;

c: caulking putty

Pat tern 1. The shape or stream of material coming from a spray gun.

2. Frequent or widespread incidence.

Pathogenic organism An organism capable of causing infection.

PCB design service bureau A business engaged in PCB design as a service for others, especially electrical engineers. The word bureau is French for desk, or office, and this service is indeed performed from an office while sitting at a desk. Also called PCB design shop.

PCB designer One who creates the artwork for printed circuit boards. For you recruiters out there who are asked to find one, and for anyone else interested, here is a plain English description for a Printed Circuit Board Designer. Hint: It is not the same as an electrical engineer.

PCB layout PCB design.

Peak Maximum value of a waveform reached during a particular cycle or operating time.

Peak output current The maximum current value delivered to a load under specified pulsed conditions.

Peak to peak The measured value of a waveform from peak in a positive direction to peak in a negative direction.

Pearl paint A type of paint that is similar to metallic paint, but instead of minute metal particles it uses mica. Mica is a kind of semi transparent, crystalline mineral that absorbs and reflects light in prismatic fashion. This gives a dramatic, multi-dimensional effect to the paint. Sometimes called "pearl coat."

PED Classification whereby equipment is required to be designed and manufactured according to 'sound engineering practice'.

Pedobarograph This is a floor mounted system for measuring pressure underneath the foot. The platform is an electronic mat, laid on top of a glass plate. When the subject walks on the mat, it is compressed onto the glass plate, which loses its reflectivity, becoming progressively darker with increased pressure. This darkening provides the means for quantitative measurement. The underside of the glass plate is usually viewed by a television camera, the monochrome image from which is processed to give a display in which different colours correspond to different levels of pressure.

Peeling A film of paint or coating lifting from the surface due to poor adhesion or to moisture that travels through or on the substrate and accumulates under the paint where it expands when the temperature rises.

Peierls stress The stress required to move a dislocation.

Pellet, fuel As used in pressurized water reactors and boiling water reactors, a pellet is a small cylinder approximately 3/8-inch in diameter and 5/8-inch in length, consisting of uranium fuel in a ceramic form—uranium dioxide, UO_2. Typical fuel pellet enrichments in nuclear power reactors range from 2.0 percent to 3.5 percent uranium-235.

Peltier effect When a current flows through a thermocouple junction, heat will either be absorbed or evolved depending on the direction of current flow. This effect is independent of joule I2 R heating.

Pelvic rotation The first determinant of gait is the way in which the pelvis twist about a vertical axis during gait. This has the effect of bringing the hip joint forward as the hip flexes, and backwards as it extends. Pelvic Rotation reduces the angle of hip flexion and extension, which in turn reduces the vertical movement of the hip.

Pelvic tilt The vertical movement of the centre of gravity accompanies pelvic tilt motion from side to side, so that when the hip of stance phase is at its highest point, the pelvis is inclined, to lower hip of the swing phase leg.

Pendentive A triangular shape that adapts the circular ring of a dome to fit onto a flat supporting wall.

Peptide A short chain of amino acids.

Perfectly balanced rotor A rotor is perfectly balanced when its mass distribution is such that it transmits no vibratory force or motion to its bearings as a result of centrifugal forces.

Performance A quantitative measure characterizing a physical or functional attribute relating to the execution of a mission/operation or function.

PERT. Program Evaluation and Review Technique

PFMEA. Process Failure Modes and Effects Analysis

PLC Product Life Cycle

PLD Programmable Logic Device

PLM Product Lifecycle Management

PM Program / Project Manager

PMI Project Management Institute

PMT Program Management Team

POF Physics of Failure

Poka-Yoke Japanese term for mistake-proofing.

Performance based regulation Required results or outcome of performance rather than a prescriptive process, technique, or procedure.

Performance based regulatory action Licensee attainment of defined objectives and results without detailed direction from the NRC on how these results are to be obtained.

Perimeter The distance around the outside of a shape.

Periodic and random deviation (PARD) The sum of all ripple and noise components measured over a specified bandwidth and stated, unless otherwise specified, in peak-to-peak values.

Periodic table An arrangement of chemical elements in order of increasing atomic number. Elements of similar properties are placed one under the other, yielding groups or families of elements. Within each group, there is a variation of chemical and physical properties, but in general, there is a similarity of chemical behavior within each group.

Peripheral A device that is external to the CPU and main memory, i.e., printer, modem or terminal, but is connected by the appropriate electrical connections.

Peritectic An isothermal reversible reaction in metals where a liquid phase reacts with a solid phase to produce a single (and different) solid phase upon cooling.

Peritectoid An isothermal reversible reaction where a solid phase reacts with a second solid phase to produce a single (and different) solid phase upon cooling.

Permanent charge The charging current which can safely be continuously supported by the battery, regardless of the state of the charge.

Permeability 1. Passage or diffusion of a gas, vapor, liquid, or solid through a material without physically or chemically affecting it;

2. Term used to express various relationships between magnetic induction and magnetizing force; either absolute permeability or specific (relative) permeability.

Personnel monitoring The use of portable survey meters to determine the amount of radioactive contamination on individuals, or the use of dosimetry to determine an individual's occupational radiation dose.

PES (Programmable Electronic System) A system for control; protection or monitoring based on one or more progamable electronic devices.

The system comprises all elements e.g. power supplies; sensors; actuators; communication paths etc.

Petroleum engineering Petroleum engineering involves sustaining the flow of oil and gas, including discovery, recovery, storage, and transportation of petroleum. Petroleum engineers integrate many engineering specializations into an efficient system of oil and gas drilling, production, and processing.

Peukert number A value that indicates how well a battery performs under heavy currents. A value close to 1 indicates that the battery performs well; the higher the number, the more capacity is lost when the battery is discharged at high currents. The Peukert number of a battery is determined empirically.

Peukert's equation An empirical formula that approximates how the available capacity of a battery changes according to the rate of discharge. The equation shows that at higher currents, there is less available energy in the battery.

Pewter Tin-base white metal containing antimony and copper; 1 to 8% Sb and 0.25 to 3% Cu.

pH (potential (of) hydrogen) A logarithmic measure of the concentration of hydrogen ions (H +) in a solution and, therefore, its acidity or alkalinity (basicity). pH = -log[H +]

The "pH" scale extends from 0 to 14 (in aqueous solutions at room temperature). A pH value of 7 indicates a neutral (neither acidic nor basic) solution. A pH value of less than 7 indicates an acidic solution, the acidity increases with decreasing pH value. A pH value of more than 7 indicates an alkaline or basic solution, the alkalinity or basicity increases with increasing pH value.

pH Junctions The Junction of a reference electrode or combination electrode is a permeable membrane through which the fill solution escapes (called the liquid junction).

pH(S) (Standard pH Scale) The conventional standard pH scale established on the basis that an individual ionic activity coefficient can be calculated from the Debye-H¸ckel law for primary buffers.

Phase A time based relationship between a periodic function and a reference. In electricity, it is expressed in angular degrees to describe the voltage or current relationship of two alternating waveforms.

Phase angle The angle that a voltage waveform leads or lags the current waveform.

Phase diagram Graphical representation of the temperature and composition limits of phase fields in an alloy or ceramic system; it can be an equilibrium diagram, approximation to an equilibrium diagram, or a representation of metastable conditions or phases.

Phase difference The time expressed in degrees between the same reference point on two periodic waveforms.

Phase proportioning A form of temperature control where the power supplied to the process is controlled by limiting the phase angle of the line voltage.

Phase rule This states that the maximum number of phases (P) that may coexist at equilibrium is two plus the number of components (C) in the mixture minus the number of degrees of freedom (F): P + F = C + 2.

Phase shift The difference between corresponding points on input and output signal wave forms (not affected by magnitude) expressed as degrees lead or lag.

Phase space A classical system of N particles can be described by its 3N position and 3N momentum coordinates. The phase space associated with the system is the 6N dimensional space defined by these coordinates.

Phase transformation Changes that can occur within a given material system; how one or more phases in an alloy change into a new phase or mixture of phases; transformation occurs because the initial state of the alloy is unstable relative to the final state; at constant temperature and pressure the relative stability of a system is determined by its Gibbs free energy.

Phenol An aromatic benzene ring with a hydroxyl substituted for one hydrogen.

Phenolic A synthetic resin used for heat or water resistance.

Phenyl A benzene ring named as a constituent group, C6H5-.

Phonon A quantum of acoustic energy, analogous to the quantum of electromagnetic radiation, the photon. Thermal excitations in a crystal or in an elastic continuum can be described as a population of phonons (analogous to blackbody electromagnetic radiation). In highly inhomogeneous solids, a description in terms of phonons breaks down and localized vibrational modes become important.

Phosphatizing Steel pretreatment by a chemical solution containing phosphates and phosphoric acid to (temporarily) inhibit corrosion.

Phosphorescence Luminescence that occurs at times greater than on the order of a second after an electron excitation event.

Photodiode Semiconductor diode that produces voltage (current) in response to illumination.

Photoelectric constant A quantity equal to h/e where h is the Planck constant, and e, the electronic charge, and which multiplied by the frequency of any radiation exciting photoemission gives the potential difference corresponding to the quantum energy absorbed by the escaping photoelectron.

$h/e = 4.1349 \times 10^{-7}$ erg.sec.emu^{-1} = 1.3793 $\times 10^{-17}$ erg.see.esu^{-1}.

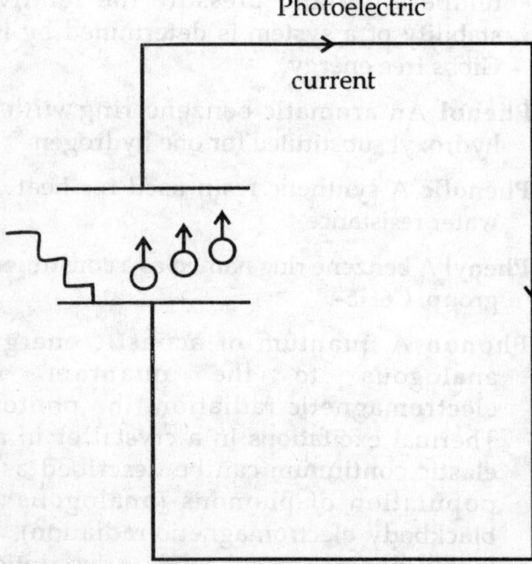

Fig. Photoelectric

Photoelectric effect An effect explained by A. Einstein that demonstrate that light seems to be made up of particles, or photons. Light can excite electrons (called photoelectrons in this context) to be ejected from a metal. Light with a frequency below a certain threshold, at any intensity, will not cause any photoelectrons to be emitted from the metal. Above that frequency, photoelectrons are emitted in proportion to the intensity of incident light.

The reason is that a photon has energy in proportion to its wavelength, and the constant of proportionality is the Planck constant. Below a certain frequency — and thus below a certain energy — the incident photons do not have enough energy to knock the photoelectrons out of the metal. Above that threshold energy, called the workfunction, photons will knock the photoelectrons out of the metal, in proportion to the number of photons (the intensity of the light). At higher frequencies and energies, the photoelectrons ejected obtain a kinetic energy corresponding to the difference between the photon's energy and the workfunction.

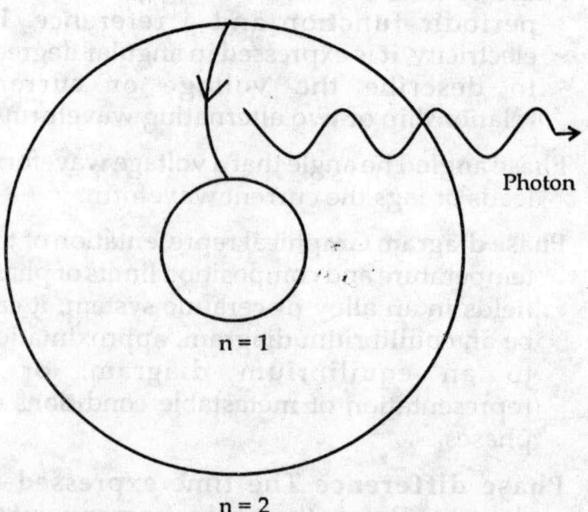

Fig. Photon emission

Photolithography Science of replicating complex circuitry onto the surface of a specimen.

Photomicrograph The picture made with a microscope.

Photon A quantum (or packet) of energy emitted in the form of electromagnetic radiation. Gamma rays and x-rays are examples of photons.

Photophosphorylation The synthesis of the energy storage compound adenosine triphosphate (ATP) from adenosine diphosphate (ADP) using solar energy.

Photoplotter Device used to generate artwork photographically by plotting objects (as opposed to copying an entire image at once as with a camera) onto film for use in manufacturing printed wiring.

Phototransistor Transistor that, when powered, produces amplified voltage (current) in response to illumination.

Phototroph Organisms which obtain energy from light using photooxidation.

Photovoltaic cell A device that directly converts the energy in light into electrical energy. Also called a photocell, a solar cell or a PV cell.

Photovoltaic effect The generation of an electromotive force as a consequence of the absorption of radiation. In practice a current which flows across the junction of two dissimilar materials when light falls upon it.

Pi bond A covalent bond formed by overlap between two p orbitals on different atoms. Pi bonds are superimposed on sigma bonds, forming double or triple bonds.

Pi filter A filter consisting of two line-to-line capacitors and a series inductance in all configuration used to attenuate noise and ripple.

Pick and dip A method of laying brick whereby the bricklayer simultaneously picks up a brick with one hand and, with the other hand, enough mortar on a trowel to lay the brick. Sometimes called the New England or Eastern method.

Pickling The chemical removal of surface oxides and other contaminants from a material by immersion in an aqueous acid solution; sulfuric and hydrochloric acids are common pickling solutions.

PID (Proportional Integral Derivative) . A three mode control action where the controller has time proportioning, integral (auto reset) and derivative rate action.

Pier 1. Intermediate support for two adjacent bridge spans.

2. Marine structure (such as a breakwater) extending into navigable water for use as a landing place, promenade, shore protection or to form an actual harbor.

3. Structural mount (as for an antena or telescope) usually of concrete, stonework, or steel.

Piezoelectric accelerometer A transducer that produces an electrical charge in direct proportion to the vibratory acceleration.

Piezoelectric material A ferroelectric material in which an electrical potential difference is created due to mechanical deformation, or conversely, in which the application of a voltage causes dimensional changes in the material.

Piezoelectricity Elastic strain caused when an electrical current is applied to a piezoelectric material and conversely, an electric current produced when pressure is applied to a piezoelectric material; these materials exhibit the Perovskite crystal structure.

Pig A colloquial term describing a container (usually lead or depleted uranium) used to ship or store radioactive materials. The thick walls of this shielding device protect the person handling the container from radiation. Large containers used for spent fuel storage are commonly called casks.

Pigment / binder ratio Ratio of total pigment to binder solids in paint.

Pile A colloquial term describing the first nuclear reactors. They are called piles because the earliest reactors were "piles" of graphite and uranium blocks.

Pile driver A noisy machine that repeatedly drops a heavy weight on top of a pile until the pile reaches solid soil or rock or cannot be pushed down any farther.

Pillar1. A : post; firm upright support for a superstructure. b : a usually ornamental column or shaft; especially : one standing alone for a monument.

2. Solid mass of rock, coal, or ore left standing to support a tunnel or mine roofrock,

3. Body part that resembles a columnrockrock.

Pilot cell A selected cell whose condition is assumed to indicate the condition of the entire battery.

Pin 1. A terminal on a through-hole component. (Derived from its physical shape on through-hole components, which predated SMT.) Also called lead.

2. In the term "pin count," pinrefers to a terminal on any component, whether through-hole or SMT.

Pin out Pin-number assignment, the relation between the logical inputs and outputs of an electronic device and their physical counterparts in the PCB package. pin-outs will involve pin numbers as a link between schematic and PCB design (both being computer generated files). In more complicated packages, they may also involve pin names. Even for devices with only two pins and no polarity, such as resistors, the netlist extracted from a schematic will have a pin 1 and pin 2 for each resistor, even though the schematic might not show a pin number label as such. (The visibility in the schematic of the pin numbers can be turned on or off at will, but the significance of the pin number assignment is still there in the schematic and subsequently, through the netlist extracted from it, the PCB database.) For CAD CAE electronics to work at all, the pin-outs for the PCB database must agree with the schematic.

Pinhole The term pinhole embraces a wide variety of oxide defects and is used in a broad sense today. Listed in this category are cracks caused by thermal contraction after oxidation or by handling, and regions of oxide with low dielectric strength caused by dust particles, inadequate masking, contamination, or poor resist adhesion.

Pinion A gear with a small number of teeth designed to mesh with a larger geared wheel or a rack. Used in rack and pinion steering and the differential ring and pinion.

Piston A partly hollow cylindrical part closed at one end, fitted to each of the engine's cylinders and attached to the crankshaft by a connecting rod. Each piston moves up and down in its cylinder, transmitting power created by the exploding fuel to the crankshaft via a connecting rod.

Pitch The up and down movement along an imaginary axis between the front and rear of a vehicle. Often during hard braking, the vehicle's nose will "dive" or pitch down in front. During acceleration the back end will "squat" or pitch down in the rear.

Pitting A form of very localized corrosion wherein small pits or holes form, usually in a vertical direction.

Pixel Picture element. Definable locations on a display screen that are used to form images on the screen. For graphic displays, screens with more pixels provide higher resolution.

pK value A measure of the strength of an acid on a logarithmic scale. The pK value is given by log10 (1/Ka), where Ka is the acid dissociation constant pK values often are used to compare the strengths of different acids.

Plane separation Of a balancing machine, is the operation of reducing the correction plane interference ratio for a particular rotor.

Planetary gears A gear set, generally found in automatic transmissions, in which all of the gears are in one plane, grouped around each other like planets around the sun. The central gear is called the "sun gear."

Plasma A highly-ionized gas containing an approximately equal number of positive ions and negative electrons. Thus, as a whole it is electrically neutral, though conductive and affected by magnetic fields.

Plastic A synthetic material made from long chains of molecules; has the capability of being molded or shaped, usually by the application of heat and pressure.

Plastic deformation The permanent (inelastic) distortion of a material under an applied stress that strains the material beyond its elastic limit; the ability of a material to be permanently deformed without fracture.

Plastic leaded chip carrier An SMT chip package that is rectangular or square-shaped with leads on all four sides. The leads are spaced at 0.050 inches, so this package is not considered fine-pitch.

Plasticizer A low molecular weight polymer additive that enhances flexibility and workability and reduces stiffness and brittleness.

Plated through hole A hole in a PWB with metal plating added after it is drilled. Its purpose it to serve either as a contact point for a through-hole component or as a via.

Platinel A non-standard, high temperature platinum thermocouple alloy whose thermoelectric voltage nearly matches a Type K thermocouple (Trademark of Englehard Industries).

Platinum 10% rhodium The platinum-rhodium alloy used as the positive wire in conjunction with pure platinum to form a Type S thermocouple.

Platinum 13% rhodium The platinum-rhodium alloy used as the positive wire in conjunction with pure platinum to form a Type R thermocouple.

Platinum 30% rhodium The platinum-rhodium alloy used as the positive wire in conjunction with platinum 6% rhodium to form a Type B thermocouple.

Platinum 6% rhodium The platinum-rhodium alloy used as the negative wire in conjunction with platinum-30% rhodium to form a Type B thermocouple.

Plausible accidents Postulated events that meet a probability test rather than the more challenging test represented by a design-basis event.

PLB(Personel Locator Beacon) Device issued to each traveler when traveling offshore on a helicopter.

It automatically emits a signal when submersed in the sea thereby giving potential for speedier rescue in event of helicopter ditching.

PLC(Programmable Logic Controller) Computer based monitoring and control package where control actions are primarily based on equipment and alarm status.

Plies The layers of cord, fiberglass, steel or structural fabric that make up the tire carcass and reinforcing belts.

Plumb rule This is a combination plumb rule and level. It is used in a horizontal position as a level and in a vertical position as a plumb rule. They are made in lengths of 42 and 48 in., and short lengths from 12 to 24 in.

Plutonium (Pu) A heavy, radioactive, manmade metallic element with atomic number 94. Its most important isotope is fissile plutonium-239, which is produced by neutron irradiation of uranium-238. It exists in only trace amounts in nature.

Ply rating A measure of the strength of tires based upon the strength of a single ply of designated construction. An eight-ply rating does not necessarily mean the tire has eight plies, but rather that the tires has the strength of eight standard plies.

PNGV(Partnership for a New Generation of Vehicles) A partnership between government, industry and academia in the USA to improve all aspects of automotive design in which batteries figure highly.

Pocket dosimeter A small ionization detection instrument that indicates ionizing radiation exposure directly. An auxiliary charging device is usually necessary.

Point defect A crystalline defect associated with one or several atomic sites.

Point of use A technique that ensures people have exactly what they need to do their job—the right work instructions, parts, tools and equipment—where and when they need them.

Pointing Finishing a mortar joint after the masonry units are laid.

Poison, neutron In reactor physics, a material other than fissionable material in the vicinity of the reactor core that will absorb neutrons. The addition of poisons, such as control rods or boron, into the reactor is said to be an addition of negative reactivity.

Poisson's ratio A bar of an isotropic, elastic material ordinarily shrinks laterally when it is stretched longitudinally. The lateral contracting strain divided by the applied tensile strain is Poisson's ratio, which varies from material to material.

Poka yoke Japanese word that refers to a mistake-proofing device or procedure used to prevent a defect during the production process.

Polarisation The change in the potential of a cell or electrode from its equilibrium value caused by the passage of an electric current through it. There are two irreversible electrochemical components, the "electrode polarisation" at the electrodes and the "concentration polarisation" in the electrolytic phase plus an ohmic loss component due to the electrical resistance of the cell. Also due to the build up of gas bubbles on the electrodes.

Polarity In electricity, the quality of having two oppositely charged poles, one positive one negative.

Polarity reversal Reversal of the polarity of a battery or cell due to over discharge.

Polarization The inability of an electrode to reproduce a reading after a small electrical current has been passed through the membrane. Glass pH electrodes are especially prone to polarization errors caused by small currents flowing from the pH meter input circuit and from static electrical charges built up as the electrodes are removed from the sample solution, or when the electrodes are wiped.

Policy A guiding principle, typically established by senior management, which is adopted by an organization or project to influence and determine decisions.

Pollution Any man made condition which adversely affects the quality of the environment.

Pollution prevention Identifying areas, processes and activities which create excessive waste products or pollutants in order to reduce or prevent them through alteration, or eliminating a process. Such activities, consistent with the Pollution Prevention Act of 1990, are conducted across all EPA programs and can involve cooperative efforts with such agencies as the Departments of Agriculture and Energy.

Polyclonal antibodies Antibodies produced by an animal's white blood cells (lymphocytes, specifically) in response to an antigen. This response occurs naturally or can purposely be created by injecting an animal, such as a rabbit or goat, with a specific antigen. More than one kind of anti-body is produced since more than one lymphocyte is producing antibodies. This is referred to as"polyclonal". The polyclonal antibodies are isolated from the animal and can be used for detection purposes. Because the antibodies are actually a mixture with different affinities (binding capability) for the antigen of interest, some variability in performance can occur from one test to another or one batch of antibodies to another.

Polycyclic A cyclic structure contains rings of bonds; a structure having many such rings is termed polycyclic. In the polycyclic structures of interest in this volume, a large fraction of the atoms are members of multiple small rings, resulting in considerable rigidity.

Polyester resin Group of synthetic resins which contain repeating ester groups. A special type of modified alkyd resin.

Polyimide Thermosetting ring chain polymer characterized by -NH group; it's increasingly used as dielectrics in high performance circuits.

Polymer Strictly it is a substance made of long repeating chains of molecules called monomers which may be identical or different. The term polymer is often used in place of plastic, rubber or elastomer. In battery technology "polymer" usually refers to a solid (plastic) ionic conductor that is an electrical insulator but passes ions.

Polymerization Chemical reaction in which two or more small molecules combine to form large molecules containing repeated structural units.

Polymorphism Different crystal structures at different temperatures or pressures for a single compound.

Polysilicon Polycrystalline silicon used as conductor in integrated circuits, and especially FETs.

Polyswitch A resettable fuse.

Polyurethane Group of thermoplastic or thermosetting polymers containing polyisocyanate; they may be hard and glossy forms; drawn into fibers; elastomeric or rubbery and flexible, or rigid foams. Used as rubber, for coatings, as fibers and foams; also used complexed in other polymers.

Polyvinyl chloride (PVC) A polyvinyl resin used extensively in the manufacture of plastics. It is also used as a rubber substitute. Residues are present in human body tissues.Chemical name: Ethene, chloro-, homopolymer

Pool reactor A reactor in which the fuel elements are suspended in a pool of water that serves as the reflector, moderator, and coolant. Popularly called a "swimming pool reactor," it is used for research and training, not for electrical generation.

Porosity The presence of numerous minute voids in a material; condition (of a solid substance) of having pores or open spaces

Port A communications connection on an electronic or computer based device.

Portland cement A hydraulic cement made by finely pulverizing the clinker produced by calcining to incipient fusion a mixture of clay and limestone or similar materials. Also defined generically as cement

Position A type of index for an aperture in an aperture list which is a number from 1 to the number of apertures in the aperture list. Position 1 is linked to D code D10, 2 is D11 and so on. Positions appear only in aperture lists, and never in a Gerber file . Cadstar aperture lists use the column heading Position to mean D code. Abbreviated "Pos" in GcPrevue.

Positive A developed image of photoplotted film, where the areas selectively exposed by the photo plotter appear black, and unexposed areas are clear. Board houses work from positives, and a photo plotter produces positives, thus one set of positives is all the film that is needed to produce a printed wiring board. (of a printed wiring image) Representing copper as black areas and absence of copper as clear areas. Typical of images of routed layers of a PWB.

Positive electrode The electrode which has a positive potential. The cathode. Electric current from this electrode flows into the external circuit.

Pot life The length of time a paint material is useful after its original package is opened or a catalyst or other curing agent is added.

Potable water Water that has does not contain harmful or objectionable impurities and is aesthetically pleasing to drink.

Potential energy The energy associated with a configuration of particles, as distinct from their motions. In macroscopic experience, potential energy can be increased (for example) by stretching a spring or by lifting a mass against a gravitational force; in molecular systems, potential energy can be increased (for example) by stretching a bond or by separating molecules against a van der Waals attraction.

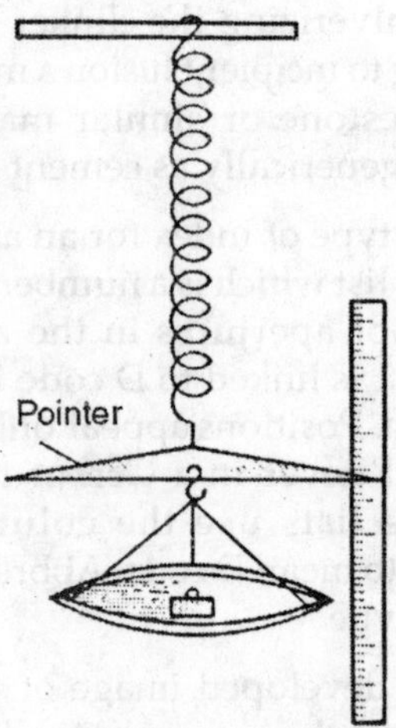

Fig. Potential Energy

Potential energy surface The potential energy of a ground-state molecular system containing N atoms is a function of its geometry, defined by 3N spatial coordinates (a configuration space). If the energy is imagined as corresponding to a height in a 3N + 1 dimensional space, the resulting landscape of hills, hollows, and valleys is the potential energy surface.

Potential well In a potential energy surface, the region surrounding a local energy minimum. Typically taken to include at least those points in configuration space such that a path of steadily declining energy can be found that leads to the minimum in question, and such that no similar path can be found to any other minimum. If the PES were a landscape, this would be the region around the minimum that could be filled with water without any flowing down and away toward another minimum.

Potentiometer 1. A variable resistor often used to control a circuit.

2. A balancing bridge used to measure voltage.

Potentiometric device Monitors the voltage between a sensing electrode and a reference electrode. A high input impedance voltmeter is used to minimize current flow. The voltage typically is proportional to the logarithm of the analyte concentration.

Potting An insulating material for encapsulating one or more circuit elements.

POTW or Publicly Owned Treatment Works Any municipally owned wastewater treatment facility.

Pouch cell A battery or cell contained in a flexible metal foil pouch.

Pound feet (LB.-FT.) Pound-feet measure twisting force or torque. Generated by the engine, torque is the "push" that sets a vehicle into motion and accelerates it. Specifications charts usually include the maximum torque the engine can develop, and the RPM at which it is generated (such as 345 lb.-ft. @ 3200 RPM).

Powdered iron core Magnetic core material that contains iron particles held together with a high resistance binder to reduce eddy currents.

Power (P) 1. Measured in watts, P=EI, I^2R or E^2/R: 1 watt = 1 joule/second, 1 joule = 1 watt-second

2. In a resistive circuit, power is the product of the in-phase components voltage and current (volt-amperes).

Power band A subjectively defined RPM range over which an engine delivers a substantial portion of its peak power. The power band usually extends from slightly below the engine's torque peak to slightly above its horsepower peak.

Power coefficient of reactivity The change in reactivity per percent change in power. The power coefficient is the summation of the moderator temperature coefficient of reactivity, the fuel temperature coefficient of reactivity, and the void coefficient of reactivity.

Power defect The total amount of reactivity added due to a given change in power. It can also be expressed as the integrated power coefficient over the range of the power change.

Power density The amount of power available from a battery. It is expressed as the power available per unit volume or per unit weight (W/L or W/kg).

Power factor The ratio of true to apparent power expressed as a decimal. It has been frequently specified as lead or lag of the current relative to voltage. However, switching power supplies have almost no phase shift but a significant difference between true and apparent power. Thus a switcher's power factor is measured with special equipment.

Power reactor A reactor designed to produce heat for electric generation (as distinguished from reactors used for research), for producing radiation or fissionable materials or for reactor component testing.

Power source Any device that furnishes electrical power, including a generator, cell, battery, power pack, power supply, solar cell, etc.

Power supply A device for the conversion of available power of one set of characteristics to another set of characteristics to meet specified requirements. Typical application of power supplies include to convert raw input power to a controlled or stabilized voltage and/or current for the operation of electronic equipment.

Power supply cord An assembly of a suitable length of flexible cord provided with an attachment plug at one end.

Power transistor A high current, bipolar transistor controlled by the current through the gate. Used in linear (series) regulators as the voltage dropper between the unregulated voltage input and the regulated output. Also used as a high current switching device in control and protection circuits. Needs a high current to turn it on and is slow to turn off and its negative temperature coefficient makes it prone to thermal runaway. For these reasons it was mostly superceded by MOSFETs in high power battery switching applications.

Powertrain A name applied to the group of components used to transmit engine power to the driving wheels. It can consist of engine, clutch, transmission, universal joints, drive shaft, differential gear, and axle shafts. Powertrain components are matched according to driver needs such as high torque, fuel economy, or convenience.

PPTC A Polymeric Positive Temperature Coefficient device. It is a non-linear thermistor, more commonly called a resettable fuse.

PQFP Plastic Quad Flat Pack.

Practical coverage The spreading rate of a paint/coating calculated at the recommended dry film thickness and assuming 15% material loss.

Pre regulation The initial regulation circuit in a system containing at least two cascade regulation loops.

Precipitation In metals, the separation of a new phase from solid or liquid solution, usually with changing conditions of temperature, pressure, or both.

Precipitation hardening Increase the hardness of a supersaturated solid solution by heat treating it to cause a second phase to precipitate out; coherency of the precipitate/matrix interface and how well the two lattices match up greatly influence the effect of precipitate.

Precipitation heat treatment Artificial aging of metals in which a constituent precipitates from a supersaturated solid solution.

Precision The degree of reproducibility among several independent measurements of the same true value under specified conditions.

Predetonation The undesirable "knock" or "ping" that occurs when the ignition of the air-fuel mixture occurs before the ignition spark. Also known as "pre-ignition".

Prefabricated brick masonry Masonry construction fabricated somewhere other than its final in-service location in the structure. Also known as preassembled, panelized and sectionalized brick masonry.

Pressure plate This instrument is used to measure pressure under foot using optoelectrical techniques. The system may use light reflection or pressure sensitive electrodes to gather data. The pattern of interference can be calibrated and viewed through a VDU. For static measurements, the scale of patterns and scale of gray can be digitised to give colour levels .

Pressure vessel A strong-walled container housing the core of most types of power reactors. It usually also contains the moderator, neutron reflector, thermal shield, and control rods.

Pressurized water reactor (PWR) A power reactor in which heat is transferred from the core to an exchanger by high temperature water kept under high pressure in the primary system. Steam is generated in a secondary circuit. Many reactors producing electric power are pressurized water reactors.

Pressurizer A tank or vessel that acts as a head tank (or surge volume) to control the pressure in a pressurized water reactor.

Primary battery A battery that is non-rechargeable.

Primary circuit A circuit electrically connected to the input or source of power to the device.

Primary device Part of a flowmeter which is mounted internally or externally to the fluid conduit and produces a signal corresponding to the flowrate and from which the flow may be determined.

Primary side control A name for an off-line switching power supply with the pulse-width modulator in the primary.

Primary standards Aqueous pH buffer solutions established by the National Bureau of Standards within the 2.5 to 11.5 pH range of ionic strength less than 0.1 and which provide stable liquid junction potential and uniformity of electrode sensitivity.

Primary system A term that may be used for referring to the reactor coolant system.

Primary treatment Process that removes the largest solid material and many microorganisms from wastewater, before it enters holding ponds for further purification.

Primitive 1. Some CAD software documentation extends this term to mean any object in a CAD database—graphics, text or otherwise; so this could be a group of graphic objects if manipulated as a unit, eg. a PCB decal . It may also mean an indivisible graphic object, i.e. a graphical object which may have component parts, but which can not have those parts separated out as individual entities. Examples of this in PCB CAD: wire segment, route, pad or padstack.

2. Any geometric shape such as a circle, polygon or square.

3. A function, operator, or type which is built into a programming language (or operating system), either for speed of execution or because it would be impossible to write it in the language. Primitives typically include the arithmetic and logical operations (plus, minus, and, or, etc.) and are implemented by a small number of machine language instructions.

Principal axes The axes of maximum and minimum normal stress.

Printed circuit board A flat plate or base of insulating material containing a pattern of conducting material. It becomes an electrical circuit when components are attached and soldered to it.

The conducting material is commonly copper which has been coated with solder or plated with tin or tin-lead alloy. The usual insulating material is epoxy laminate. But there are many other kinds of materials used in more exotic technologies.

Single-sided boards, the most common style in mass-produced consumer electronic products, have all conductors on one side of the board. With two-sided boards, the conductors, or copper traces, can travel from one side of the board to the other through plated-thru holes called vias , or feed-throughs. In multilayer boards, the vias can connect to internal layers as well as either side.

Prism Small masonry assemblage made with masonry units and mortar. Primarily used to predict the compressive and flexural strengths of full scale masonry members.

Prismatic cell A slim rectangular sealed cell in a metal case. The positive and negative plates are stacked usually in a rectangular shape rather than rolled in a spiral as done in a cylindrical cell.

Probabilistic risk analysis A systematic method for addressing the risk triplet as it relates to the performance of a complex system to understand likely outcomes, sensitivities, areas of importance, system interactions, and areas of uncertainty. The risk triplet is the set of three questions that the NRC uses to define "risk": 1.

Probability density function Consider an uncertain physical property and a corresponding space describing the range of values that the property can have (e.g., the configuration of a thermally excited N particle system and the corresponding 3N dimensional configuration space). The probability density function associated with a property is defined over the corresponding space; its value at a particular point is the probability per unit volume that the property has a value in an infinitesimal region around that point.

Probe A generic term that is used to describe many types of temperature sensors.

Process analysis Simply a method of mathematically describing a complex phenomenon as the net result of a number of different processes. The rate concept more specifically assumes that the changes in the state of a system with time can be systematically treated as the sum of the rates of the individual processes acting on the system.

Process engineering Engineering a sequence of individual processes or conversions to efficiently convert a raw material to a finished product. The processes involved can comprise physical, chemical, or biological changes in the raw material as it becomes a product. Physical processes change the shape, size, color, temperature, or concentration of a compound but do not alter its molecular structure. Chemical or biological reaction processes alter the very nature of a molecule, changing it to something different.

Process meter A panel meter with sizeable zero and span adjustment capabilities, which can be scaled for readout in engineering units for signals such as 4-20 mA, 10-50 mA and 1-5 V.

Producers Autotrophic organisms which produce protoplasm using inorganic carbon and energy from the sun.

Product A product is a given set of items. The set could comprise of system, subsystem, hardware, or software items and their documentation.

Product lifecycle management Product Lifecycle Management (PLM) generally refers to the process of managing all data related to the product over its life cycle. It broadens the concept of product data management (PDM) to address managing the product configuration and the availability of product data into its later lifecycle stages of production, operation, support and disposal and to address managing the product's related process data. ARC Advisory Group defines six components of PLM: innovation and portfolio management, project and program management, collaborative design, product data management, manufacturing process planning, and service and support management. Michigan's PLM Development Consortium defines it as "an integrated, information driven approach to all aspects of a product's life from its design inception, through its manufacture, deployment and maintenance and culminating in its removal from service and final disposal. PLM is the integration of business systems to manage a product's lifecycle." CIMdata Inc., says that PLM represents "a business approach to solving the problem of managing the complete set of product and plant definition information and the processes through which it passes. The PLM process includes creating and changing that information, managing it through its life and disseminating and using it throughout the lifecycle of the product."

Product roadmap A tabular or graphic representation of product plans mapped against time to show the evolution of product models, their capabilities and the relationships to one another. Product roadmaps may be based on an individual product line or a product platform, and it may also show the relationship to supporting technologies. The product roadmap represents the long-term product plan to meet the needs of a defined market.

Production expense Production expenses are a component of generation expenses that includes costs associated with operation, maintenance, and fuel.

Profibus Vendor-independent open fieldbus standard increasingly used in process control.

Standardised in IEC 61158 and EN50170.

Profile The term used to describe the anchor pattern of a surface produced by sandblasting, acid etching or similar method.

Program An undertaking requiring concerted effort, which is focused on developing and/ or maintaining a specific product. The product may include hardware, software, and other components. Typically a project has its own funding cost accounting, and delivery schedule with the acquirer (Customer).

Program manager The role at a high enough level in an organization that the primary focus is the long-term vitality of the organization, rather than the short-term project and contractual concerns and pressures. In general, a senior manager for engineering would have responsibility for multiple projects. Perceptual Map A Perceptual Map is a visual method for comparing customer perceptions of different products considering two different characteristics of those products. It is used to show relationships between marketplace competitors and the criteria used by buyers in making purchase decisions and recommendations. Perceptual maps may be used for market segmentation, concept development and evaluation, and tracking changes in marketplace perceptions.

Programmable power supply A power supply with an output controlled by an applied voltage, current, resistance or digital code.

Programming The control of a power supply parameter, such as output voltage, by means of a control element or signal.

Project management The management process, tools, and techniques used to define the project's goal; plan, schedule, and budget all the work necessary to reach that goal; lead the project; monitor progress; and ensure that the project is completed in a satisfactory way.

Project office A designated location where the administrative work of the project is conducted and the project management skills (resources) such as cost accounting, estimating, scheduling, training, etc. are retained. In the past, this was usually only economically possible on large projects. Recently some companies have established this type of facility to support a pool of smaller projects in an effort to develop and promote improved project management capabilities.

Prom Programmable read-only memory. A semiconductor memory whose contents cannot be changed by the computer after it has been programmed.

Pronation A triplane motion which requires movement in all three planes Movement is described as abduction, dorsiflexion and eversion.

Proof pressure The specified pressure which may be applied to the sensing element of a transducer without causing a permanent change in the output characteristics.

Proportional counter A radiation instrument in which an electronic detection system receives pulses that are proportional to the number of ions formed in a gas-filled tube by ionizing radiation.

Proportioning band A temperature band expressed in degrees within which a temperature controller's time proportioning function is active.

Proportioning control mode A time proportioning controller where the amount of time that the relay is energized is dependent upon the system's temperature.

Proportioning control plus derivative function A time proportioning controller with derivative function. The derivative function senses the rate at which a system's temperature is either increasing or decreasing and adjusts the cycle time of the controller to minimize overshoot or undershoot.

Proportioning control plus integral A two-mode controller with time proportioning and integral (auto reset) action. The integral function automatically adjusts the temperature at which a system has stabilized back to the setpoint temperature, thereby eliminating droop in the system.

Proportioning control with integral and derivative functions Three mode PID controller. A time proportioning controller with integral and derivative functions. The integral function automatically adjusts the system temperature to the set point temperature to eliminate droop due to the time proportioning function. The derivative function senses the rate of rise or fall of the system temperature and automatically adjusts the cycle time of the controller to minimize overshoot or undershoot.

Proprietary information Privately owned knowledge or data, such as that protected by a registered patent, copyright, or trademark.

Prosthesis (pl. Prostheses) Artificial substitute for a body part.

Protection A facility incorporated into battery packs to protect the cells from out of tolerance working conditions or misuse.

Protection head An enclosure usually made out of metal at the end of a heater or probe where connections are made.

Protection tube A metal or ceramic tube, closed at one end into which a temperature sensor is inserted. The tube protects the sensor from the medium into which it is inserted.

Protein Living cells contain many molecules that consist of amino acid polymers folded to form more-or-less definite three-dimensional structures; these are termed proteins. Short polymers lacking definite three-dimensional structures are termed peptides. Many proteins incorporate structures other than amino acids, either as covalently attached side chains or as bound ligands. Molecular objects made of protein form much of the molecular machinery of living cells.

Pull production The opposite of push production. It means products are made only when the customer has requested or "pulled" it, and not before. Doing so prevents building products that are not needed.

Pulse A step rise, a level, and a step fall of voltage or current. Characteristics of a pulse are: rise time, duration and fall time.

Pulse charger Versatile, hybrid charger having some of the advantages of both switch-mode and linear chargers. More costly than both.

Pulse discharge A high rate discharge, usually of 1 second or less.

Pulse width modulation (PWM) A method of regulating the output voltage of a switching power supply by varying the duration, but not the frequency, of a train of pulses that drives a power switch.

Push pull converter A power switching circuit that uses two or more power switches driven alternately on and off.

Pushrod A general term for any rod that transfers force in compression. In a conventional overhead valve layout, pushrods are used to transfer reciprocating motion from the cam followers to a more distant part of a valve train, typically the rocker arms. Pushrods are eliminated in overhead camshaft designs.

PWA Printed Wiring Assembly; same as PCB.

Pyroclastic flows High-density fluid mixtures of hot, dry rock fragments and hot gases that move away from their source (volcanic vents) at high speeds. They may result from the explosive eruption of molten or solid rock fragments, or both, or from the collapse of vertical eruption columns of ash and larger rock fragments. Pyroclastic flows may also result from a laterally directed explosion, or the fall of hot rock debris from a dome or thick lava flow.

Pyroelectricity The property of certain crystals, such as tourmaline, of acquiring opposite electrical charges on opposite faces when heated.

Pyrometer Instrument used to measure surface temperatures.

Q The energy-storing characteristic of an electronic circuit, system or device, equal to reactance divided by resistance. Q determines rate of decay of stored energy; the higher the 0, the longer it takes for the energy to be released.

Q factor A rating, applied to coils, capacitors, and resonant circuits, equal to the reactance divided by the resistance. The ratio of energy stored to energy dissipated per cycle in an electrical or mechanical system.

QFP Quad Flat Pack, a fine-pitch SMT package that is rectangular or square with gull-wing shaped leads on all four sides. The lead pitch of a QFP is typically either 0.8mm or 0.65mm, although there are variations on this theme with smaller lead pitches: TQFP also 0.8mm; PQFP tooled at either 0.65mm (0.026") or 0.025" and SQFP at 0.5mm (0.020"). Any of these packages can have a wide variety of lead counts from 44 leads on up to 240 or more. Although these terms are descriptive, there are no industry- wide standards for sizes. Any printed circuit designer will need a spec sheet for the particular manufacturer's part, as a brief descrition like "PQFP-160" is inadequate to define the mechanical size and lead pitch of the part.

QRA Quantitative Risk Assessment. A method for quantifying major accident hazards (estimation of scenario frequencies) and their potential effects.

QS-9000 QS-9000 is a quality management standard developed by the Big Three Automakers for the automotive sector. Currently largely replaced by Technical Specification 16949 (ISO/TS 16949).

Quadro riportato The simulation of a wall painting for a ceiling design in which painted scenes are arranged in panels resembling frames on the surface of a shallow, curved vault.

Qualification testing Testing performed to demonstrate to the acquirer that a CSCI, HWCI, system or subsystem meets its specified requirements.

Quality An inherent or distinguishing characteristic of a material, product or assembly. Or, having a high degree of excellence. The qualitiy of a thing tends to be better the more care, thought (and often money) its maker puts into its making.

Quality assurance A planned and systematic pattern of all actions necessary to provide adequate confidence that management and technical planning and controls are adequate to: ·Establish correct technical requirements for design and manufacturing; · Manage and design activity standards, drawings, specifications, or other documents referenced on drawings, lists, or technical documents.

Quality control Techniques and systems ensuring that a high (or at least a predetermined) level of quality is maintained through various stages of a process at the provider rather than at the owner end of the business.

Quality factor The factor by which the absorbed dose (rad or gray) is to be multiplied to obtain a quantity that expresses, on a common scale for all ionizing radiation, the biological damage (rem or sievert) to an exposed individual. It is used because some types of radiation, such as alpha particles, are more biologically damaging internally than other types.

Quality function deployment A structured planning and decision-making methodology for capturing customer requirements (voice of the customer) and translating those requirements into product characteristics, part characteristics, process plans, and quality/process control requirements using a series of matrices.

Quality of life The degree of emotional, intellectual, or visual/cultural satisfaction in a person's everyday life as distinct from the degree of material comfort.

Quality of motion Assessment of quality of motion depends upon the range of motion; direction of motion and symmetry of motion. When a joint which should give 60-70% of movement only offers 20%, this can be classified as abnormal. Qualitative assessment is based on appearance, feel and sound, as the joint is passively and actively examined. Diagnosis is enhanced by radiographic changes and inter articulor changes by use of an arthrogram. A smooth uninterrupted excursion suggests a joint is normal. At either end the movement should cease abruptly.

Quantization The concept that energy can occur only in discrete units called quanta.

Quantum mechanics Quantum mechanics describes a system of particles in terms of a wave function defined over the configuration space of the system. Although the concept of particles having distinct locations is implicit in the potential energy function that determines the wave function (e.g., of a ground-state system), the observable dynamics of the system cannot be described in terms of the motion of such particles from point to point. In describing the energies, distributions, and behaviors of electrons in nanometer-scale structures, quantum mechanical methods are necessary. Electron wave functions help determine the potential energy surface of a molecular system, which in turn is the basis for classical descriptions of molecular motion. Nanomechanical systems can almost always be described in terms of classical mechanics, with occasional quantum mechanical corrections applied within the framework of a classical model.

Quantum theory The concept that energy is radiated intermittently in units of definite magnitude, called quanta, and absorbed in a like manner.

Quarry An open pit mine from which stone is taken by cutting, digging or blasting. The most famous quarries for white marble are around Carrara, Italy. Others are Colombara, Polvaccio, Sponda, and Campiglia. In other countries, quarries for white marble include: Paros, Proconnesos, and Dokimeion in Greece, Aphrodysias in Turkey. In the US, marble is mined in Colorado and Georgia. Reknowned quarries for other stones: Seravezza in Italy for breccia; and Monsummano in Italy for limestone. US granite and limestone deposit are mostly on the eastern seaboard.

Quarter panel A sheet metal panel that covers the area from the rear-door opening to the taillight area, and from the bottom of the surface to the base of the roof, or from the headlamp area to the front-door opening, and from the bottom of the surface to the base of the hood.

Quartz halogen headlamps A headlamp bulb having a quartz envelope holding the tungsten filament and filled with an inert gas containing iodine or another of the five halogen gases. The gas serves to remove the tungsten deposits from the bulb wall and redeposit them on the filament, preventing blackening of the bulb surface and reduction of light output. This kind of cycle requires very high filament operation temperatures which necessitates the use of quartz instead of glass. These lamps produce more lighting power per watt of electrical power than standard sealed beam headlamps.

Quatrefoil In architecture, an ornament having four leaves, lobes, or foils.

Queen closure (closer) Cut brick having a nominal 2 in. horizontal face dimension; half brick used in a masonry course to prevent vertical joints from falling one above another

Quench Rapidly cool a material; typically done to retain a structure at room temperature that otherwise is only stable at high temperature.

Quick charge Charging in three to six hours at about 0.3C rate. Needs special charger.

Quick test A shear test of a cohesive soil without allowing the sample to drain.

Quicklime Lime, burnt lime, caustic lime.

Quiescent current The current which continues to be drawn from the battery when the application it powers is in standby or hibernation mode.

Quoin 1. Projecting right angle masonry corner.

2. A : Corner stones lending either strength or emphasis, distinguished from the rest of the surface by greater size, different color, rustication, or the imitation of same in brick or paint; b: A large square ashlar or stone at the angle of a wall to limit the rubble and make the corner true and strong; an exterior masonry corner.

3. The keystone of an arch.

4. A wedge to support or steady a stone.

Quoin header Corner header in the face wall which also serves as a strecher for the side wall; a large, sometimes rusticated, usually slightly projecting stone (or stones) that often form the corners of the exterior walls of masonry buildings.

Quoin post Heel post.

QUV An accelerated testing device designed to evaluate the aging and color fading properties of a coating by exposure to high intensity, ultraviolet light.

R

R Chart A statistical process control (SPC) chart that monitors the range (variability) of the process. A sample of parts is collected from the process periodically. The range (maximum minus minimum) of the sample is plotted on the control chart and a determination is made if the process is "under control" or not.

R&D Research and Development

RDMP Rapid Decision-Making Practices.

R Symbol for Resistance.

Rack and pinion steering A steering gear in which a pinion on the end of the steering shaft merges with a rack of gear teeth on the major cross member of the steering linkage. When the steering wheel is turned, the pinion gear turns, moving the rack to the left or right, thus steering the wheels.

Racking A method entailing stepping back successive courses of masonry.

Rad The special unit for radiation absorbed dose, which is the amount of energy from any type of ionizing radiation (e.g., alpha, beta, gamma, neutrons, etc.) deposited in any medium (e.g., water, tissue, air). A dose of one rad means the absorption of 100 ergs (a small but measurable amount of energy) per gram of absorbing tissue (100 rad = 1 gray).

Radiation (ionizing radiation) Alpha particles, beta particles, gamma rays, x-rays, neutrons, high-speed electrons, high-speed protons, and other particles capable of producing ions. Radiation, as used in 10 CFR Part 20, does not include non-ionizing radiation, such as radio- or microwaves, or visible, infrared, or ultraviolet light.

Radiation area Any area with radiation levels greater than 5 millirems (0.05 millisievert) in one hour at 30 centimeters from the source or from any surface through which the radiation penetrates.

Radiation damage Chemical changes (bond breakage, ionization) caused by high-energy radiation (e.g., x-rays, gamma rays, high-speed electrons, protons, etc.).

Radiation shielding Reduction of radiation by interposing a shield of absorbing material between any radioactive source and a person, work area, or radiation-sensitive device.

Radiation sickness (syndrome) The complex of symptoms characterizing the disease known as radiation injury, resulting from excessive exposure (greater than 200 rads or 2 gray) of the whole body (or large part) to ionizing radiation. The earliest of these symptoms are nausea, fatigue, vomiting, and diarrhea, which may be followed by loss of hair (epilation), hemorrhage, inflammation of the mouth and throat, and general loss of energy. In severe cases, where the radiation exposure has been approximately 1000 rad (10 gray) or more, death may occur within two to four weeks. Those who survive six weeks after the receipt of a single large dose of radiation to the whole body may generally be expected to recover.

Radiation source Usually a sealed source of radiation used in teletherapy and industrial radiography, as a power source for batteries (as in use in space craft), or in various types of industrial gauges. Machines, such as accelerators and radioisotope generators, and natural radionuclides may be considered sources.

Radiation standards Exposure standards, permissible concentrations, rules for safe handling, regulations for transportation, regulations for industrial control of radiation, and control of radioactive material by legislative means.

Radiation warning symbol An officially prescribed symbol (a magenta or black trefoil) on a yellow background that must be displayed where certain quantities of radioactive materials are present or where certain doses of radiation could be received.

Radiation, nuclear Particles (alpha, beta, neutrons) or photons (gamma) emitted from the nucleus of unstable radioactive atoms as a result of radioactive decay.

Radical A structure with an unpaired electron (but excluding certain metal ions). In organic molecules, a radical is often associated with a highly reactive site of reduced valence. The term radical is sometimes used to describe a substructure within a molecule; the term free radical then describes a radical in this sense, viewed as the result of cleaving the bond linking the substructure to the rest of the molecule.

Radioactive decay Large unstable atoms can become more stable by emitting radiation. This process is called radioactive decay. This radiation can be emitted in the form of a positively charged alpha particle, a negatively charged beta particle, or gamma rays or x-rays.

Radioactive series A succession of nuclides, each of which transforms by radioactive disintegration into the next until a stable nuclide results. The first member is called the parent, the intermediate members are called daughters, and the final stable member is called the end product.

Radioactivity The spontaneous emission of radiation, generally alpha or beta particles, often accompanied by gamma rays, from the nucleus of an unstable isotope. Also, the rate at which radioactive material emits radiation. Measured in units of becquerels or disintegrations per second.

Radiography Nondestructive method of internal examination in which metal objects are exposed to a beam of X-ray or gamma radiation; differences in thickness, density, or absorption caused by internal defects or inclusions are apparent in the shadow image produced on a fluorescent screen or photographic film placed behind the object.

Radioisotope An unstable isotope of an element that decays or disintegrates spontaneously, emitting radiation. Approximately 5,000 natural and artificial radioisotopes have been identified.

Radiological sabotage Any deliberate act directed against a plant or transport in which an activity licensed pursuant to 10 CFR Part 73 of NRC's regulations is conducted or against a component of such a plant or transport that could directly or indirectly endanger the public health and safety by exposure to radiation.

Radiological survey The evaluation of the radiation hazards accompanying the production, use, or existence of radioactive materials under a specific set of conditions. Such evaluation customarily includes a physical survey of the disposition of materials and equipment, measurements or estimates of the levels of radiation that may be involved, and a sufficient knowledge of processes affecting these materials to predict hazards resulting from expected or possible changes in materials or equipment.

Radiology That branch of medicine dealing with the diagnostic and therapeutic applications of radiant energy, including x-rays and radioisotopes.

Radionuclide A radioisotope.

Radiosensitivity The relative susceptibility of cells, tissues, organs, organisms, or other substances to the injurious action of radiation.

Radium (Ra) A radioactive metallic element with atomic number 88. As found in nature, the most common isotope has a mass number of 226. It occurs in minute quantities associated with uranium in pitchblende, camotite, and other minerals.

Radon (Rn) A radioactive element that is one of the heaviest gases known. Its atomic number is 86. It is a daughter of radium.

Ragone plot The graphical illustration of the specific energy of a cell as a function of its specific power.

Rainbow Means the splitting of sunlight into a spectrum of colours by droplets of water suspended in the air due to refraction and internal reflection of sunlight by them.

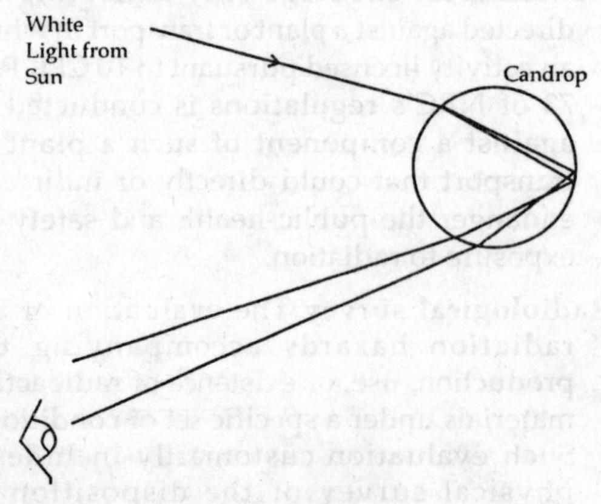

Fig. Rainbow

RAL Number The RAL catalogue is an international color standard that defines colours according to numbers and names e.g. RAL7032 is a flinty grey colour.

RAM cells Rechargeable Alkaline Manganese cells.

Random access memory (RAM) Memory that can be both read and changed during computer operation. Unlike other semi-conductor memories, RAM is volatile-if power to the RAM is disrupted or lost, all the data stored is lost.

Range The difference between the minimum and maximum values of sensor output in the intended operating range. Defines the overall operating limits of a sensor.

Range of gyration The range of gyration is a measure of the distribution of mass in a body segment. This is the rotational equivalent to mass, and is determined by the size, shape and density of the segment. The range of gyration is often converted to the Moment of Inertia by the Parallel Axis Theorem.

Range of motion This refers to the available movement provided by the joint and is determined by the type of joint and its articular surfaces as well as the physiological control of movement across the joint.

Rangeability The ratio of the maximum flowrate to the minimum flowrate of a meter.

Rankine (°R) An absolute temperature scale based upon the Fahrenheit scale with 180° between the ice point and boiling point of water. 459.67°R = 0°F.

Rapid manufacturing Rapid Manufacturing refers to the use rapid prototyping technologies to directly manufacture low volumes of parts.

Rapid prototyping 1. Rapid Prototyping refers to various technologies such as stereolithography and selective laser sintering that can rapidly create parts for visualization, product mock-ups, or functional product prototypes or produce rapid tooling to manufacture small to medium volumes of parts. Rapid prototyping processes involve devices, ranging from office modelers to four-ton machines, that accept 3D CAD files, slice the data into cross-sections, and construct layers from the bottom up, bonding one on top of the other, to produce physical prototypes.

2. More generally,it is the process of quickly generating prototypes or mockups of what a product system will look like. Rapid prototyping may be done with paper prototypes such as sketches, low-fidelity physical prototypes, CAD visualization, rapid application development, or video prototyping.

Rapid tooling Rapid Tooling refers to the use rapid prototyping technologies to fabricate tooling in a mucher shorter period of time than conventional tooling. Rapid tooling technologies include methods such as RTV molds, high-speed milling, centrifugal casting, etc.

RCCA Root Cause and Corrective Action.

RAPS (Remote area power supplies) Power systems deriving their energy from local solar or wind sources using a battery for energy storage and supplying the load through DC-DC converters or AC inverters.

Rate When applied to cells it usually means the cells current carrying capacity.

Rate action The derivative function of a temperature controller.

Rate time the time interval over which the system temperature is sampled for the derivative function.

Rated capacity The specified capacity of a battery.

Rated output current The maximum continuous load current a power supply is designed to provide under specified operating conditions.

Ratiometric measurement A measurement technique where an external signal is used to provide the voltage reference for the dual-slope A/D converter. The external signal can be derived from the voltage excitation applied to a bridge circuit or pick-off supply, thereby eliminating errors due to power supply fluctuations.

Ratsnest A bunch of straight lines (unrouted connections) between pins which represents graphically the connectivity of a PCB CAD database. (Derived from the pattern of the lines: as they crisscross the board, the lines form a seemingly haphazard and confusing mess similar to a rat 's nest.)

Reactance Portion of impedance that characterizes non-dissipative, energy storage effects.

Reaction A process that transforms one or more chemical species into others. Typical reactions make or break bonds; others change the state of ionization or other properties taken to distinguish chemical species.

Reactive A component that exhibits the property of either capacitance or inductance.

Reactive waste A waste which; 1. reacts violently with water,

2. Forms potentially explosive mixtures with water,

3. Is normally unstable,

4. Contains cyanide or sulfide in sufficient quantity to evolve toxic fumes at high or low pH,

5. Is capable of exploding if heated under pressure, or

6. Is an explosive compound listed in Department of Transportation (DoT) regulations. One of EPA's four hazardous waste properties.

Reactivity A term expressing the departure of a reactor system from criticality. A positive reactivity addition indicates a move toward supercriticality (power increase). A negative reactivity addition indicates a move toward subcriticality (power decrease).

Reactor coolant system The system used to remove energy from the reactor core and transfer that energy either directly or indirectly to the steam turbine.

Reactor engineering Concerned with the exploitation of a chemical, biological, or enzymatic reaction under a controlled set of conditions to affect either information about the reaction or to produce a product or products efficiently. Reactor design is that activity specifically directed toward setting the size and operating parameters of a reactor to affect some desired goal.

Reactor, nuclear A device in which nuclear fission may be sustained and controlled in a self-supporting nuclear reaction. The varieties are many, but all incorporate certain features, including fissionable material or fuel, a moderating material (unless the reactor is operated on fast neutrons), a reflector to conserve escaping neutrons, provisions of removal of heat, measuring and controlling instruments, and protective devices. The reactor is the heart of a nuclear power plant.

Read only memory (ROM) Memory that contains fixed data. The computer can read the data, but cannot change it in any way.

Reaeration The dissolving of molecular oxygen from the atmosphere into the water.

Reagent A chemical species that undergoes change as a result of a chemical reaction.

Reagent device A large reagent structure (or a large structure that binds a smaller reagent) serving as a component of a mechanochemical system. A reagent device exists chiefly to hold, position, and manipulate the environment of a reagent moiety.

Real time monitoring Monitoring and measuring environmental developments with technology and communications systems that provide time-relevant information to the public in an easily understood format people can use in day-to-day decision-making about their health and the environment.

Receiving water A water which receives wastewater (treated or otherwise) discharges.

Receiving water quality standards Standards which require a discharger to maintain a certain quality level in the receiving water.

Receptor A structure that can capture a molecule (often of a specific type in a specific orientation) owing to complementary surface shapes, charge distributions, and so forth, without forming a covalent bond.

Rechargeable Capable of being recharged; refers to secondary cells or batteries.

Reciprocal lattice A group of points arranged about a center in such a way that the line joining each point of the center is perpendicular to a family of planes in the crystal, and the length of this line is inversely proportional to their interplanar distance.

Recombinant system Sealed secondary cells in which gaseous products of the electrochemical charging cycle are made to recombine to recover the active chemicals. A closed cycle system preventing loss of active chemicals. Used in Nicads and SLA batteries.

Reconditioning One or more deep discharges below 1.0 V/cell with a very low controlled current, causing a change to the molecular structure of the cell and a rebuilding of its chemical composition. Reconditioning helps break down large crystals to a more desirable small size, often restoring the battery to its full capacity. Applies to nickel-based batteries.

Reconstruction A crystal consists of a regular array of atoms, and the simplest model of a crystal surface would be generated by simply discarding all atoms to one side of a surface without changing the positions of the rest. In reality, however, the positions of the remaining atoms do change. A pattern of displacements that lowers the symmetry of the surface (relative to the ideally terminated crystal) is termed a surface reconstruction; some reconstructions alter the pattern of bonds.

Record Throughout the PMS, the requirements to "record" information are interpreted to mean "set down in a manner that can be retrieved and viewed." The result can take many forms including, but not limited to, hand-written notes, hard-copy, or electronic documents, and data recorded in computer-aided software engineering (CASE) and project management tools.

Redox battery A battery in which the chemical energy is stored in two dissolved ionic reactants separated by a membrane.

Redox potential The potential developed by a metallic electrode when placed in a solution containing a species in two different oxidation states.

Reduced mass Many dynamical properties of a system consisting of two interacting masses, m1 and m2, are equivalent to those of a system in which one mass is fixed in space and the other has a mass (the reduced mass) with the value m1m2/(m1 + m2). The reduced mass description has fewer dynamical variables.

Redundancy Power supplies connected in parallel operation so that if one fails, the others will continue delivering enough current to supply the maximum load. This method is used in applications where power supply failure cannot be tolerated.

Reengineering The process of examining and altering an existing system to reconstitute it in a new form. May include reverse engineering (analyzing a system and producing a representation at a higher level of abstraction, such as design from code), restructuring (transforming a system from one representation to another at the same level of abstraction), recommendation (analyzing a system and producing user and support documentation), forward engineering (using software products derived from an existing system, together with new requirements, to produce a new system), and translation (transforming source code from one language to another or from one version of a language to another).

Reference designator ("ref des") The name of a component on a printed circuit by convention beginning with one or two letters followed by a numeric value. The letter designates the class of component; eg. "Q" is commonly used as a prefix for transistors. Reference designators appear as usually white or yellow epoxy ink (the "silkscreen") on a circuit board. They are placed close to their respective components but not underneath them, so that they are visible on the assembled board. By contrast, on an assembly drawing a reference designator is often placed within the boundaries of a footprint —a very useful technique for eliminating ambiguity on a crowded board where reference designators in the silkscreeen may be near more than one component.

Reference junction The cold junction in a thermocouple circuit which is held at a stable known temperature. The standard reference temperature is 0°C (32°F). However, other temperatures can be used.

Reference man A person with the anatomical and physiological characteristics of an average individual that is used in calculations assessing internal dose (also may be called "standard man").

Reference mark Any diagnostic point or mark which can be used to relate a position during rotation of a part to its location when stopped.

Reference plane Any plane perpendicular to the shaft axis to which an amount of unbalance is referred.

Reference voltage The defined or specified voltage to which other voltages are compared.

Reflection coefficient For plane waves or transmission lines, the ratio of the reflected wave to the incident wave. By extension, the concept is applied to networks to express the effect of an impedance mismatch. From the general conception, three specific definitions follow: 1. The acoustic reflection coefficient is the ratio of the flow of reflected sound energy to the flow of incident sound energy.

2. For a transition or discontinuity between two transmission media, the reflection coefficient is that which would be observed at a specified point in one medium if the other medium were match-terminated.

3. The reflection coefficient in a transmission medium is defined as follows: At a given point, and for a given mode of transmission, the ratio of some quantity associated with the reflected wave to the corresponding quantity in the incident wave. The reflection coefficient may be different for different associated quantities, and the chosen quantity should be specified. The "voltage reflection coefficient" is most commonly used, and is defined as the ratio of the complex electric field strength (or voltage) of the reflected wave to that of the incident wave.

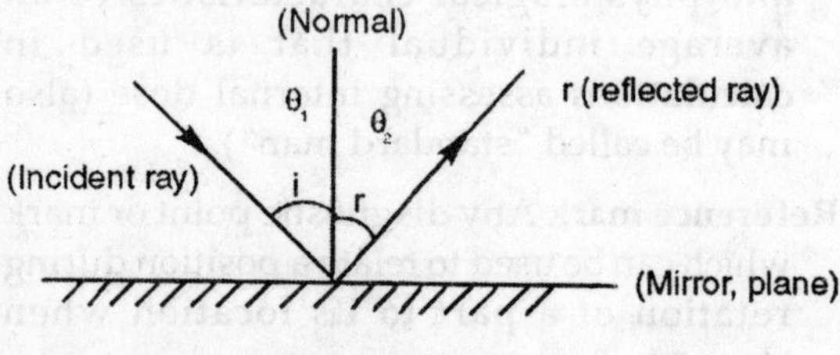

Fig: Reflection

Reflector A layer of material immediately surrounding a reactor core that scatters back (or reflects) into the core many neutrons that would otherwise escape. The returned neutrons can then cause more fissions and improve the neutron economy of the reactor. Common reflector materials are graphite, beryllium, water, and natural uranium.

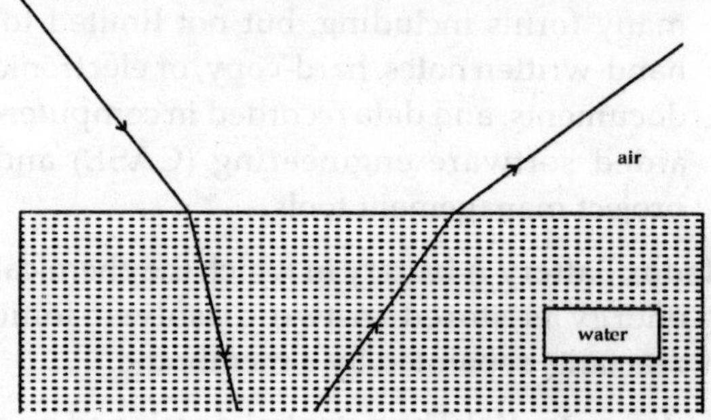

Fig. Refraction of light

Reflow soldering Process of joining metallic surfaces through the mass heating of solder/ solder paste to form solder fillets at metallized areas; it creates a mechanical and electrical connection between components and a PCB.

Refraction Bending of a light beam when passing from one medium to another, at different velocities of light.

Refractory A metal or ceramic that does not deteriorate rapidly or does not melt when exposed to extremely high temperatures.

Refractory metal thermocouple A class of thermocouples with melting points above 3600°F. The most common are made from tungsten and tungsten/rhenium alloys Types G and C. They can be used for measuring high temperatures up to 4000°F (2200°C) in non-oxidizing, inert, or vacuum environments.

Refractivity 1. In general, the property of refraction, or a quantitative relationship by which it is expressed, which is commonly some function of the index of refraction.

2. The quantity (n - 1) which enters many optical formulas is sometimes called "refractivity." Here n is the refractive index.

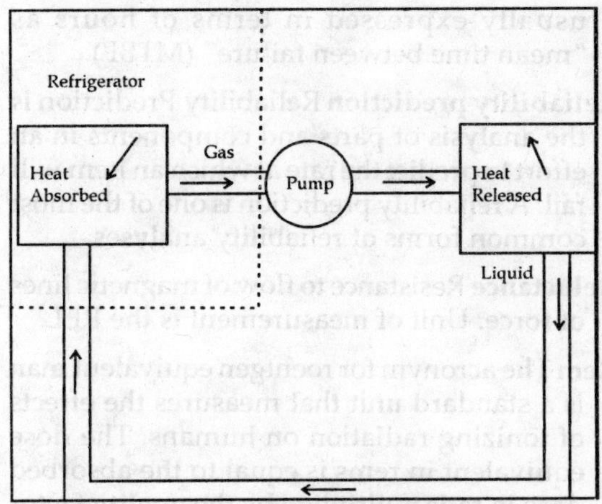

Fig. Refrigerator

Refurbishing The repair of worn out or damaged batteries. This is not the same as reconditioning.

Refuse All forms of solid waste.

Refuse derived fuel (RDF) A fuel derived from the combustible portion of municipal solid waste. The fuel is often processed into small briquettes, similar in size to charcoal.

Regenerative braking This uses the electrical drive motor in an electric vehicle to act as a generator returning energy to the battery when overdriven mechanically by the vehicle wheels. This provides a powerful braking effect and at the same time captures energy which would otherwise be wasted or dissipated in the brakes.

Regenesys A high power Sodium Polysulfide Bromine "Flow Battery".

Register In printed board manufacture, many terms are borrowed from the subject of printing. :

Proper alignment of various plates, stones, or screens to assure clear and accurate reproduction, as of color. Examples: in register, off register.

In printed circuit design, the designer gets his photoplot files in register before he views them with his Gerber file viewer. The board manufacturer produces film from the Gerber files and uses them in register with respect to the panels of material from which he will build the boards. He is going to want the pads on both sides and on internal layers to be in register before he drills holes in the panel.

Regulated power supply A device that maintains within specified limits a constant output voltage or current for specified changes in line, load, temperature or time.

Rehabilitation The process of returning a property to a state of utility, through repair or alteration, which makes possible an efficient contemporary use while preserving those portions and features of the property which are significant to its historic, architectural, and cultural values.

Rehabilitation not only encourages the repair of historic buildings, it allows appropriate alterations to ensure their efficient contemporary use. Examples include the continued use of hotels, stores, and private residences, as well as the adaptation of vacant schools into apartments, warehouses into offices, and industrial buildings into commercial space.

Because rehabilitation focuses on how buildings can be successfully used for contemporary purposes, it may be considered somewhat more flexible than more traditional treatments, such as preservation and restoration. But even though rehabilitation allows for more change, a historic building's distinctive materials, features, and spaces still must be preserved.

Reimbursable A type of contract where contractor is re-imbursed for all time taken to carry out and complete the assignment.

Reinforced concrete Concrete with steel bars or mesh embedded in it for increased strength in tension; in pre-tensioned concrete, the embedded steel bars or cables are stretched into tension before the concrete hardens; in post-tensioned concrete, the embedded steel bars or cables are stretched into tension after the concrete hardens.

Reinforced insulation As defined by Underwriters Laboratories, Inc., an improved basic insulation with such mechanical and electrical qualities that it, in itself, provides the same degree of protection against electrical shock as double insulation. It may consist of one or more layers of insulating material. It is acceptable in place of double insulation.

Reinforced masonry Masonry units, reinforcing steel, grout and/or mortar combined to act together in resisting forces.

Relative magnetic permeability (μr) The ratio of the magnetic permeability of some medium to that of a vacuum (unitless), or: $\mu r = \mu/\mu 0$, where μ0 is the permeability of vacuum, a universal constant, which has a value of 1.257×10^{-6} H/m.

Relaxation time A measure of the rate at which a disequilibrium distribution decays toward an equilibrium distribution. The electron relaxation time in a metal, for example, describes the time required for a disequilibrium distribution of electron momenta (e.g., in a flowing current) to decay toward equilibrium in the absence of an ongoing driving force and can be interpreted as the mean time between scattering events for a given electron.

Relay (Mechanical) An electromechanical device that completes or interrupts a circuit by physically moving electrical contacts into contact with each other.

Relay (Solid State) A solid state switching device which completes or interrupts a circuit electrically with no moving parts.

Reliability (life, multi-use vs. single, calibration vs. accuracy drift) How well a sensor maintains both precision and accuracy over its expected lifetime. Also includes the robustness of the sensor.

Reliability analysis A predictive tool used to estimate the "life" of a product. This is usually expressed in terms of hours as "mean time between failure" (MTBF).

Reliability prediction Reliability Prediction is the analysis of parts and components in an effort to predict the rate at which an item will fail. A reliability prediction is one of the most common forms of reliability analyses.

Reluctance Resistance to flow of magnetic lines of force. Unit of measurement is the REL.

Rem The acronym for roentgen equivalent man is a standard unit that measures the effects of ionizing radiation on humans. The dose equivalent in rems is equal to the absorbed dose in rads multiplied by the quality factor of the type of radiation.

Remedial design A phase of remedial action that follows the remedial investigation/ feasibility study and includes development of engineering drawings and specifications for a site cleanup.

Remedial investigation An in-depth study designed to gather data needed to determine the nature and extent of contamination at a Superfund site; establish site cleanup criteria; identify preliminary alternatives for remedial action; and support technical and cost analyses of alternatives. The remedial investigation is usually done with the feasibility study. Together they are usually referred to as the "RI/FS".

Remedial project manager (RPM) The EPA or state official responsible for overseeing on-site remedial action.

Remedial response Long-term action that stops or substantially reduces a release or threat of a release of hazardous substances that is serious but not an immediate threat to public health.

Remediation Cleanup or other methods used to remove or contain a toxic spill or hazardous materials from a Superfund site.

Remote Not hard-wired; communicating via switched lines, such as telephone lines. Usually refers to peripheral devices that are located a site away from the CPU.

Remote sensing A technique for regulating the output voltage of a power supply at the load by connecting the error-sensing leads to the load. Remote sensing compensates for voltage drops in cables and connectors between the power supply and the load. Since the remote sense leads usually directly control the power supply's internal feedback loop, the system should not be operated with the remote sense leads disconnected. Also, when connecting remote sense leads, make sure the + lead goes to the + end of the load. A very common mistake is to reverse the leads, which can damage the system.

Remote voltage adjustment Adjustments of the output voltage made remotely over a limited range by a variable resistor. This type of control is often found on high density DC/DC converters.

Repeatability The ability of a transducer to reproduce output readings when the same measurand value is applied to it consecutively, under the same conditions, and in the same direction. Repeatability is expressed as the maximum difference between output readings.

Representative point The point in a configuration space that represents the geometry of a system.

Reproducibility The closeness of agreement among repeated measurements of an output for the same value of input made under the same operating conditions over a period of time.

Requirement A function, feature or capability that a product must provide or meet to satisfy the customer's needs and enterprise's objectives for a new product.

Requirements allocation matrix Requirements Allocation Matrix is a matrix showing the allocation of a requirement (e.g., reliability, weight, cost) to various subsystems or subassemblies so that requirement can be accurately flowed-down and the satisfaction of the overall requirement can be tracked and managed.

Resealable safety vent The resealable vent internal to a cell to release excessive internal pressure.

Reserve battery Batteries which are stored in an inactive state without their electrolyte. They are only activated when needed by the introduction of the electrolyte.

Reserve capacity The number of minutes at which the battery can be discharged at 25 Amps and maintain a terminal voltage higher than 1.75 volts per cell, on a new, fully charged battery at 80degrees Fahrenheit(27C). Defines a battery's ability to power a vehicle with an inoperative alternator or fan belt. Used for comparing automotive SLI batteries.

Reserved word A word that has a defined function in the language, and cannot be used as a variable name.

Reset current A current injected into the winding of a magnetic component to reset the core.

Reset signal A signal used to return a circuit to a desired state.

Resettable fuse A fuse which protects against excessive current and temperature by interrupting the flow of current. After opening it will reset after the fault conditions have been removed but only after it has cooled. It requires no manual resetting or replacement. The "Polyswitch" is an example of this.

Residual A conservative projection of the market value of a vehicle at the end of a lease. Residual values are provided by major auto manufacturers or independent companies that specialize in auto valuation. A higher residual value will lower the monthly payment. However, the vehicle must be resalable at the residual amount or the lessor will lose money. Conversely, adjusting the residual too low will increase the monthly payment and may make the lease payment unattractive. Adjustments to residual may be made for excess mileage or wear and tear.

Residual flux The flux that exists in a magnetic core after the H field is returned to zero.

Residual stress Internal stress in a material often resulting from thermal or mechanical straining.

Resilience The tendency of a material to return to its original shape after the removal of a stress that has produced elastic strain.

Resin An organic polymer that crosslinks to form a thermosetting plastic when mixed with a curing agent.

Resistance (Ù-ohm) Characteristic of a resistor: in a 1-ohm resistance a current of 1 A produces a voltage drop of 1 V.

Resistance ratio characteristic For thermistors, the ratio of the resistance of the thermistor at 25°C to the resistance at 125°C.

Resistance welding Type of welding process in which the work pieces are heated by the passage of an electric current through the contact; this includes spot welding, seam or line welding, and percussion welding; flash and butt welding are sometimes considered as resistance welding processes.

Resistivity (ñ) The reciprocal of electrical conductivity, and a measure of a material's resistance to passing electric current.

Resistor Energy dissipative element consisting of a poor conductor in series with connecting wires.

Resolution The smallest detectable increment of measurement. Resolution is usually limited by the number of bits used to quantize the input signal. For example, a 12-bit A/D can resolve to one part in 4096 (2 to the 12 power equals 4096).

Resolution of forces This is a technique used to convert a single force into two forces acting in different directions. When resolving a force into components, it is not necessary that both components be at right angles with each other, although it makes the calculations much easier, since the right angled triangle lends itself to simple geometrical and trigomentrical methods.

Resonance 1. The state in which the natural response frequency of a circuit coincides with the frequency of an applied signal, or vice versa, yielding intensified response.

Fig. Resonance

2. The state in which the vibration frequency of a body coincides with an applied vibration force, or vice versa, yielding rein-forced vibration of the body.

Resonant circuit A circuit in which inductive and capacitive elements are in resonance at an operating frequency.

Resonant converter A class of converters that uses a resonant circuit as part of the regulation loop.

Resonant frequency The natural frequency at which a circuit oscillates or a device vibrates. In an L circuit, inductive and capacitive reactances are equal at the resonant frequency.

Resonator A small auxiliary muffler that assists the main muffler in reducing exhaust noise.

Respiration Energy production in which oxygen is the terminal electron acceptor, i.e. oxidation to produce energy where oxygen is the oxidizing agent.

Response time The time it takes for the sensor's output to reach its final value. A measure of how quickly the sensor will respond to changes in the environment. In general, this parameter is a measure of the speed of the sensor and must be compared with the speed of the process.

Responsiveness to change Changes to system and program requirements in response to directed changes by the procuring activity, or problem solutions identified shall be evaluated for total program impact with respect to performance, cost, and schedules.

Rest periods Interruptions to the charging process to allow the chemical reactions in the battery to stabilise.

Restricted area Any area to which access is controlled for the protection of individuals from exposure to radiation and radioactive materials.

Return The name for the common terminal of the output of a power supply; it carries the return currents for the outputs.

Reveal That portion of a jamb or recess which is visible from the face of a wall.

Reverse bias The insulating bias for a p-n junction rectifier; electrons flow into the p side of the junction.

Reverse engineering 1. Reverse Engineering is the process of capturing the geometry of existing physical objects and then using the data obtained as a foundation for designing a duplicate of the original or an entirely new adaptation.

2. Reverse engineering refers to the procedure of carefully dismantling and inspecting a competitor's product to look for design features that can be incorporated into one's own product.

Reverse polarity A connection that is opposite to that which is specified or intended.

Reverse recovery time (Diode) The time required to remove charge carriers from the junction of rectifier when reverse voltage is applied.

Reverse voltage protection A circuit or circuit element that protects a power supply from damage by a voltage of reverse polarity applied at the input or output terminals.

Reversible reaction A reaction in which the reactant(s) proceed to product(s), but the products react at an appreciable rate to reform reactant(s).

ReWheel To change out the internals (bundle) of a compressor to alter its throughput capabilities.

Reynolds number The Reynolds number is the ratio of inertial forces, as described by Newton's second law of motion, to viscous forces. If the Reynolds number is high, inertial forces dominate and turbulent flow exists. If it is low, viscous forces prevail, and laminar flow results.

RFI Radio Frequency Interference. Noise induced upon signal wires by ambient radio-frequency electromagnetic radiation with the effect of obscuring an instrument signal.

RFID Radio Frequency Identification. Small tags incorporating a radio transmitter which can be used to identify or track items of value.

RFQ Request for Quotation.

RGIT RGIT Montrose are a provider of safety training. The term - RGIT - has become synonymous with the offshore safety course - BOSIET.

Rheology Study of flow characteristics.

Rheostat A variable resistor.

RHS Rectangular Hollow Section. Term used in structural steel work.

Richter scale Used to measure the magnitude of an earthquake; introduced in 1935 by the seismologists Beno Gutenberg and Charles Francis Richter.

Right sizing A process that challenges the complexity of equipment. It examines how equipment fits into an overall vision for how work will flow through the factory. When possible, right sizing favors smaller, dedicated machines rather than large, multipurpose, batch-processing machines.

Rigid Ability to resist deformation when subjected to a load; rigidity (n.) the measure of a structure's ability not to change shape when subjected to a load.

Rigid rotor A rotor is considered rigid when it can be corrected in any two (arbitrarily selected) planes and after that correction, its unbalance does not significantly exceed the balancing tolerances (relative to the shaft axis) at any speed up to maximum operating speed and when running under conditions which approximate closely to those of the final supporting system.

Rigid structure As used in this volume, a covalent structure that is reasonably stiff. In a typical rigid structure, all modes of deformation encounter first-order restoring forces resulting from some combination of bond stretching and angle bending; such a structure cannot undergo deformation by bond torsion alone. Meeting this condition usually requires a polycyclic diamondoid structure.

Ring and pinion gear Any gear set consisting of a small gear (the pinion gear) which turns a large-diameter annular gear (the ring gear). Used in rear-drive differentials (rear ends) to transfer power from the driveshaft to the axle and wheels.

Ripple The periodic ac component at the power source output harmonically related to source or switching frequencies. Unless specified otherwise, it is expressed in peak-to-peak units over a specified bandwidth.

Ripple voltage The periodic ac component of the dc output of a power supply.

Rise time The time required for an output voltage of a digital cirucit to change from low voltage level (0) to high voltage leve 1., after the change has started. Very short rise times, not high clock speeds, are the primary cause of cross-talk in PCBs. Rise times are characteristc of the technology being used in a circuit. Gallium Arsenide components can have rise times around 100-picoseconds (millionths of millionths of seconds), 30 to 50 times faster than some CMOS components.

Riser Steel or flexible pipe which transfer well fluids from the seabed to the surface.

Risk based decisionmaking An approach to regulatory decisionmaking in which such decisions are made solely based on the results of a probabilistic risk analysis.

Risk informed decisionmaking An approach to decisionmaking in which insights from probabilistic risk analyses are considered with other engineering insights.

Risk informed regulation Incorporating an assessment of safety significance or relative risk in NRC regulatory actions. Making sure that the regulatory burden imposed by individual regulations or processes is commensurate with the importance of that regulation or process to protecting public health and safety and the environment.

Risk management An organized process to identify what can go wrong, to quantify and access associated risks, and to implement/ control the appropriate approach for preventing or handling each risk identified. A management process consisting of identification, assessment, mitigation, and management of all project, technical and market risks using formal tools and methods.

Risk significant When used to qualify an object, such as a system, structure, component, accident sequence, or cut set, this term identifies that object as exceeding a predetermined criterion related to its contribution to the risk from the facility being addressed. One that is associated with a level of risk that exceeds a predetermined significance criterion.

Robust design Design of the product in a manner to desensitize the product to variation including misuse and increase the probability that it will perform as intended.

Robustness The condition of a product or process where its operating parameters remain relatively stable with a minimum of variation even though factors which influence operation or usage , such as wear or environment, change.

ROCE Return On Capital Employed. The ratio of operating profits generated to the amount of operating capital invested. In effect it is a measure of how productively a company manages its assets.

Rock tunnel A passage constructed through solid rock.

Rocking chair cell A lithium ion cell.

Roentgen (R) A unit of exposure to ionizing radiation. It is the amount of gamma or x-rays required to produce ions resulting in a charge of 0.000258 coulombs/kilogram of air under standard conditions. Named after Wilhelm Roentgen, the German scientist who discovered x-rays in 1895.

Rolling radius Tire-rolling radius is the distance from the center of the wheel to the road. Static radium applies when the vehicle is standing still. Dynamic rolling-radius described wheels in motion. The latter is used to measure tire revolutions per mile and is usually slightly higher than static radius.

Rolling resistance This is motion resisting force that is present from the instant the wheels begin to turn. On normal road surfaces, rolling resistance decreases with increased tire pressure and increases with vehicle weight. Rolling resistance can also be affected by tire construction and tread design.

Room conditions Ambient environmental conditions under which transducers must commonly operate.

Root cause analysis Root Cause Analysis – Study of original reason for nonconformance with a process. When the root cause is removed or corrected, the nonconformance will be eliminated.

Root mean square (RMS) value The equivalent DC voltage that produces exactly the same amount of heating as the AC waveform in a simple (non-inductive) resistor. The RMS value of a sine wave is 0.707 x Peak Value.

Rosin A hard, natural resin, consisting of abietic acid and pimaric acids and their isomers, some fatty acids, and terepene hydrocarbons; the resin is extracted from pine trees and subsequently refined.

Rotary wing landing pad A paved or unpaved surface for takeoffs and landings of rotary wing aircraft. It is physically smaller than a rotary wing runway, typically 100 by 100 feet square, and is normally located at a site that is remote from an airfield or heliport. For inventory purposes, only the prepared surface is included.

Rotor A rotor is a rotating body whose journals are supported by bearings.

Route 1. n. A layout or wiring of a connection.

2. v. The action of creating such a wiring.

ROV Remote Operated Vehicle. An underwater robot.

Rowlock A brick laid on its face edge so that the normal bedding area is visible in the wall face. Frequently spelled rolok.

RS485 connection A standard for serial transmission of data between multiple devices.

RSI Repetitive Strain Injury. Injury occurring from repeated physical movements which causes damage to tendons; nerves; muscles and other soft body tissues.

Repeated movements can be simple and not necessarily strenuous e.g. keyboard use etc.

RTD Resistance Temperature Detector. Resistance Temperature Detectors (RTD's) provide temperature measurement through changes in electrical resistance.

RTJ Ring Type Joint. Description of the mating face of a flange where the mating face has a machined groove into which the gasket is inserted.

Rubblization A decommissioning technique involving demolition and burial of formerly operating nuclear facilities. All equipment from buildings is removed and the surfaces are decontaminated. Above-grade structures are demolished into rubble and buried in the structure's foundation below ground. The site surface is then covered, regraded and, landscaped for unrestricted use.

Runoff The water that flows overland to lakes or streams during and shortly after a precipitation event.

Runway overrun A cleared area extending beyond the ends of a runway. These are not normal traffic areas and are intended only to minimize the probability of serious damage to aircraft using these areas accidentally or in cases of emergency.

Runway, surfaced A flexible or rigid paved airfield surface used for normal takeoffs and landings of fixed or rotary wing aircraft. It can also accommodate rotary wing aircraft. For inventory purposes, only the prepared runway surface is included.

Runway, unsurfaced An unpaved, prepared surface for training, emergency, and other special takeoff and landing operations of fixed or rotary wing aircraft. It can also accommodate rotary wing aircraft. For inventory purposes, only the prepared runway surface is included.

S

Sacrificial anode An active metal or alloy that corrodes and protects another metal or alloy to which it is electrically coupled.

Sacrificial layer A thin film that is later removed to release a microstructure from its substrate.

Safe extra low voltage (SELV) International safety standards as described in IEC 380 and VDE 0806.

Safe shutdown earthquake Is the maximum earthquake potential for which certain structures, systems, and components, important to safety, are designed to sustain and remain functional.

Safeguards As used in regulation of domestic nuclear facilities and materials, the use of material control and accounting programs verify that all special nuclear material is properly controlled and accounted for, and the physical protection (also referred to as physical security) equipment and security forces.As used by the International Atomic Energy Agency (IAEA), verifying that the "peaceful use" commitments made in binding non-proliferation agreements, both bilateral and multilateral, are honored.

Safety The expectation that a system does not, under defined conditions lead to a state in which human life is endangered.

Safety ground A conductive path from a chassis, panel or case to earth to help prevent injury or damage to personnel and equipment.

Safety injection The rapid insertion of a chemically soluble neutron poison (such as boric acid) into the reactor coolant system to ensure reactor shutdown.

Safety limit A restriction or range placed upon important process variables that are necessary to reasonably protect the integrity of the physical barriers that guard against the uncontrolled release of radioactivity.

Safety related In the regulatory arena, this term applies to systems, structures, components, procedures, and controls of a facility or process that are relied upon to remain functional during and following design-basis events. Their functionality ensures that key regulatory criteria, such as levels of radioactivity released, are met. Examples of safety related functions include shutting down a nuclear reactor and maintaining it in a safe shutdown condition.

Safety significant When used to qualify an object, such as a system, structure, component, accident sequence, or cut set, this term identifies that object as having an impact on safety, whether determined through risk analysis or other means, that exceeds a predetermined significance criterion.

Safety vent A safety mechanism that is activated when the internal gas pressure rises above a normal level.

SAFSTOR A method of decommissioning in which the nuclear facility is placed and maintained in such condition that the nuclear facility can be safely stored and subsequently decontaminated to levels that permit release for unrestricted use.

Sagittal plane This is a flat vertical plane passing through the body from front to back, dividing the body into a right and left half. The Cardinal Sagittal Plane passes through the centre of gravity and divides the body into equal symmetrical halves. All other sagittal planes are parallel to it.

Salt bridge The salt bridge of a reference electrode is that part of the electrode which contains the filling solution to establish the electrolytic connection between reference internal cell and the test solution. Auxiliary Salt Bridge: A glass tube open at oneend to receive intermediate electrolyte filling solution, and the reference electrode tip and a junction at the other end to make contact with the sample.

Salt effect (fx) The effect on the activity coefficient due to salts in the solution.

Saltwater intrusion The gradual replacement of freshwater by saltwater in coastal areas where excessive pumping of groundwater occurs.

SAMA (Scientific Apparatus Makers Association) An association that has issued standards covering platinum, nickel, and copper resistance elements (RTDs).

Sampling rate The repetition frequency at which digital samples are taken of an analogue quantity.

Sandstone Sandstones usually can be identified by their coarse, granular, sandy texture. They are sedimentary rocks that consist of consolidated sand grains (mainly quartz and feldspar) cemented together with a variety of minerals (silicates, iron oxides, limonite, calcite and clays). These cementing materials make one sandstone behave very differently from another. Sandstones containing silica are quite hard, strong and decay resistant, whereas those containing calcite resemble limestone in their susceptibility to acid damage, and those containing clay absorb water and deteriorate more easily. Because they have a granular texture throughout, sandstone surfaces stay matte even when worked. Most sandstones have a tendency to absorb moisture and do not withstand frost action well, so they should not be used as foundation stones and should be protected from excessive moisture.

Sanitary engineering Sanitary engineering is a branch of civil and environmental engineering that deals with sanitary issues affecting public health, such as safe drinking water and sewage disposal. Among other things, sanitary engineers deal with preventing toxins and dangerous microorganisms from endangering the public in such systems.

Sanitizing One of the 5S's used for workplace organization. Sanitizing is the act of cleaning the work area. Dirt is often the root cause of premature equipment wear, safety problems and defects.

SAP (Systems Analysis and Program Development) Software company based in Germany perhaps best known for their purchasing software suite.

Saponifier An alkaline chemical added to water to improve its ability to dissolve rosin flux residues.

SAT (Site Acceptance Test) A test of equipment carried out at site following installation of equipment but prior to commissioning.

Satellite design lab A laboratory founded by Dr. Glenn Lightsey at the University of Texas

Department of Aerospace Engineering and Engineering Mechanics. This facility is used for student projects related to the conception and fabrication of space vehicles including satellites, sounding rocket payloads, and near-space balloon flights. It is located on the fourth floor of the W.R. Woolrich Laboratories building on the UT Austin campus.

Satellite formations When several satellites are flown together in orbit, they are arranged in a way that prevents collisions, promotes communication, and allows them to complete their mission successfully. These arrangements are referred to as satellite formations, and are commonly used with small satellites like FASTRAC.

Saturated An organic molecule is described as saturated if it is a closed shell species lacking double or triple bonds; forming a new bond to a saturated molecule requires the cleavage of an existing bond.

Saturation 1. The operating condition of a transistor when an increase in base current produces no further increase in collector current.

2. A circuit condition whereby an increase in the driving or input signal no longer produces a change in the output.

3. The condition when a transistor is driven so hard that it becomes biased in the forward direction. In a switching application, the charge stored in the base region prevents the transistor from turning off quickly under saturation conditions.

4. Generally, that state in which a semiconductor device is conducting most heavily for a given applied voltage. In many devices it is also a state in which the normal amplification mechanisms have become "swamped" and inoperative. (Graf).

SCADA (Supervisory Control and Data Acquisition) A term commonly used to describe a PC based software package that allows operator control of a process from a PC and has the PC collecting and storing process information.

Often used with PLCs.

Scale deposit (Corrosion Eng.) Partially adherent layers of corrosion products on metals which generally develop in hard water and at high temperature.

Scanning electron microscope (SEM) A microscope producing an image by using reflected electron beams that scan the surface of a specimen.

Scanning tunneling microscope A device in which a sharp conductive tip is moved across a conductive surface close enough to permit a substantial tunneling current (typically a nanometer or less). In a common mode of operation, the voltage is kept constant and the current is monitored and kept constant by controlling the height of the tip above the surface; the result, under favorable conditions, is an atomic-resolution map of the surface reflecting a combination of topography and electronic properties. The STM has been used to manipulate atoms and molecules on surfaces.

Scattered radiation Radiation that, during its passage through a substance, has been changed in direction. It may also have been modified by a decrease in energy. It is one form of secondary radiation.

Schematic A diagram which shows, by means of graphic symbols, the electrical connections, components and functions of an electrical system. The components are represented by agreed-upon symbols, and the conductors connecting them by lines. If two lines cross each other, a large dot represents a junction, whereas no dot represents no connection.

Schottky diode A diode device that exhibits a low forward voltage drop and fast recovery time relative to a standard silicon diode.

Scientific law A relationship between quantities, usually described by an equation in the physical sciences; is more important and describes a wider range of phenomena than a scientific principle.

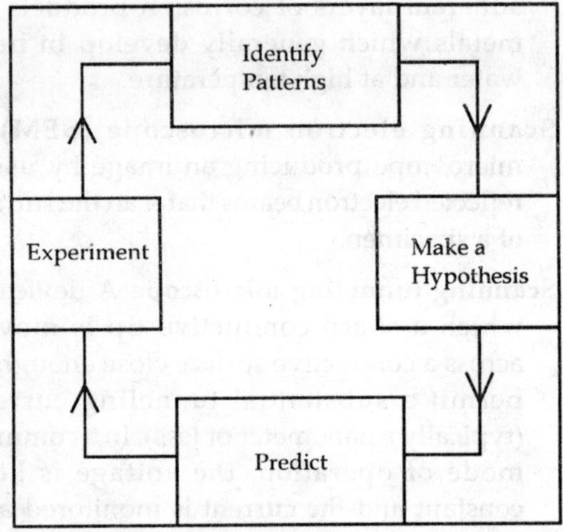

Fig. Scientific law

Scintillation detector The combination of phosphor, photomultiplier tube, and associated electronic circuits for counting light emissions produced in the phosphor by ionizing radiation.

Scoliosis Lateral curvature of the spine, named in accordance with the location and deviation of the convexity, as right thoracic.

Scram The sudden shutting down of a nuclear reactor, usually by rapid insertion of control rods, either automatically or manually by the reactor operator. May also be called a reactor trip. It is actually an acronym for "safety control rod axe man," the worker assigned to insert the emergency rod on the first reactor .

Scroll To move all or part of the screen material up to down, left or right, to allow new information to appear.

SD&P (Simultaneous Drilling & Production) A technique of drilling a new well on a platform while it continues to produce oil or gas.

Sealed beam headlamp A one piece, hermetically sealed headlamp in which the filament is an integral part of the unit and the lens itself is the bulb. Sealed beams are relatively inexpensive and when one burns out or the lens cracks, the whole unit is replaced.

Sealed cells A cell which remains closed and does not release gas or liquid when operated within the limits of charge and temperature specified by the manufacturer. An essential component in recombinant cells.

Sealed source Any radioactive material or byproduct encased in a capsule designed to prevent leakage or escape of the material.

Secondary battery A battery which can be recharged and used repeatedly.

Secondary circuit A circuit electrically isolated from the input or source of power to the device.

Secondary circuit protection Overcurrent protection located in the secondary circuit.

Secondary device A part of the flowmeter which receives a signal proportional to the flowrate, from the primary device, and displays, records and/or transmits the signal.

Secondary output An output of a switching power supply that is not sensed by the control loop.

Secondary radiation Radiation originating as the result of absorption of other radiation in matter. It may be either electromagnetic or particulate in nature.

Secondary standard pH buffer solutions which do not meet the requirements of primary standard solutions but provide coverage of the pH range not covered by primary standards. Used when the pH value of the primary standard is not close to the sample pH value.

Secondary system The steam generator tubes, steam turbine, condenser, and associated pipes, pumps, and heaters used to convert the heat energy of the reactor coolant system into mechanical energy for electrical generation. Most commonly used in reference to pressurized water reactors.

Secondary treatment In wastewater treatment, the conversion of the suspended, colloidal and dissolved organics remaining after primary treatment into a microbial mass with is then removed in a second sedimentation process. Secondary treatment included both the biological process and the associated sedimentation process.

Secondary winding A coil that receives energy from the primary winding by mutual induction and delivers energy to the load is the secondary winding.

Secured landfill A landfill which has containment measures such as liners and a leachate collection system so that materials placed in the landfill will not migrate into the surrounding soil, air and water.

Sedimentation The gravity settling, and thus removal, of materials more dense than the suspending fluid.

Seebeck coefficient The derivative (rate of change) of thermal EMF with respect to temperature normally expressed as millivolts per degree.

Seebeck effect When a circuit is formed by a junction of two dissimilar metals and the junctions are held at different temperatures, a current will flow in the circuit caused by the difference in temperature between the two junctions.

Seebeck EMF The open circuit voltage caused by the difference in temperature between the hot and cold junctions of a circuit made from two dissimilar metals.

Selectivity The ability of a sensor to measure only one metric or, in the case of a chemical sensor, to measure only a single chemical species.

Self discharge Capacity loss during storage due to the internal current leakage between the positive and negative plates.

Self heating Internal heating of a transducer as a result of power dissipation.

Self inductance Inductance that produces an induced voltage in itself as the result of a change in current flow.

SEM (Scanning Electron Microscope) Apparatus used to investigate the physical structure of cell components and surfaces. They typically cost about $500,000 or more.

Semiconductor An insulator whose conductivity can be manipulated by the addition of impurities (doping), by introduction of an electric field, by exposure to light , or by other means.

Semiregulated output A subjective term indicating partial regulation.

Sense line The conductor which routes output voltage to the control loop.

Sense line return The conductor which routes the voltage on the output return to the control loop.

Sensing element That part of the transducer which reacts directly in response to the input.

Sensitivity shift A change in slope of the calibration curve due to a change in sensitivity.

Sensitivity The amount of change in a sensor's output in response to a change at a sensor's input over the sensor's entire range. Provides an indication of a sensor's ability to detect changes. For some sensors, the sensitivity is defined as the input parameter change required to produce a standardized output change.

SEPA (Scottish Environment Protection Agency) Public body responsible for the protection of the environment in Scotland.

Separator An ionic, permeable, electronically nonconductive spacer which prevents electronic contact between electrodes of opposite polarity in the same cell.

Sequential access An access mode in which records are retrieved in the same order in which they were written. Each successive access to the file refers to the next record in the file.

Serial transmission Sending one bit at a time on a single transmission line. Compare with parallel transmission.

Series 1. The interconnection of two or more power sources such that alternate polarity terminals are connected so their voltages sum at a load.

2. The connection of circuit components end to end to form a single current path.

Series connection The connection of positive to negative of two or more cells to form a battery of higher voltage.

Series pass A controlled active element, such as a transistor, in series with a load that is used to regulate voltage.

Series regulator A regulator in which the active control element is in series with the dc source and the load.

Service maintenance The percent of rated capacity remaining after a specified period of time.

Set point The temperature at which a controller is set to control a system.

Setting range The range over which the value of the stabilized output quantity may be adjusted.

Settling time The time for a power supply to stabilize within specifications after an excursion outside the input/output design parameters. Related topic: Transient Response

Severe accident A type of accident that may challenge safety systems at a level much higher than expected.

Sewage Water used by industry and households. Sewage typically contains everything from soap to solid waste, and must be purified before it can be safely returned to the ecosystem.

Shear A shear deformation is one that displaces successive layers of a material transversely with respect to one another, like a crooked stack of cards. Shear is a dimensionless quantity measured by the ratio of the transverse displacement to the thickness over which it occurs.

Shelf life The length of time that a given item can remain in a saleable condition on a supplier's shelf; this is highly relevant for most sealants as they degrade in time even when stored in sealed, unopened containers.

Shielding Any material or obstruction that absorbs radiation and thus tends to protect personnel or materials from the effects of ionizing radiation.

Shock load Influent wastewater entering the plant which has an unusually high organic content and/or high flow rate.

Shock phase During stance phase there are two high shock phases; the first occurs just after heel strike and the second as the heel lifts off the ground. The twin peaks of force are considered to represent normal walking.

Short Short circuit. 1. An abnormal connection of relatively low resistance between two points of a circuit. The result is excess (often damaging) current between these points. Such a connection is considered to have occurred in a printed wiring CAD database or artwork anytime conductors from different nets either touch or come closer than the minimum spacing allowed for the design rules being use.

Short circuit A direct connection that provides a virtually zero resistance path for current.

Short circuit current (SCC) The initial value of the current obtained from a power source in a circuit of negligible resistance.

Short circuit test A test in which the output is shorted to ensure that the short circuit current is within its specified limits.

Shoved joints Vertical joints filled by shoving a brick against the next brick when it is being laid in a bed of mortar.

Shrinkage 1. Decrease in size which may result in cracking, stresses, strain.

2. If scratches or substrate imperfections have not been properly filled, they will show up as the paint shrinks into them.

3. Decrease in wood dimensions due to water loss the wood cell walls. Shrinkage across the grain of wood occurs when the moisture content falls below the fiber saturation point (28-30 percent). Below the fiber saturation point, shrinkage is generally proportional to moisture content. Shrinkage is expressed as a percentage of the green wood dimensions.

4. A: decrease in volume of a soil (or fill) material through the reduction of voids by mechanical compaction, superimposed loads, or natural consolidation; b: settling or reduction in volume of earthen fills, cement slurries, or concrete on setting; c: decrease in volume of clayey soil or sediment due to reduction of void volume, principally by drying.

5 Cracking due to drying

Shrinkage crack A crack produced in fine-grained sediment or rock by the loss of contained water during drying or dehydration; e.g., a desiccation crack or a mud crack. AGI

Shrinkage index The numerical difference between the plastic and shrinkage limits. ASCE

Shrouded terminals Terminals surrounded by an insulating shroud which prevents accidental contact with the terminal.

Shuffling gait During this type of gait the foot is still moving forward at the time of initial contact. In some subjects contact is made by the foot flat, in others there is heel strike. Some people also experience shuffling during midswing. This condition is the direct result of degeneration of the basal ganglia in a disordered extra pyramidal system and is called Parkinsonism.

Shunt 1. A parallel conducting path in a circuit.

2. A low value precision resistor used to monitor current.

Shunt regulator A voltage regulator which uses a transistor or FET, in parallel with the load, which shorts out the excess voltage when the applied input voltage exceeds a specified limit producing a regulated output voltage. It is a simple but lossy design.

Shutdown A decrease in the rate of fission (and heat production) in a reactor (usually by the insertion of control rods into the core).

Shutdown margin The instantaneous amount of reactivity by which the reactor is subcritical or would be subcritical from its present condition assuming all full-length rod cluster assemblies (shutdown and control) are fully inserted except for the single rod cluster assembly of highest reactivity worth that is assumed to be fully withdrawn.

Shuttlecock cell A lithium ion cell.

SI System Internationale. The name given to the standard metric system of units.

SI units International System of units based on the metric system and units derived from the metric system.

Sievert (Sv) The international system (SI) unit for dose equivalent equal to 1 Joule/kilogram. 1 sievert = 100 rem. Named for physicist Rolf Sievert.

Sifting One of the 5S's used for workplace organization. Sifting involves screening through unnecessary materials and simplifying the work environment. Sifting is the separating of the essential from the nonessential.

Sigma bond A covalent bond in which overlap between two atomic orbitals (e.g., of sp, sp^2, or sp^3 hybridization) produces a single bonding orbital in which the distribution of shared electrons has a roughly cylindrical symmetry about the axis linking the two atoms. By themselves, sigma bonds present little barrier to rotation of one substructure with respect to another, although steric effects and cyclic structures may hinder or block rotation.

Signal An electrical transmittance (either input or output) that conveys information.

Signal conditioning To process the form or mode of a signal so as to make it intelligible to, or compatible with, a given device, including such manipulation as pulse shaping, pulse clipping, compensating, digitizing, and linearizing.

Silicon controlled rectifier (SCR) A unidirectional, four-layer (PNPN) junction device in which conduction is initiated by the application of a gate current. Conduction will continue until the current is reduced to some minimum value.

Silicon wafer A thin, iridescent, silvery disk of silicon which contains a set of integrated circuits, prior to their being cut free and packaged. A silicon wafer will diffract reflected light into rainbow patterns and, being a similar size, looks so much like a music CD that it could be mistaken for one (except that it has no label or hole in the middle). On closer inspection, one can see the individual (usually rectangular- or square-shaped) integrated circuits which form a uniform patchwork quite unlike the surface of a music CD. When cut or etched from the wafer these circuits are then called chips or dice.

Silkscreen (Also called "silkscreen legend") 1. The decals and reference designators in epoxy ink on a printed wiring board, so called because of the method of application—the ink is "squeegeed" through a silk screen, the same technique used in the printing of T-shirts.

A silk mesh size commonly used is 6 mils. With this mesh size, the absolute minimum line width of any silkscreen legend artwork is 6 mils, which leaves a very faint line. 7 mils works better for a practical minimum line width.

Newer silkscreening methods allow for sikscreen draws of 5 mils, which come out very clear. A good reference designator size to use is 35 mils high with a 5 mil draw.

2. A Gerber file controlling the photoplotting of this legend.

Silt Sediment particles ranging from 0.004 to 0.06 mm (0.00016 to 0.0024 inch) in diameter.

Simulation Modeling or representation of hardware, software or systems to determine or verify their behavior, operation or fit. Simulation is used to provide confidence that the hardware, software or systems will operate as intended without investing the time or expense of physically constructing the object to verify its operation. In electronic design, simulation is a technique in which the properties of a circuit are represented indirectly by test vectors.

Sine wave A wave form of a single frequency alternating current whose displacement is the sine of an angle proportional to time or distance.

Single bond A sigma bond having no associated pi bonds.

Single crystal A crystalline solid for which the periodic and repeated atomic pattern extends throughout its entirety without interruption.

Single ended input A signal-input circuit where SIG LO (or sometimes SIG HI) is tied to METER GND. Ground loops are normally not a problem in AC-powered meters, since METER GND is transformer-isolated from AC GND.

Single piece flow A process in which products proceed, one complete product at a time, through various operations in design, order-taking and production without interruptions, backflows or scrap.

Single Plane (Static) balancing machine A single plane balancing machine is a gravitational or centrifugal balancing machine that provides information for accomplishing single plane balancing.

Single point ground The one point in a system that connects multiple grounds and returns. Also known as star ground, or holy point ground.

Single precision The degree of numeric accuracy that requires the use of one computer word. In single precision, seven digits are stored, and up to seven digits are printed. Contrast with double precision.

Single support This describes the stage in stance phase, where only one leg is in ground contact. During single support the centre of gravity of the body sits at its highest point.

Single track PCB design with only one route between adjacent DIP pins.

Singlet An electronic state of a molecule in which all spins are paired.

Sintering Densification of a particulate ceramic compact involving a removal of the pores between the starting particles (accompanied by equivalent shrinkage) combined with coalescence and strong bonding between adjacent particles.

Site remediation The process of cleaning up a hazardous waste disposal site that has either been abandoned or that those responsible either refuse to cleanup or are financially unable to cleanup.

Six sigma 1. Six Sigma is a statistical measurement of processes that produce less than 3.4 defects or mistakes per million opportunities (or 99.99966% good).

2. Six Sigma is a bottom-line oriented, data-driven quality program. It is based on achieving a level of quality which equates with only 3.4 defects per million opportunities for each product. The six sigma process includes five steps a) define, b) measure, c) analyze, d) improve, and e) control.

SLA Stereolithography Apparatus.

SLS Selective Laser Sintering

SM Solids Modeling

SLA Battery Sealed Lead Acid battery. In sealed batteries the generated oxygen combines chemically with the lead and then the hydrogen at the negative electrode, and then again with reactive agents in the electrolyte, to recreate water. A recombinant system. The net result is no significant loss of water from the cell.

SLD (Single Line Diagram) A schematic diagram showing the electrical connection from the main service entrance to all of the facility's electrical distribution panels and electrical loads.

Slenderness ratio Ratio of the effective height of a member to its effective thickness.

Slewing rate The maximum rate of change a power supply output can produce when subjected to a large step response or specified step change.

Slip Plastic deformation resulting from dislocation motion; also, the shear displacement of 2 adjacent planes of atoms.

Slip casting A forming technique used to shape ceramic materials. A slip or suspension of solid particles in water is poured into a porous mold. A solid layer forms on the inside wall as water is absorbed by the mold, leaving a shell (or a solid piece) in the shape of the mold.

Slow charge Charging overnight in 14 to 16 hours at about 0.1C rate. Safe and simple.

Slow start A feature that ensures the smooth, controlled rise of the output voltage, and protects the switching transistors from transients when the power supply is turned on.

Sludge A mixture of solid waste material and water. Sludges result from the concentration of contaminants in water and wastewater treatment processes. Typical wastewater sludges contain from 0.5 to 10 percent solid matter. Typical water treatment sludges contain 8 to 10 percent solids.

Slushed joints Vertical joints filled, after units are laid, by "throwing" mortar in with the edge of a trowel. (Generally, not recommended.)

Smallest bending radius The smallest radius that a strain gage can withstand in one direction, without special treatment, without suffering visible damage.

Smart battery An intelligent battery which contains information about its specification, its status and its usage profile which can be read by its charger or the application in which it is used.

Smart sensor A sensor in which the electronics that process the output from the sensor, and forms the modifier, are partially or fully integrated on a single chip.

SMBus System Management Bus. A two wire, 100 KHz, serial bus for interconnecting Smart Batteries which have built in intelligence, with their associated chargers or applications.

Snubber An RC network used to reduce the rate of rise of voltage in switching applications.

Snubber network A circuit that uses a RC network and a diode in unipolar switching applications.

Soft Pertaining to or consisting of software.

Soft copy An electronic form of a document; a data file in computer memory or stored on storage media. When one is looking at a soft copy he is viewing the document as displayed on a computer monitor.

Soft-burned Clay products which have been fired at low temperature ranges, producing relatively high absorptions and low compressive strengths.

Softening The removal of divalent cations by precipitation or ion exchange.

Soft-ground tunnels A passage constructed through loose, unstable, or wet ground, requiring supports to keep the walls from collapsing.

Software Computer programs and computer databases.

Software architecture Software Architecture refers to the high-level structure of software systems. The architecture of a software system identifies a set of components that collaborate to achieve the system goals. The architecture specifies the "externally visible" properties of the component, i.e., those assumptions other components can make of a component, such as its provided services, performance characteristics, fault handling, shared resource usage, etc. It also specifies the relationships among the components and how they interact.

Software configuration management Software Configuration Management (SCM) is the specialization of Configuration Management.

Software development A set of activities that results in software products. Software development may include new development, modification, reuse, re-engineering, maintenance, or any other activities that result in software products.

Software management group A group of specialists who establish, maintain, and improve the software management process used during software development.

Software system A system consisting solely of software and possibly the computer equipment on which the software operates.

SOLAS Safety of Life at Sea (SOLAS) requirements stipulate minimum equipment to be provided in liferafts.

Liferafts are provided on offshore installations, helicopters etc.

Solder A low melting point alloy, usually of lead (Pb) and tin (Sn), that can wet copper, conduct current, and mechanically join conductors.

Solder ball The round solder balls bonded to a transistor contact area and used to make connection to a conductor by face-down bonding techniques. In IBM's ceramic BGA's, the solder used in solder balls has a higher melting point than that used in soldering the ball to the chip substrate and the BGA to a board. IBM uses 10/90 tin/lead for the solder ball and eutectic solder for the assembly. The high melt balls of the BGA do not melt during PCB assembly and thus create a pre-determined standoff height for the component. PBGA packages use a eutectic solder ball which provides a collapsible joint similar to C4 .

Solder bridging When solder paste or solder on two or more adjacent pads come into contact to form a conductive path or bridge.

Solder bumps Solder balls.

Solder mask A dielectric material used to cover the entire surface (except where the joints are to be formed) of a PCB primarily to protect the circuitry from environmental damage; it also helps to reduce bridging.

Solenoid A cylindrical coil of wire which generates a force when an electrical current is applied. Commonly used with actuators to activate a process.

Solid state battery Cells with solid electrolytes. Lithium polymer cells are examples of this technology.

Solidification The change from liquid state to solid state upon cooling through the melting temperature or melting range.

Solidification range The temperature between the liquidus and solidus.

Solidification shrinkage crack A crack that forms due to internal stresses developed from shrinkage during solidification of a metal casting; also called a hot crack.

Solidstate switch A switch that uses no moving parts.

Solidus The temperature in a phase equilibrium diagram below which no liquids are present; the highest temperature at which a metal or alloy is completely solid.

Solution heat treatment Heating an alloy to a suitable temperature, holding at that temperature for enough time to allow one or more constituents to enter into solid solution, and then rapidly cooling to hold the consituents in solution.

Solvation Ions in solution are normally combined with at least one molecule of solvent. This phenomenon is termed solvation.

Somatic effects of radiation Effects of radiation limited to the exposed individual, as distinguished from genetic effects, that may also affect subsequent unexposed generations.

Sorting One of the 5S's used for workplace organization. Sorting involves organizing essential materials. It allows the operator to find materials when needed because they are in the correct location.

Sour A sour fluid is a fluid that contains hydrogen sulphide (H_2S) at 10 ppm or more.

Source Origin of the input power, e.g., generator, utility lines, mains, batteries, etc.

Source code A non-executable program written in a high-level language. A compiler or assembler must translate the source code into object code (machine language) that the computer can understand and process.

Source material Uranium or thorium, or any combination thereof, in any physical or chemical form or ores that contain by weight 1/20 of one percent (0.05 percent) or more of 1. Uranium,

2. Thorium, or

3. Any combination thereof. Source material does not include special nuclear material.

Source reduction The elimination or reduction of the waste at the source by modification of the actual process which produces the waste.

Souring The result of the conversion of sulfate or other oxidized forms of sulfur-containing ions to hydrogen sulfide gas and sulfide ions through the respiratory activities of sulfur-reducing bacteria in an oil reservoir. In producing operations, the most significant detrimental effect of microbial souring is the generation of toxic and corrosive hydrogen sulfide that attacks vessels and pipeline and decreases the quality of the associated natural gas. This may require expensive production equipment (rated for sour service) and gas sweetening operations to bring the gas back to export pipeline standards.

SOx (Sulphur Oxides) Sulphur oxide gases (principally Sulphur dioxide - SO_2) are formed when fuel e.g. coal and oil; containing sulphur is burned.

sp, sp2, sp3 An isolated carbon atom has four valence orbitals: three mutually perpendicular p orbitals, each with a single nodal plane, and one spherically symmetric s orbital. A carbon atom in a typical molecule can be regarded as bonding with four orbitals consisting of weighted sums (termed hybrids) of these s and p orbitals. One common pattern has four equivalent orbitals, each formed by combining the three p orbitals with the s orbital; this is sp3 hybridization. An sp3 carbon atom forms four sigma bonds, usually in a roughly tetrahedral arrangement. Another common pattern has three equivalent orbitals formed by combining two p orbitals with the s orbital; this is termed sp2 hybridization. An sp2 carbon atom forms three roughly coplanar sigma bonds, usually separated by ~120 , and one pi bond (or several fractional pi bonds). If a single p orbital is combined with the s orbital, the result is sp hybridization, forming two sigma bonds and two pi bonds (usually in a straight line). Atoms of other kinds (e.g., N and O) can hybridize in an analogous manner.

Space reddening The observed reddening, or absorption of shorter wavelengths, of the light from distant celestial bodies due to scattering by small particles in interstellar space. Compare red shift.

Space simulator 1. Any device used to simulate one or more parameters of the space environment used for testing space systems or components.

2. Specifically, a closed chamber capable of approximately the vacuum and normal environments of space.

Space suit A pressure suit for wear in space or at very low ambient pressures within the atmosphere, designed to permit the wearer to leave the protection of a pressurized cabin.

Space transformer A major component of certain high-density probe cards . It provides pitch reduction, high routing density and localized mid-frequency decoupling. A major developer of ATE systems which use space transformers is Wentworth Labs. .

Span 1. The dimension of a craft measured between lateral extremities; the measure of this dimension.

2. Specifically, the dimension of an airfoil from tip to tip measured in a straight line.

Span is not usually applied to vertical airfoils.

Span adjustment The ability to adjust the gain of a process or strain meter so that a specified display span in engineering units corresponds to a specified signal span. For instance, a display span of 200°F may correspond to the 16 mA span of a 4-20 mA transmitter signal.

Spare A connector point reserved for options, specials, or other configurations. The point is identified by an (E#) for location on the electrical schematic.

Spark discharge That type of gaseous electrical discharge in which the charge transfer occurs intermittently along a relatively constricted path of high ion density, resulting in high luminosity. It is of short duration and to be contrasted with the nonluminous point discharge, with the diffuse corona discharge, and also with the continuous arc discharge.

The exact meaning to be attached to the term spark discharge varies somewhat in the literature. It is frequently applied to just the transient phase of the establishment of any arc discharge. A lightning discharge is a large-scale spark discharge, though its very length introduces certain details not found in laboratory short-spark processes.

Spark spectrum The spectrum of an ion. The degree of ionization, or order of the spectrum, is indicated by a Roman numeral following the symbol for the element. The first spark spectrum is indicated by II, the second by III, and so on. Thus Fe IV indicates the spectrum of an iron atom which has lost three electrons.

Spastic diplegia A common manifestation of cerebral palsy. It affects both legs although there may be considerable asymmetry between the two sides. The commonest pattern consists of hip flexion and internal rotation, due to the over activity by the iliopsoas, rectus femorus, and hip adductors. Knee extension, due to over activity of the hamstrings, especially on the medial side; and equinus deformity of the foot and eversion of the rearfoot, due to over activity by the triceps surae and peronei.

Special causes (of variation) Special Causes are causes of variation in output from a manufacturing process or system of procedures that is not due to the inherent operation of the process or system itself (common causes), but is due to the intrusion into the system of a one-time or external cause of variation. One-time or external causes do not spring from the system, and so are preventable - i.e., their occurrence can be prevented. Consequently, the reason for each special cause must be investigated and steps then taken to see that it does not occur again. The presence of a special cause of variation must be determined statistically. This is done by knowing that variation in output due to common causes follows a regular pattern corresponding to the Normal curve - i.e., with an average and a deviation on either side of the average within three standard deviations. Variation due to a special cause results in a performance outside these statistical limits.

Special perturbations A method of orbit determination by numerical integration which takes into account the perturbing forces which are causing the orbit to depart from the orbit as calculated by Kepler laws.

Specific gravity SG The ratio of the weight of a solution compared with the weight of an equal volume of water at a specified temperature. It is used to determine the charge condition in lead acid batteries.

Specific heat The ratio of thermal energy required to raise the temperature of a body 1° to the thermal energy required to raise an equal mass of water 1°.

Specific propellant consumption The reciprocal of the specific impulse, i.e., the required propellant flow to produce one pound of thrust in an equivalent rocket.

Specification A document intended primarily for use in procurement, which describes the essential technical requirements for items, materiels or services including the procedures for determining whether or not the requirements have been met.

Spectral 1. Of or pertaining to a spectrum.

2. Referring to thermal radiation properties, for ratios such as emittance, reflectance, and transmittance, at a specified wavelength; for powers, such as emissive power, within a narrow wavelength band centered on a specified wavelength.

Spectral filter A filter which allows only a specific band width of the electromagnetic spectrum to pass, i.e., 4 to 8 micron infrared radiation.

Spectral function The Fourier representation of a given function; that is, the Fourier transform if the given function is aperiodic or the set of coefficients of the Fourier series if the given function is periodic. Also called spectrum.

Spectral line A bright, or dark, line found in the spectrum of some radiant source.

Spectroheliograph An instrument for taking photographs (spectroheliograms) of the image of the sun monochromatic light. The wavelength of light chosen for this purpose corresponds to one of the Fraunhofer lines, usually the light of hydrogen or ionized calcium. A similar instrument used for visual, instead of photographic, observations in a spectrohelioscope.

Spectrophotometer A photometer which measures the intensity of radiation as a function of the frequency (or wavelength) of the radiation. Also called spectroradiometer. In one design, radiation enters the spectrophotometer through a slit and is dispersed by means of a prism. A bolometer having a fixed aperture scans the dispersed radiation so that the intensity over a narrow wave band is obtained as a function of frequency.

Spectroscope An apparatus to effect dispersion of radiation and visual display of the spectrum obtained.

A spectroscope with a photographic recording device is called a spectrograph.

Spectrum 1. In physics, any series of energies arranged according to wavelength (or frequency).

2. The series of images produced when a beam of radiant energy is subject to dispersion.

3. Short for electromagnetic spectrum or for any part of it used for a specific purpose as the radio spectrum (10 kilocycles to 300,000 megacycles).

4. In mathematics, = function.

5. In acoustics, the distribution of effective sound pressures or intensities measured as a function of frequency in specified frequency bands.

Spectrum analysis Utilizing frequency components of a vibration signal to determine the source and cause of vibration.

Specular reflection Reflection in which the reflected radiation is not diffused; reflection as from a mirror. Also called regular reflection, simple reflection. Compare diffuse reflection.

The angle between the normal to the surface and the incident beam is equal to the angle between the normal to the surface and the reflected beam. Any surface irregularities on a specular reflector must be small compared to the wavelength of the incident radiation.

Speed Rate of motion.

Rate of motion in a straight line is called linear speed, whereas change of direction per unit time is called angular speed. Speed and velocity are often used interchangeably although some authorities maintain that velocity should be used only for the vector quantity.

Speed of light The speed of propagation of electromagnetic radiation through a perfect vacuum; a universal dimensional constant equal to 299,792.5 +/- 0.4 kilometers per second. Also called velocity of light.

Spent (depleted) fuel Nuclear reactor fuel that has been used to the extent that it can no longer effectively sustain a chain reaction.

Spent fuel pool An underwater storage and cooling facility for spent (used) fuel elements that have been removed from a reactor.

Spent nuclear fuel Fuel that has been removed from a nuclear reactor because it can no longer sustain power production for economic or other reasons.

Sphere of influence The surface in space about a planet where the ratio of the force with which the sun perturbs the motion of a particle about the planet, to the force of attraction of the planet equals the ratio of the force with which the planet perturbs the motion of a particle about the sun, to the force of attraction of the sun on the particle.

The volume inside this surface defines the region where the attracting body exerts the primary influence on a particle.

Spherical coordinates 1. A system of coordinates defining a point on a sphere or spheroid by its angular distances from a primary great circle and from a reference secondary great circle, as latitude and longitude.

2. Space polar coordinates.

Spherical excess The amount by which the sum of the three angles of a spherical triangle exceeds 180 degrees.

Spherical system A trajectory measuring system, whose locus of the measured range is a sphere with the ground equipment at the center.

A unique point in space is determined by the intersection of three or more spheres. The term spherical system has been applied to systems using three or more slant ranges to determine space position.

Spherical triangle A closed figure having arcs of three great circles as sides.

Spherical wave A wave whose phase front surfaces are spheres. Such waves propagate from a point source.

Spherics Variant spelling of sferics.

Spheroid An ellipsoid; a figure resembling a sphere. Also called ellipsoid or ellipsoid of revolution from the fact that it can be formed by revolving an ellipse about one of its axes. If the shorter axis is used as the axis of revolution, an oblate spheroid results, and if the longer axis is used, a prolate spheroid results. The earth is approximately an oblate spheroid.

Spheroidal excess The amount by which the sum of the three angles of a triangle on the surface of a spheroid exceeds 180 degrees.

Spicules Bright spikes extending into the chromosphere of the sun from below.

They are several hundred miles in diameter and extend outward 5000 to 10,000 miles. Spicules have a lifetime of several minutes and may be related to granules.

Spillway An overflow channel that allows dam operators to release lake water when it gets high enough to threaten the safety of a dam.

Spin stabilization Directional stability of a spacecraft obtained by the action of gyroscopic forces which result from spinning the body about its axis of symmetry.

Spin table A flat round platform on which human and animal subjects can be placed in various positions and rapidly rotated, much as on a phonograph record, in order to simulate and study the effects of prolonged tumbling at high rates.

Complex types of tumbling can be simulated by mounting the spin table on the arm of a centrifuge.

Spinal kyphosis Angular curvature of the spine, the convexity of the curve being posterior, usually situated in the thoracic region and involving few or many of the vertebrae; the result of such diseases as tuberculosis, osteoarthrosis, or rheumatoid arthritis of the spine, or an improper posture habit.

Spinal lordosis Forward curvature of the lumbar spine. It may result from increased weight of abdominal contents as in pregnancy or extreme obesity. Other causes include poor posture, flexion contracture of the hip, rickets and tuberculosis of the spine.

Spinner Rotating part of a radar antenna used to impact any subsidiary motion in addition to the primary slewing of the beam.

Spinodal structure A fine homogeneous mixture of two phases that form by growth of composition waves in a solid solution during suitable heat treatment; the phases differ in composition from each other and the parent phase, but have the same crystal structure as the parent phase.

Spiral scanning Scanning in which the direction of maximum radiation describes a portion of a spiral. The rotation is always in one direction.

Spiral wound Battery construction in which the electrodes with the electrolyte and separator between them are rolled into a spiral like a jelly roll (Swiss roll).

Spire An architectural or decorative feature of a skyscraper; the Council on Tall Buildings and Urban Habitat includes spires but not antennae when calculating the official height of a skyscraper.

Split bobbin winding The method of winding a transformer whereby the primary and secondary are wound side-by-side on a bobbin with an insulation barrier between the two windings.

Spoiler A plate, series of plates, comb, tube, bar, or other device that projects into the airstream about a body to break up or spoil the smoothness of the flow, especially such a device that projects from the upper surface of an airfoil, giving an increased drag and a decreased lift.

Spoilers are normally movable and consist of two basic types: the flap spoiler, which is hinged along one edge and lies flush with the airfoil or body when not in use, and the retractable spoiler, which retracts edgewise into the body.

Spontaneous emission The decay of an atom or ion in an excited energy state Ej to a lower state Ei without the influence of any external perturbation. This process results in the emission of a photon of energy.

Spontaneous-ignition temperature In testing fuels, the lowest temperature of a plate or other solid surface adequate to cause ignition in air of a fuel upon the surface.

Spot size The diameter of the circle formed by the cross section of the field of view of an optical instrument at a given distance.

Spread reflection Reflection from a rough surface with large irregularities. Also called mixed reflection.

Springs, torsion bar A long straight bar that is fastened to the frame at one end and to a control arm at the other. Spring action is produced by a twisting of the bar.

Spurious disk The round image of perceptible diameter of a star as seen through a telescope, due to diffraction of light in the telescope.

Spurious radiation 1. Any undesired emission from a radio transmitter.

2. Any electromagnetic radiation from a radio receiver. Also called spurious emission.

Spurious response Output from a receiver due to a signal or signals having frequencies other than that to which the receiver is tuned.

Spurious transmitter output Any part of the radio frequency output of a transmitter which is not a component of the theoretical output as determined by the type of modulation and specified bandwidth limitations.

Spurious tube counts In radiation-counter tubes, counts other than background counts and those caused directly by the radiation to be measured.

Spurious counts are caused by failure of the quenching process, electrical leakage, and the like. Spurious counts may seriously affect measurement of background counts.

Sputtering A deposition process wherein a surface, or target, is immersed in an inert-gas plasma and is bombarded by ionized molecules that eject surface atoms. The process is based on the disintegration of the target material under ion bombardment. Atoms broken away from the target material by gas ions deposit on the part (substrate), forming a thin film. (Graf).

Square wave 1. An oscillation, the amplitude of which shows periodic discontinuities between two values, remaining constant between jumps.

2. Specifically, in radar a pulse initiated by a rapid rise to peak power, maintained at a constant peak power over the finite pulse length, and terminated by rapid decrease from peak power.

Squeeze film damping Effect of ambient fluid and spacing on the vertical movement of a structural member with respect to a substrate.

Squib 1. Any of various small explosive devices.

2. An explosive device used in the ignition of a rocket. Usually called an igniter.

SS loran Sky-wave synchronized loran, or loran in which the sky wave rather than the ground wave from the master controls the slave. SS loran is used with unusually long baselines.

Stability 1. The property of a body, as an aircraft or rocket, to maintain its attitude or to resist displacement, and, if displaced, to develop forces and moments tending to restore the original condition.

2. Of a fuel, the capability of a fuel to retain its characteristics in an adverse environment, e.g. Extreme temperature.

Stability augmentation system An auxiliary system to the basic manual vehicle control system whereby response of the control surfaces to inputs by the pilot can be adjusted to give a preselected vehicle response by selection of certain fixed gains in a standard feedback loop on control-surface output.

Stabilized data Radar data output corrected for tilt or roll of an unstabilized radar antenna, such as shipboard installations, etc.

Stable Ability to resist collapse and deformation; stability (n.) characteristic of a structure that is able to carry a realistic load without collapsing or deforming significantly.

Stable datum A datum along which all other data align. From any confusion, order and sanity can emerge providing one merely selects a datum, assigns it importance or seniority and then begins to align other data against it. (Read this essay by L. Ron Hubbard: Confusion and the Stable Datum).

The stable datum for any PCB layout could be stated this way: The schematic is the "Bible." In other word, the schematic says the circuit is this way, and the PCB design must follow that pattern perfectly.

Stable isotope An isotope that does not undergo radioactive decay.

Stable platforms A gyroscopic device so designed as to maintain a plane of reference in space regardless of the movement of the vehicle carrying the stable platform.

Stable strictly speaking, A system is termed stable if no rearrangement of its parts can form a system of lower free energy. In practice, the term is used with an implicit proviso regarding the transformations to be considered. Hydrogen is not considered unstable merely because it is subject to nuclear fusion at extreme temperatures. A system is usually regarded as stable (more precisely, as kinetically stable) if its rate of transformation to a state of lower free energy is negligible (by some standard) under the ambient conditions. In nanomechanical systems, a structure can commonly be regarded as stable if it has an extremely low rate of transformations when subjected to its intended operating conditions.

Stack effect Results from the difference in air temperature between indoor and outdoor air during the heating season. Warm air, being lighter than cold air, rises in a building, creating a suction at the base and exerting an outward pressure at the top. The higher the building, the greater the pressure difference across the walls and roof. The suction is greatest at the base, decreasing as the building rises to a neutral pressure plane somewhere between the ground floor and the roof. Above the neutral pressure plane the pressure becomes positive (active outwards) and increases with height, reaching its highest value at the roof. The quantity of air entering the building below the neutral pressure plane is equal to the quantity of air leaving the building above that level

Stadimeter An instrument for determining the distance to an object of know dimension by measuring the angle subtended at the observer by the object. The instrument is graduated directly in distance.

Stage 1. A self-propelled separable element of a rocket vehicle.

2. A step or process through which a fluid passes, especially in compression or expansion.

3. A set of stator blades and a set of rotor blades in an axial-flow compressor or in a turbine; an impeller wheel in a radial-flow compressor.

Stage and a half A liquid-propellant rocket of which only part of the propulsion unit falls away from the rocket vehicle during flight, as in the case of booster rockets falling away to leave the sustainer engine to consume remaining fuel.

Stagnation pressure 1. The pressure at a stagnation point.

2. In compressible flow, the pressure exhibited by a moving gas or liquid brought to zero velocity by an isentropic process.

3. total pressure.

4. impact pressure.

Because of the lack of a standard meaning, stagnation pressure should be defined when it is used.

Stagnation region Specifically, the region at the front of a body moving through a fluid where the fluid has negligible relative velocity.

Stainless steel Any steel containing at least 10.5% Cr as the principal alloying element.

Stance phase During the gait cycle stance phase is when the foot is on the ground, and the body passes over the top of it. It lasts from heel strike to toe off. This section of the gait cycle requires the foot to act as a shock absorber; mobile adapter; rigid lever and pedestal. The stance phase can be sub divided into contact phase and support phase.

Stance time During the gait cycle, stance phase is when the foot is on the ground and the body passes over the top of it. Referred to as support phase or contact phase lasts from heel contact to toe off.

Standard 1. An exact value, or a concept, that has been established by authority or agreement, to serve as a model or rule in the measurement of a quantity or in the establishment of a practice or procedure.

2. A document that establishes engineering and technical limitations and applications for items, materials, processes, methods, design, or engineering practices.

Standard artillery atmosphere A set of values describing atmospheric conditions on which ballistic computations are based: namely, no wind, a surface temperature of 15 degrees C, a surface pressure of 1000 millibars, a surface relative humidity of 78 percent, and a lapse rate which yields a prescribed density-altitude relation.

Standard artillery zone A vertical subdivision of the standard artillery atmosphere. It may be considered a layer of air of prescribed thickness and density.

Standard atmosphere A hypothetical vertical distribution of atmospheric temperature, pressure, and density which, by international agreement, is taken to be representative of the atmosphere for purposes of pressure altimeter calibrations, aircraft performance calculations, aircraft and rocket design, ballistic tables, etc. The air is assumed to be devoid of dust, moisture, and water vapor and to obey the perfect gas law and the hydrostatic equation (the air is static with respect to the earth).

Standard charge The normal C/10 charge used to recharge a cell or battery in 10 hours. Other definitions (charging periods) also apply.

Standard electrode potential (E0) The standard potential E0 of an electrode is the reversible emf between the normal hydrogen electrode and the electrode with all components at unit activity.

Standard error of estimate A measure of the dispersion (scatter) of data points with respect to a curve of regression. S is the positive square root of the arithmetic mean of the squares of the deviations from a curve of regression:

Where d is deviation from R ; R is curve of regression; and n is number of points.

S is a measure of the variation to be expected in making predictions from the regression equation.

Standard hydrogen electrode (SHE) A platinum conductor in contact with a 1 M H+ ions and bathed by hydrogen gas at one atmosphere.

Standard pressure 1. In meteorology, usually a pressure of 1000 millibars, but other pressures may be used as standard for specific purposes.

2. In physics, a pressure of 1 standard atmosphere.

Standard propagation The propagation of radio energy over a smooth spherical earth of uniform dielectric constant and conductivity under conditions of standard refraction in the atmosphere, i.e., an atmosphere in which the index of refraction decreases uniformly with height at a rate of 12 N-units per 1000 feet.

Standard propagation results in a ray curvature due to refection which has a value approximately one-fourth that of the earth's curvature, giving a radio horizon which is about 15 percent greater than the distance to the geometrical horizon. This is equivalent to straight-line propagation over a fictitious earth whose radius is four-thirds the radius of the actual earth.

Standard refraction The refraction which would occur in an idealized atmosphere in which the index of refraction decreases uniformly with height at the rate of 39 X 10^{-6} per kilometer.

Standard refraction may be included in ground-wave calculations by use of an effective earth radius of 8.5 X 10^6 meters, or four-thirds the geometrical radius of the earth.

Standard review plan A document that provides guidance to the staff for reviewing an application to obtain an NRC license to construct or operate a nuclear facility or to possess or use nuclear materials.

Standard temperature 1. A temperature that depends upon some characteristic of some substance, such as the melting, boiling, or freezing point, that is used as a reference standard of temperature.

2. In physics, usually the ice point (0 degrees C); less frequently, the temperature of maximum water density (4 degrees C).

3. In meteorology, this has no generally accepted meaning, except that it may refer to the temperature at zero altitude in the standard atmosphere (15 degrees C).

Standard temperature and pressure (STP). Usually a temperature of 0 degrees C but also used to designate a temperature of 15 degrees C and 1 standard atmosphere.

Standard work instructions A lean tool that enables operators to observe the production process with an understanding of how assembly tasks are to be performed. It ensures that the quality level is understood and serves as an excellent training aid. It enables absentee replacement individuals to easily adapt and perform the assembly operation.

Standardization 1. The act or process of reducing something to, or comparing it with, a standard.

2. A measure of uniformity.

3. A special case of calibration whereby a known input is applied to a device or system for the purpose of verifying the output or adjusting the output to a desired level or scale factor.

Applied to transducers, standardization indicates adjustments of the output to a standard value within specified limits of error.

Standardized work A precise description of each work activity specifying cycle time, takt time, the work sequence of specific tasks and the minimum inventory of parts on hand needed to conduct the activity.

Standby state A state in which the main functions of a circuit have been powered down to save energy, but power remains applied to the circuit ready to make a rapid restart.

Standby vessel Rescue vessel that is always within 5 miles of the rig or platform.

Close standby i.e. closer than 5 miles is provided if activities on the asset increase the risk of man overboard.

Standing wave A periodic wave having a fixed distribution in space which is the result of interference of progressive waves of the same frequency and kind. Such waves are characterized by the existence of nodes or partial nodes and antinodes that are fixed in space.

Stanton number (symbol nst). A number expressing the ratio of the heat transmission perpendicular and parallel to the flow direction, defined as h/cp ñv where h is the heat transfer coefficient, cp is the specific heat, ñ is the density, and v is the flow velocity.

Star 1. A self-luminous celestial body exclusive of nebulas, comets, and meteors; any one of the suns seen in the heavens. Distinguished from planets or planet satellites that shine by reflected light.

2. Any luminous body seen in the heavens.

Star catalogue A listing of stars giving positions for a specified mean equinox and equator. Stars are often identified by catalogue numbers.

Star classification Stars are classified by their spectra, designated by letters, sometimes with numerical subdivisions, as the sun is a G1-type star. The seven main types with their principal spectral characteristics are, in order of decreasing temperature:

O - He II absorption;

B - He I absorption;

A - H absorption;

F - Ca II absorption;

G - strong metallic lines;

K -bands developing;
M - very red.

Also, the letters, P, W, Q, R, N, and S are used to designate comparatively rare types of stars which do not fall into the main series.

Star tracker A telescopic instrument on a rocket or other flight borne vehicle that locks onto a celestial body and gives guidance reference to the vehicle during flight.

Stark effect The broadening or splitting of a spectral line observed when a luminous gas is acted upon by a strong electric field.

Starting pressure In rocketry, the minimum chamber pressure required to establish shock-free flow in the exit plane of a supersonic nozzle.

State A physical system is said to be in a particular state when its physical properties fall within some particular range; the boundaries of the range defining a state depend on the problem under consideration. In a classical world, each point in phase space could be said to correspond to a distinct state. In the real world, time-invariant systems in quantum mechanics have a set of discrete states, particular superpositions of which constitute complete descriptions of the system. In practice, broader boundaries are usually drawn. A molecule is often said to be in a particular excited electronic state, regardless of its state of mechanical vibration. In nanomechanical systems, the PES often corresponds to a set of distinct potential wells, and all points in configuration space within a particular well can be regarded as one state. Definitions of state in the thermodynamics of bulk matter are analogous, but extremely coarse by these standards.

State of health SOH A measurement that reflects the general condition of a battery and its ability to deliver the specified performance compared with a fresh battery. It takes into account such factors as charge acceptance, internal resistance, voltage and self-discharge. It is not as precise as the SOC determination.

State of the art The level to which technology and science have at any designated cutoff time been developed in a given industry or group of industries.

State variable Any independent variable in a problem which must be specified to define a condition of state, as for example a component of position.

Statement of Work (SOW) A document primarily for use in procurement, which specifies the work requirements for a project or program. It is used in conjunction with specifications and standards as a basis for a contract. The SOW will be used to determine whether the contractor meets stated performance requirements.

Static 1. Involving no variation with time.

2. Involving no movement, as in static test.

3. Any radio interference detectable as noise in the audio stage of a receiver.

Static pressure (symbol p) 1. The pressure with respect to a stationary surface tangent to the mass-flow velocity vector.

2. The pressure with respect to a surface at rest in relation to the surrounding fluid.

Static testing The testing of a rocket or other device in a stationary or hold-down position, either to verify structural design criteria, structural integrity, and the effects of limit loads or to measure the thrust of a rocket engine.

Static unbalance Static unbalance is that condition of unbalance for which the central principal axis is displayed only parallel to the shaft axis.

Station constants In tracking and telemetry, constants usually associated with instrumentation sites, e.g., survey coordinates, zeroing correction, etc.

Station error In geodesy and surveying, the difference, usually negligible, between the astronomical and geodetic latitudes, due to local gravitational anomalies.

Station pressure The atmospheric pressure computed for the level of the station elevation.

This may or may not be the same as either the climatological station pressure or the actual pressure, the difference being attributable to the difference in reference elevations. Station pressure usually is the base value from which sea-level pressure and altimeter setting are determined.

Stationary orbit An orbit in which the satellite revolves about the primary at the angular rate at which the primary rotates on its axis. From the primary, the satellite thus appears to be stationary over a point on the primary.

A stationary orbit with respect to the earth is commonly called a 24-hour orbit.

Stationary wave A standing wave in which the net energy flux is zero at all points.

Stationary waves can only be approximated in practice.

Statistical mechanics Statistical mechanics treats the detailed state of a system (its quantum state or, in classical models, its position in phase space) as unknown and subject to statistical uncertainties; entropy is a measure of this uncertainty. Statistical mechanics describes the distribution of states in an equilibrium system at a given temperature (describing either the distribution of probabilities of quantum states or the probability density function in phase space), and can be used to derive thermodynamic properties from properties at the molecular level. These equilibrium results are useful in nanomechanical design.

Stator In machinery, a part or assembly that remains stationary with respect to a rotating or moving part or assembly such as the field frame of an electric motor or generator, or the stationary casing and blades surrounding an axial-flow-compressor rotor or turbine wheel; a stator blade.

Status accounting The process of documenting the correct, approved status of the system, including a historical record of the development of specifications and all approved changes.

Status signals Logic signals that indicate normal or abnormal conditions of operation, including: ac low, dc low, ac ok, dc ok, overtemperature, overvoltage, under temperature, overcurrent.

Steady state 1. The condition of a substance or system whose local physical and chemical properties do not vary with time.

2. Specifically, the stable operating condition of a reactor in which the neutron inventory remains constant; that is, the effective multiplication factor ke is equal to 1.

Steady state diffusion The diffusion condition for which there is no net accumulation or depletion of diffusing species. The diffusion flux is independent of time.

Steady-state vibration A condition that exists in a system if the velocity of each particle is a continuing periodic quantity.

Steam generator The heat exchanger used in some reactor designs to transfer heat from the primary (reactor coolant) system to the secondary (steam) system. This design permits heat exchange with little or no contamination of the secondary system equipment.

Steering function An empirical relation based on the relative distance and velocity of the target, used in guidance of rockets and spacecraft.

Steering ratio A predetermined ratio of the steering gears. Usually, the lower the steering ratio, the quicker the response.

Stellar inertial guidance The guidance of a flight-borne vehicle by a combination of celestial and inertial guidance; the equipment which accomplishes the guidance.

Stellar map matching A process during the flight of a vehicle by which a chart of the stars set into the guidance system is automatically matched with the position of the stars observed through telescopes so as to give guidance to the vehicle.

Stellarator machine An experimental thermonuclear device where containment in a magnetic field is achieved by closing the field upon itself and thus allowing the particles to perform endless spiral motion.

Stellite Cobalt based alloy commonly applied as coating to valve trims to provide high resistance to wear; corrosion and high temperatures.

Step down transformer A transformer with a turns ratio more than one. The output voltage is less than the input voltage.

Step length This is the distance by which the normal foot moves forward in front of the other one. In pathological gait, it is perfectly possible for the two lengths to be different.

Step response The response of a system to an instantaneous jump in the input signal.

Steppage This is a very simple swing phase modification, consisting of exaggerated knee and hip flexion to lift the foot higher than usual for increased ground clearance. It is particularly used to compensate for a plantar flexed ankle, where the dorsiflexors are inadequate to control foot drop.

Steradian The unit solid angle which cuts unit area from the surface of a sphere of unit radius centered at the vertex of the solid angle. There are 4 steradians in a sphere.

Stereolithography A rapid prototyping (RP) process, introduced in 1987 by 3D Systems Inc. which launched the RP industry. A Stereolithography Apparatus (SLA) machine builds physical models in this manner: it focuses an ultraviolet (UV) light onto the surface of a vat filled with liquid photopolymer. The light beam, moving under computer control, draws each layer of an object onto the surface of the liquid. Wherever the beam strikes the surface, liquid changes to solid. 3D parts are built from the bottom up, one layer at a time; when the part is finished, it is exposed to UV light for curing.

Steric Pertaining to the spatial relationships of atoms in a molecular structure, and in particular, to the space-filling properties of a molecule. If molecules were rigid and had hard surfaces, steric properties would merely be an opaque way of saying "shape"; a flexible side-chain, however, has definite steric properties but no fixed shape. Nanomechanical systems make extensive use of the steric properties of relatively rigid molecules, for which the term "shape" has essentially its conventional meaning so long as one remembers that the surface interactions are soft on small length-scales.

Steric hindrance Slowing of the rate of a chemical reaction owing to the presence of structures on the reagents that mechanically interfere with the motions associated with the reaction, typically by obstructing the reaction site.

Stiffness The stiffness of a system with respect to a deformation (e.g., the stiffness of a spring with respect to stretching) is the second derivative of the energy with respect to the corresponding displacement; this measures the curvature of the potential energy surface along a particular direction. Positive stiffness is associated with stability, and a large stiffness can result in a small positional uncertainty in the presence of thermal excitation. Negative stiffnesses correspond to unstable locations on the potential energy surface. Alternative terms for stiffness include force gradient and rigidity.

Stirling cycle A theoretical heat engine cycle in which heat is added at constant volume, followed by isothermal expansion with heat addition. The heat is then rejected at constant volume, followed by isothermal compression with heat rejection.

If a regenerator is used so that heat rejected during the constant-volume process is recovered during heat addition at constant volume, the thermal efficiency of the Stirling cycle is the same as for the Carnot cycle, with less compressive work needed.

Stochastic effects Effects that occur by chance, generally occurring without a threshold level of dose, whose probability is proportional to the dose and whose severity is independent of the dose. In the context of radiation protection, the main stochastic effects are cancer and genetic effects.

Stochastic process An ordered set of observations in one or more dimensions, each being considered as a sample of one item from a probability distribution.

Stoichiometric Of a mixture of chemicals, having the exact proportions required for complete chemical combination, applied especially to combustible mixtures used as propellants.

Stoichiometry For ionic compounds, the state of having exactly the ratio of cations to anions specified by the chemical formula. Stoichiometric quantities refers to quantities of reactants mixed in exactly the correct amounts so that all are used up at the same time.

Stonewall The maximum stable flow and maximum head condition for a centrifugal compressor.

Stooping An atmospheric refraction phenomenon, a mirage; a special case of sinking in which the curvature of light rays due to atmospheric refraction decreases with elevation so that the visual image of a distant object is foreshortened in the vertical.

The opposite of stooping is towering.

Stop bit A signal following a character or block that prepares the receiving device to receive the next character or block.

STOP Safety Training and Observation Programme. DuPont initiated scheme designed to create a culture of spotting potential hazards before they cause an accident.

Stopping point The end of the highly luminous path of a visual meteor. Also called Hemmungspunkt.

Storable Of a liquid; subject to being placed and kept in a tank without benefit of special measures for temperature or pressure control, as in storable propellant.

Storage 1. The act of storing information.

2. Any device in which information can be stored. Also called a memory device.

3. In a computer, a section used primarily for storing information. Such a section is sometimes called a memory or a store.

The physical means of storing information may be electrostatic, ferroelectric, magnetic, acoustic, optical, chemical, electronic, electrical, mechanical, etc., in nature.

Storage capacity The amount of information, usually expressed in bits (i.e., the log2 of the number of distinguishable states in which the storage can exist), that can be retained in storage. Also called memory capacity.

Storage life The length of time a cell or battery can be stored on open circuit without permanent deterioration of its performance.

Storage temperature The range of ambient temperatures through which an inoperative power supply can remain in storage without degrading its subsequent operation.

Store 1. To retain information in a device from which it can later be withdrawn.

2. To introduce information into such a device.

3. A container, rocket, bomb, or vehicle carried externally in a craft.

STP (Standard Pressure and Temperature) STP is used in many thermodynamic calculations and tabulations and is defined as 0 degrees Celsius and 1 atmosphere of pressure.

Straddle carrier A ground vehicle that carries its load suspended between its wheels.

Strain In mechanical engineering, strain is a measure of the deformation resulting from stress (that is, force per unit area); the displacement of one point with respect to another, divided by their equilibrium separation in the absence of stress. In chemistry, a molecular fragment generally has some equilibrium geometry (bond lengths, interbond angles, etc.) when the rest of the molecular structure does not impose special constraints (e.g., bending bonds to form a small ring). Deviations from this equilibrium geometry are described as strain, and increase the energy of the molecule. Strain in the mechanical engineering sense causes strain in the chemical sense.

Strain gauge An element (wire or foil) that measures a strain based on electrical resistance changes of the gauge that result from a change in length or dimension strain of the wire or foil.

Stratosphere radiation Any infrared radiation involved in the complex infrared exchange continually proceeding within the stratosphere.

Stratosphere The atmosphere from approximately 12 km to 70 km. The temperature of the atmosphere increases in this region.

Streamline A line whose tangent at any point in a fluid is parallel to the instantaneous velocity vector of the fluid at that point. The differential equations of the streamlines may be written dr X v = 0, where dr is an element of the streamline and v is the velocity vector; or in Cartesian coordinates, dx / u = dy /v = dz /w, where u, v, w, are the fluid velocities along the orthogonal X, Y, Z axes, respectively.

In steady-state flow the streamlines coincide with the trajectories of the fluid particles; otherwise, the streamline pattern changes with time.

Stress Force per unit area applied by one part of an object to another. Pressure is an isotropic compressive stress. Suspending a mass from a fiber places it in tensile stress. Gluing a layer of rubber between two plates and then sliding one over the other (while holding their separation constant) places the rubber in shear stress.

Stress corrosion (cracking) A form of failure resulting from the combined action of a tensile stress and a corrosion environment, occurring at lower stress levels than required when the corrosion environment is absent.

Stress intensity factor (K) A scale factor to define the magnitude of the crack-tip stress field; the functionality depends on the configuration of the cracked component and the manner in which the loads are applied.

Stress tensor The complete set of stress components in a solid or fluid medium, which are written as a tensor ôij. It has nine components, one for each of the coordinate faces of an imaginary element upon which the stress acts (j = x, y, z) and for each direction in which the stress is directed (i = x, y, z).

By definition, an inviscid fluid is one in which the six tangential stresses (i +/- j) are zero, and the three normal stresses (i = j) are equal to the negative of the pressure.

Stretcher A masonry unit laid with its greatest dimension horizontal and its face parallel to the wall face.

Stretchout An action whereby the time for completing an action, especially a contract, is extended beyond the time originally programmed or contracted for.

Strictly One Ampere hour is the charge transferred by one amp flowing for one hour. 1Ah = 3600 Coulombs.

Stride length This is the distance between two successive placements of the same foot. It consists of two step lengths. The stride length measured between successive positions of the left foot must always be the same as that measured by the right foot, unless the subject is walking in a curve.

Stringer A slender, lightweight, lengthwise fill-in structural member in a rocket body, or the like, serving to reinforce and give shape to the skin.

Strong acid An acid that, for practical purposes, ionizes completely under the conditions of interest. Common strong acids are hydrochloric, nitric, and sulfuric.

Strouhal number A nondimensional parameter important in vortex meter design defined as: s = Fh/V where f = frequency, V = velocity, and h = a reference length.

Struck joint Any mortar joint which has been finished with a trowel.

Structural engineer An engineer who investigates the behavior and design of all kinds of structures, including dams, domes, tunnels, bridges, and skyscrapers, to make sure they are safe and sound for human use.

Structural integrity The component must itself be capable of resisting the imposed load or must be supported by one that can. It must be capable of resisting the strongest loads or combinations acting directly or indirectly on it without rupturing or detaching itself away from its supported and supporting elements, or fail in creep. The component must be sufficiently rigid to resist displacement.

Structural volume The interior of a diamondoid structure typically consists of a dense network of covalent bonds; a larger excluded volume, however, is determined by nonbonded repulsions at the surface. The structural volume corresponds to a region smaller than the excluded volume, chosen to make properties such as the strength and modulus nearly size independent by correcting for surface effects.

Strut The main support member in a MacPherson suspension system. The strut also serves as the shock absorber.

Sub panel A group of printed circuits (called modules) arrayed in a panel and handled by both the board house and the assembly house as though it were a single printed wiring board. The sub-panel is usually prepared at the board house by routing most of the material separating individual modules, leaving small tabs. The tabs are strong enough so that the sub-panel can be assembled as a unit, and weak enough so that final separation of assembled modules is easily done.

Subatomic particle Any particle of less than atomic mass, e.g., the electron, proton, and neutron, also called atomic particle.

Subatomic particles are classified by relative mass into four groups: leptons, mesons, nucleons, and hyperons, from lowest to highest masses, respectively.

Subaudio frequency A frequency below the audiofrequency range, below about 15 cycles per second.

Subcarrier A carrier which is applied as a modulating wave to modulate another carrier or an intermediate subcarrier.

Subcarrier oscillator In a telemetry system, the oscillator which is directly modulated by the measurand or by the equivalent of the measurand in terms of changes in the transfer elements of a transducer.

Subcommutation In telemetry, commutation of additional channels with output applied to individual channels of the primary commutator.

Subcommutation is called synchronous if its rate is a submultiple of that of the primary commutator. Unique identification must be provided for the subcommutation frame pulse.

Subcontractor An individual, partnership, corporation, or an association that contracts with an organization (i.e., the prime contractor) to design, develop, and/or manufacture one or more products.

Subcritical mass An amount of fissionable material insufficient in quantity or of improper geometrical configuration to sustain a fission chain reaction.

Subcriticality The condition of a nuclear reactor system when the rate of production of fission neutrons is lower than the rate of production in the previous generation owing to increased neutron leakage and poisons.

Subframe In telemetry, a complete sequence of frames during which all subchannels of a specific channel are sampled once.

Subharmonic A sinusoidal quantity having a frequency that is an integral submultiple of the fundamental frequency of a periodic quantity to which it is related.

Sublimation The transmission of a substance directly from the solid state to the vapor state, or vice versa, without passing through the intermediate liquid state.

Sublunar point The geographical position of the moon; that point on the earth at which the moon is in the zenith at a specified time.

Subluxation An incomplete dislocation of a joint. A gradual displacement or partial dislocation produced within a joint, that is contrary to its plane or motion, or that exceeds the range of motion of that joint.

Subrefraction Less-than-normal refraction, particularly as related to atmospheric refraction.

Greater-than-normal refraction is called super refraction.

Subroutine A set of instructions necessary to direct a computer to carry out a well-defined mathematical or logical operation; a submit of a routine, usually coded in such a manner that it can be treated as a black box by the routine using it.

Subsolar point The geographical position of the sun; that point on the earth at which the sun is in the zenith at a specified time.

Subsonic In aerodynamics, of or pertaining to, or dealing with speeds less than acoustic velocity as in subsonic aerodynamics.

Subsonic flow Flow of a fluid, as air over an airfoil, at speeds less than acoustic velocity.

Aerodynamic problems of subsonic flow are treated with the assumption that air acts as an incompressible fluid.

Substandard propagation The propagation of radio energy under conditions of substandard refraction in the atmosphere, that is, refraction by an atmosphere or section of the atmosphere in which the index of refraction decreases with height at a rate of less than 12 N-units per 1000 feet.

Substandard propagation produces a less-than-normal downward bending or even an upward bending of radio waves as they travel through the atmosphere, giving closer radio horizons and decreased radar and radio coverage. It results primarily when propagation takes place through a layer in which moisture remains constant or increases with height.

Substellar point The geographical position of a star; that point on the earth at which the star is in the zenith at a specified time. Also called subastral point.

Substrate level phosphorylation The synthesis of the energy storage compound adenosine triphosphate (ATP) from adenosine diphosphate (ADP) using organic substrates without molecular oxygen.

Substrate The supporting material on or in which the parts of an integrated circuit are attached or made. The substrate may be passive (thin film , hybrid) or active (monolithic compatible).

Subtend To be opposite, as an arc of a circle subtends an angle at the center of the circle , the angle being formed by the radii joining the ends of the arc with the center.

Sudden commencement magnetic storm A magnetic storm characterized by a world-wide sudden commencement which takes place in a matter of minutes and shows no recurrence after 27 days, the period of rotation of the sun.

Sudden ionospheric disturbance (SID). A complex combination of sudden changes in the condition of the ionosphere and the effects of these changes.

The following are the most important effects accompanying a sudden ionospheric disturbance: (a) radio fadeout, a condition in which there is a marked and abrupt increase in absorption in the D-region for high frequency radio waves (2 to 3 megacycles) and a consequent loss of long-distance radio reception in this range of frequencies; (b) magnetic crotchet, a sudden change in the earth's magnetic field due to an increase in the conductivity of the lower ionosphere, the change being in the nature of an augmentation of the normal quiet-day magnetic change; (c) sudden enhancements of long-wave atmospherics recorded in the frequency range between 10 and 100 kilocycles due to the improved reflectivity at oblique incidence of the D-region for such low-frequency radio waves; (d) sudden phase anomalies of discrete low-frequency radio waves (10 to 100 kilocycles) due to descent of the D-layer; and (e) sudden field-strength anomalies of distant low-frequency radio signals (10 to 100 kilocycles) due to interference between the groundwave and the skywave. A sudden ionospheric disturbance usually occurs a few minutes after a solar flare and is noted only on the sunlit side of the earth. The return of the ionosphere to its normal condition following a pronounced sudden ionospheric disturbance usually takes from half an hour to an hour, sometimes longer.

Sulphation Growth of lead sulphate crystals in Lead-Acid batteries which inhibits current flow. Sulphation is caused by storage at low state of charge.

Summer solstice 1. That point on the ecliptic occupied by the sun at maximum northerly declination. Sometimes called June solstice, first point of Cancer.

2. That instant at which the sun reaches the point of maximum northerly declination, about June 21.

Sun The star at the center of the solar system, around which the planets, planetoids, and comets revolve. It is a G-type star.

The sun visible in the sky is called apparent or true sun. A fictitious sun conceived to move eastward along the celestial equator at a rate that provides a uniform measure of time equal to the average apparent time is called mean sun; a fictitious sun conceived to move eastward along the ecliptic at the average rate of the apparent sun is called dynamical mean sun.

Sun tracker A species of star tracker designed to lock onto the sun to afford guidance to a rocket or other flight-borne object.

Sunspot A relatively dark area on the surface of the sun consisting of a dark central umbra surrounded by a penumbra which is intermediate in brightness between the umbra and the surrounding photosphere.

Sunspots usually occur in pairs with opposite magnetic polarities. They have a lifetime ranging from a few days to several months. Their occurrence exhibits approximately an 11-year period (the sunspot cycle).

Sunspot cycle A cycle with an average length of 11.1 years but varying between about 7 and 17 years in the number and area of sunspots, as given by the relative sunspot number. This number rises from a minimum of 0 to 10 to a maximum of 50 to 140 about 4 years later, and then declines more slowly.

An approximate 11-year cycle has been found or suggested in geomagnetism, frequency of aurora, and other ionospheric characteristics. The u-index of geomagnetic intensity variation shows one of the strongest known correlations to solar activity. Eleven-year cycles have been suggested for various tropospheric phenomena, but none of these has been substantiated.

Super heating 1. The heating of a liquid above its boiling temperature without the formation of the gaseous phase.

2. The heating of the gaseous phase considerably above the boiling-point temperature to improve the thermodynamic efficiency of a system.

Superadiabatic lapse rate An environmental lapse rate greater than the dry-adiabatic lapse rate, such that potential temperature decreases with height.

Superalloy An alloy developed for very high temperature service where relatively high stresses (tensile, thermal, vibratory, and shock) are encountered and where oxidation resistance is frequently required.

Supercharger Supercharging is the compression of an engine's intake charge above atmospheric pressure by means of an air pump driven by a crankshaft. This is not to be confused with a turbocharger which is an air pump that is exhaust driven. A supercharger can provide boost faster than a turbo and over a much broader engine rpm range. The disadvantages of supercharging are higher power demands, more mechanical noise and more complex control requirements.

Supercommutation Commutation by connection of single data input source to equally spaced contacts of the commutator (cross-patching).

Corresponding crosspatching is required at the decommutator.

Superconductivity A phenomenon occurring below a very low, characteristic critical temperature in certain materials (superconductors), characterised by the complete absence of electrical resistance and the damping of the interior magnetic field (the Meissner effect). Superconductors can carry currents that will not decay.

Supercriticality The condition for increasing the level of operation of a reactor. The rate of fission neutron production exceeds all neutron losses, and the overall neutron population increases.

Superficial Confined to or pertaining to the surface

Superheating The heating of a vapor, particularly steam, to a temperature much higher than the boiling point at the existing pressure. This is done in some power plants to improve efficiency and to reduce water damage to the turbine.

Superior mirage A spurious image of an object formed above its true position by abnormal atmospheric refraction conditions; opposite to an inferior mirage. Compare towering, looming, inferior mirage.

Superior mirages occur when the temperature lapse rate near the earth's surface is less than its normal value or, especially, when the temperature actually increases with height. Under these conditions the velocity of light increases upward in such a way that light rays are bent downward as they propagate through the layer in quasi-horizontal directions. The downward curvature gives the impression that the position of the object viewed is well above its true position in space. The object also appears inverted. Complex combination of superior and inferior mirages may occur with unusual density stratifications.

Supersonic compressor A compressor in which a supersonic velocity is imparted to the fluid relative to the rotor blades, the stator blades, or to both the rotor and stator blades, producing oblique shock waves over the blades to obtain a high pressure rise.

Supersonic diffuser A diffuser designed to reduce the velocity and increase the pressure of fluid moving at supersonic velocities.

Supersonic flow In aerodynamics, flow of a fluid over a body at speeds greater than the acoustic velocity and in which the shock waves start at the surface of the body. Compare hypersonic flow.

Superstandard propagation The propagation of radio waves under conditions of superstandard refraction (superrefraction) in the atmosphere, that is, refraction by an atmosphere or section of the atmosphere in which the index of refraction decreases with height at a rate of greater than 12 N-units per 1000 feet.

Superstandard propagation produces a greater-than-normal downward bending of radio waves as they travel through the atmosphere, giving extended radio horizons and increased radar coverage. It results primarily from propagation through layers near the earth's surface in which the moisture lapse rate is greater than normal, or the temperature lapse rate less than normal, or both. A condition in which warn dry air moves out over a cool water surface is an example of superrefraction. A layer in which the downward bending is greater than the curvature of the earth is called a radio duct. Frequently, the general term, anomalous propagation, is used for superstandard propagation.

Superstandard refraction Refraction by an atmosphere or section of the atmosphere in which the index of refraction decreases with height at a rate greater than 12 N-units per 1000 feet. Also called superrefraction.

Supination It is a component of motion in all three cardinal planes, none of the three component motions can take place independently of the other two. The motion of supination is a complex triplanar motion consisting of plantar flexion, adduction and inversion, all three motions occur simultaneously. The axis of motion is identical to that of pronation.

Supplement An angle equal to 180 degrees minus a given angle. Thus 110 degrees is the supplement of 70 degrees and the two are said to be supplementary.

Supplementary angles Two angles whose sum is 180 degrees.

Supplementary insulation As defined by Underwriters Laboratories, Inc., an independent insulation provided in addition to the basic insulation to protect against electric shock in case of mechanical rupture or electrical breakdown of the basic insulation. An enclosure of insulating material may form a part of the whole of the supplementary insulation.

Suppliers The term 'suppliers' includes contractors, sub-contractors, vendors, developers, sellers or any other term used to identify the source from which products or services are obtained.

Supply chain management The procurement, stocking and distribution of components, subassemblies and products throughout the design, manufacturing, and distribution stages, ensuring that the correct components, subassemblies and products are delivered to their appropriate destination at the proper time, the lowest overall cost, and acceptable quality levels.

Surface 1. A two-dimensional extent; the outside or superficies of any body; especially, the surface of the earth, either land or water, used in combinations as surface-to-air, etc.

2. A wing, rudder, propeller blade, vane, hydrofoil, or the like - applied in this sense to the entire structure or body.

Surface boundary layer That thin layer of air adjacent to the earth's surface, extending up to the so-called anamometer level (the base of the Ekman layer). Within this layer the wind distribution is determined largely by the vertical temperature gradient and the nature and contours of the underlying surface; shearing stresses are approximate constant. Also called surface layer, friction layer, atmospheric boundary layer, ground layer.

Surface insulation resistance The electrical resistance of an insulating material between a pair of contacts or conductors; SIR is determined under specified environmental and electrical conditions.

Surface modeling A 3D modeling technique to describe geometry by its surfaces. This is typically used where surface shape is critical such as the design of auto body panels and aerostructures and industrial design.

Surface mount Surface mount technology. The technology of creating printed wiring wherein components are soldered to the board without using holes. The result is higher component density, allowing smaller PWB 's. Abbreviated SMT.

Surface mount technology Method of assembling printed circuit boards where the components are mounted onto the surface of the board rather than being inserted into holes in the board.

Surface mounted devices (SMD) A family of components intended to be mounted directly upon the surface of a substrate or circuit board.

Surface plasmon A collective motion of electrons in the surface of a metal conductor, excited by the impact of light of appropriate wavelength at a particular angle.

Surface water Water which is contained in lakes, rivers, and oceans.

Surge An unstable operating condition when the flow through a compressor is decreased to the point that momentary flow reversals occur.

Can lead to major damage of compressor.

Surge current A current of short duration that occurs when power is first applied to capacitive loads or temperature dependent resistive loads such as tungsten or molybdenum heaters-usually lasting no more than several cycles.

Survey The process of determining accurately the position, extent, contour, etc., of an area, usually for the purpose of preparing a chart.

Survey meter Any portable radiation detection instrument especially adapted for inspecting an area or individual to establish the existence and amount of radioactive material present.

Suspended growth reactor A reactor in which the microorganisms are suspended in the wastewater. Examples of suspended growth reactors are activated sludge reactors and anaerobic digesters.

Suspension In physical chemistry, a system composed of one substance (suspended phase, suspensoid) dispersed throughout another substance (suspending phase) in a moderately finely divided state, but not so finely divided as to acquire the stability of a colloidal system.

Given sufficient time, a suspension will, be definition, separate itself by gravitational action into two visibly distinct portions, whereas a colloidal system, by definition, is stable. Dust in the atmosphere is an example of a suspension of a solid in a gas.

Suspension bridge A bridge in which the roadway deck is suspended from cables that pass over two towers; the cables are anchored in housings at either end of the bridge.

Suspension effect The source of error due to varied reference liquid junction potential depending upon whether the electrodes are immersed in the supernatant fluid or deeper in the sediment. Normally encountered with solutions containing resins or charged colloids.

Suspension system Includes springs, shock absorbers/struts, and linkage used to suspend a vehicle's frame, body, engine and drivetrain above the wheels.

Sustainer Anything that acts to sustain an action or movement already begun; specifically, a sustainer engine.

Sustainer engine A rocket engine that maintains the velocity of a rocket vehicle once it has achieved its programmed velocity by use of booster or other engine.

This term is applied, for example, to the remaining engine of the Atlas after the two booster engines have been jettisoned. The term is also applied to a rocket engine used on an orbital flider to provide the small amount of thrust now and then required to compensate for the drag imparted by air particles in the upper atmosphere.

Sustaining One of the 5S's used for workplace organization. Sustaining is the continuation of sifting, sweeping, sorting and sanitizing. It is the most important and the most difficult, because it addresses the need to perform the 5S's on an on-going and systematic basis.

Sustentaculum tali A process of the calcaneum supporting the talus

Sweep The motion of the visible dot across the face of a cathode-ray tube, as a result of deflections of the electron beam.

A linear time-base sweep has a constant sweep speed before retrace. An expanded time-base sweep is produced if the sweep speeds is increased during a selected part of the cycle; a delayed time-base sweep if the start of the sweep is delayed, usually to provide an expanded scale for a particular part. A sweep intended primarily for measurement of range may be called a range sweep.

Swing phase The swing phase occupies about 40% of the gait cycle. Divided into acceleration phase and deceleration phase separated by mid swing. Swing phase of the right foot corresponds with stance phase of the left foot. The walking velocity depends to a large extent on the efficiency of the swing phase, since stride length is the amount by which the foot can be moved during this time.

Swing time This accounts for 40% of the gait cycle and includes toe off, acceleration through to mid swing then deceleration, ending with heel contact of the opposite foot. As cadence alters swing phase time alters directly.

Swing-around trajectory A planetary round trip trajectory which requires no propulsion at the destination planet, but uses the planet's gravitational field to effect the necessary orbit change to return to earth.

Switch mode charger Charger which uses a switch mode regulator. More efficient but more costly than a Linear charger.

Switch mode regulator A switching regulator is a voltage regulator which uses an output stage, switched repetitively on and off, together with energy storage components (capacitors and inductors) to generate a DC output voltage. Regulation is achieved through Pulse Width Modulation (PWM). Output voltages can be generated that are greater than or less than the input voltage, and multiple output voltages can be generated with a single regulator.

Switching power supply "Switchers" are typically smaller (by about 10X), lighter (by about 10X) and generate less heat (by about 5X) compared to an equivalent linear supply. At output power ratings above about 25W they are usually less expensive. However, their transient response is slower, output regulation less precise, and have higher PARD (noise). These latter issues aren't usually a problem for most digital circuits, but must be considered when powering sensitive analog circuits.

Switching regulator A switching circuit that operates in a closed loop system to regulate the power supply output.

Synchrocyclotron A cyclotron in which the frequency of the electric field is frequency modulated to permit the acceleration of particles to relativistic energies. Also called FM cyclotron.

Synchronism The relationship between two or more periodic quantities of the same frequency when the phase difference between them is zero or constant at a predetermined value.

Synchronous computer A computer in which the starting time of every ordinary operational cycle is controlled by signals which occur at regular intervals. Contrast with asynchronous computer.

Synchronous rectification A rectification scheme in a switching power supply in which a FET or bipolar transistor is substituted for the rectifier diode to improve efficiency.

Synchronous satellite An equatorial west-to-east satellite orbiting the earth at an altitude of approximately 35,900 kilometers at which altitude it makes one revolution in 24 hours, synchronous with the earth's rotation.

Synchrotron A device for accelerating particles, ordinarily electrons, in a circular orbit in an increasing magnetic field by means of an alternating electric field applied in synchronism with the orbital motion.

Synergic curve A curve plotted for the ascent of a rocket vehicle calculated to give the vehicle an optimum economy in fuel with an optimum velocity.

This curve, plotted to minimize air resistance, starts off vertically, but bends towards the horizontal between 20 and 60 miles altitude to minimize the thrust required for vertical ascent.

Synergism Is the act of working together. Two chemicals which are synergistic have a greater effect together than the sum of their individual effects. The effect can be either positive or negative.

Synodic period The interval of time between and planetary configuration of a celestial body, with respect to the sun, and the next successive same configuration of that body, as from inferior conjunction to inferior conjunction.

Synodic satellite A hypothetical earth satellite, situated 0.84 of the distance to the moon on a line joining the centers of the earth and moon and having the same period of revolution as the moon.

Synodical month The average period of revolution of the moon about the earth with respect to the sun, a period of 29 days 12 hours 44 minutes 2.8 seconds.

This is sometimes called the month of the phases, since it extends from new moon to the next new moon. Also called lunation.

Synoptic meteorology The study and analysis of weather information gathered at the same time.

Synoptic Pertaining to or affording an overall view:

In meteorology, this term refers to meteorological data obtained simultaneously over a wide area for the purpose of presenting a comprehensive and nearly instantaneous picture of the state of the atmosphere. Thus, to a meteorologist, synoptic takes on the additional connotation of simultaneity.

Synthesis The translation of input requirements (including performance, function, and interface) into possible solutions (resources and techniques) satisfying those inputs. Defines a physical architecture of people, product and process solutions for logical groupings of requirements (performance, functions, and interface) and their designs for those solutions.

System design The process of designing a system that comprises the interaction and integration of subsystems and subassemblies into a single system that performs an intended function. The sub-assemblies can consist of electrical, mechanical, optical, software, and other components to achieve overall functionality.

System design authority The part of the developing contractors organization which is responsible for the overall design of the CIs under development.

System elements A system element is a balanced solution to a functional requirement or a set of functional requirements and must satisfy the performance requirements of the associated item. A system element is part of the system (hardware, software, facilities, personnel, data, material, services, and techniques) that, individually or in combination, satisfies a function (task) the system must perform.

System Integration The successive combining and testing of hardware and software system components in a prescribed manner to prove compatibility and performance.

System Integration Team A System Integration Team is a higher-level IPT Integrated Product Team) used in a larger program which flows down requirements and workscope in individual IPT's, monitors and coordinates their activities from a technical perspective, resolves interface and integration issues, and redirects technical activities when required to assure that the development work is accomplished to meet the overall system requirements.

System of astronomical constants An interrelated group of values constituting a model of the earth and the motions which together with the theory of celestial mechanics serves for the calculations of ephemerides.

System requirement - A condition or capability that must be met or possessed by a system or system component to satisfy a condition or capability needed by a user to solve a problem.

Systematic error An error that is always a function of the magnitude of the quantity observed.

When the error is constant it is called a bias error. Systematic errors are often caused by false elements in an instrument. An example is an eccentrically mounted azimuth circle or an azimuth circle with graduation errors.

Systems engineering Systems engineering is the process of specifying the system requirements, allocating the system requirements to the hardware and software components, specifying the interfaces between the hardware and software components, and monitoring the design and development of these components to ensure conformance with their specifications. Systems engineering transforms an operational need into a description of system performance parameters and a system configuration through the use of an iterative process (e.g., definition, syntheses, analysis, design, test and evaluation, etc.); integrates related technical parameters and assure compatibility of all physical, functional, and program interfaces in a manner which optimizes the total system definition and design; and integrates reliability, maintainability, safety, human, and other such factors into the total engineering effort.

Systems engineering management process group A group of specialists who facilitate the definition, maintenance, and improvement of the systems engineering process used by the organization. In the key practices, this group is generically referred to as "the group responsible for the organization's systems and software engineering process activities".

Systems security engineering An element of system engineering that applies scientific and engineering principles to identify security vulnerabilities and minimize or contain risks associated with these vulnerabilities. It uses mathematical, physical, and related scientific disciplines, and principles and methods of engineering design and analysis to specify, predict, and evaluate the vulnerability of the system to security threats.

Systems specification A system level requirements specification. A system specification may be a System/subsystem specification, Prime Item Development Specification, or a Critical Item Development Specification.

Syzygy A point of the orbit or a planet or satellite at which it is in conjunction or opposition.

The term is used chiefly in connection with the moon, when it refers to the points occupied by the moon at new and full phase.

T

T Symbol for temperature in 0C.

T Variously, the symbol for transformer, absolute temperature. The abbreviation for Tesla

TAB bonding Tape automated bonding; semiconductor packaging technique that uses a tiny lead-frame to connect circuitry on the surface of the chip to a substrate instead of wire bonds.

Tabs Flat connectors used on pouch cells.

Tachometer An instrument for measuring the speed of the engine crankshaft in revolutions per minute (RPM).

Tafel equation The relationship between the internal electrode potentials in a battery and the current which flows. This is an exponential relationship based on empirical results which quantifies the elecrochemical reactions. It is analogous to the Arrhenius equation which quantifies the thermochemical process relating the temperature to the rate at which a chemical action progresses.

Taguchi methods A quality engineering methodology developed by Genichi Taguchi that includes off-line quality control, on-line quality control, and system of experimental design to reduce costs and improve quality. Taguchi methods are not just a statistical application of design of experiments. Taguchi methods include the integration of statistical design of experiments into a powerful engineering process. The goal is not just to optimize an arbitrary objective function, but also to reduce the sensitivity of engineering designs to uncontrollable factors or noise. This moves design targets toward the middle of the design space so that external variation affects the behavior of the design as little as possible. This permits large reductions in both part and assembly tolerances, which are major drivers of manufacturing cost.

Tailings dam A dam, usually made of earth and rock, used to contain mining waste.

Tailoring The tailoring of requirements is the responsibility of the acquirer (customer), suggested tailoring may be provided by prospective and selected developers (prior to contract agreement).

Takt time A reference number that is used to help match the rate of production to the rate of sales. In other words, the rate at which customers require finished units. It is determined by dividing the total available production time per shift by the customer demand rate per shift. "Takt" is a German word for pace or beat.

Talipes This refers to a number of deformities especially those of congenital origin, such as clubfoot. Combinations of various types called talipes equinovalgus; equinovarus, talipes calcaeovalgus; and talipes calcaneovarus.

Talipes equinus Talipes in which the person elevates the heel and weight bearing is thrown onto the anterior of the foot.

Talipes valgus Talipes in which the outer border of the foot is everted, with inward rotation of the tarsus and flattening of the plantar arch.

Tape A recording media for data or computer programs. Tape can be in permanent form, such as perforated paper tape, or erasable, such as magnetic tape. Generally, tape is used as a mass storage medium, in magnetic form, and has a much higher storage capacity than disk storage, but it takes much longer to write or recover data from tape than from a disk.

Taper charge In quick chargers the charging current is is progressively reduced in a controlled way by controlling the supply voltage. In slow chargers the voltage is fixed and the charging current reduces in an uncontrolled way due to increase in the cell voltage as the charge builds up.

TAR Turnaround. Term given to a planned outage of plant for maintenance; modification etc.

Target costing A market-driven strategy and process that begins with what price a product can sell for in the marketplace to achieve a desired sales volumes. Target cost is then calculated by subtracting the desired profit margin from this target price. The target cost is treated as an independent variable that must be satisfied along with other customer requirements rather than the result of design decisions (dependent variable). This cost would be considered the unit production cost that is expected to be achieved during a mature production stage. Depending on the definition, it may or may not include warranty costs and selling, general and administrative costs.

Target standard The target design standards for prototype Systems, toward which work during the development phase is directed in regard to prototypes.

Task 1. A sequence of instructions treated as a basic unit of work.

2. A well-defined unit of work in the software process that provides management with a visible checkpoint into the status of the project. Tasks have readiness criteria (preconditions) and completion criteria (postconditions).

Taxiways Pavements used for the powered ground movement of aircraft between runway systems and other airfield facilities.

TCP/IP Transmission Control Protocol/ Internet Protocol. The suite of communications protocols used to connect devices on the internet or similar network.

TDR Time Domain Reflectometer, a device which a board house can use for measuring characteristic impedance of a conductor on a printed board, thus insuring an accurate build for controlled impedance.

Team A collection of people, often drawn from diverse but related groups, assigned to perform a well-defined function for an organization or a project. Team members may be part-time participants of the team and have other primary responsibilities.

Technical (Formal) reviews A series of system engineering activities by which the technical progress on a project is assessed relative to its technical or contractual requirements. The formal reviews are conducted at logical transition points in the development effort to identify and correct problems resulting from the work completed thus far before the problem can disrupt or delay the technical progress. The reviews provide a method for the contractor and procuring activity to determine that the development of a CI and its identification have met contract requirements.

Technical data Recorded information, regardless of medium or characteristics, of a scientific or technical nature. It may, for example document research, experimental, developmental, or engineering work; or be usable or used to define a design or process or to acquire, support, maintain, or operate materiel. Technical data does not include computer software or financial, administrative, cost and pricing, and management data, or other information incidental to contract administration.

Technical data package A collection of product related engineering data comprised of the Engineering Drawing Package (EDP) and non-EDP data related to the design and manufacture of the item or system. The EDP contains all the descriptive documentation needed to ensure the competitive re-procurement of an item or system. The non-EDP consists of data such as system and development specifications, product specifications, concurrent repair list packaging data sheets, special production tool data, acceptance inspection, equipment data, project standards, repair manuals, supplementary quality assurance provisions, preparation for delivery requirements and other data as required.

Technical effort A technical effort is any activity that influences system performance by defining, designing, or executing a task, requirement, or procedure. All the activities required to implement and execute the systems engineering process are technical efforts.

Technical objectives Technical objectives shall be established for each program so that meaningful relationships among need, urgency, risks, and worth can be established.

Technical performance measurement The continuing prediction and demonstration of the degree of anticipated or actual achievement of selected technical objectives. It includes an analysis of any differences among the "achievement to date", "current estimate", and the specification requirement. "Achievement to date" is the value of a technical parameter estimated or measured in a particular test and/or analysis. "Current Estimate" is the value of a technical parameter predicted to be achieved at the end of the contract within existing resources.

Technical program planning and control The management of those design, development, test, and evaluation tasks required to progress from an operational need to the deployment and operation of the system by the user.

Technical specifications Part of an NRC license authorizing the operation of a nuclear production or utilization facility. A Technical Specification establishes requirements for items such as safety limits, limiting safety system settings, limiting control settings, limiting conditions for operation, surveillance requirements, design features, and administrative controls.

Technique A method of performing an analysis.

Technology Specification requirements shall be delineated in light of acceptable technological risks defined by risk assessment.

Technology roadmap A tabular or graphic representation of technology plans mapped against time to guide the selection and use of technology in new product development or represent the technology embodied in future products.

Technology transfer Technology Transfer is the process of transferring research and technology from laboratories, government and outside organizations into the enterprise for practical application in new products.

TEG (Triethylene Glycol) A colourless odourless non volatile hygroscopic liquid whose major application is as a drying agent for natural gas.

Telecommunication Synonym for data communication. The transmission of information from one point to another.

TEMPCO Abbreviation for "temperature coefficient": the error introduced by a change in temperature. Normally expressed in %/°C or ppm/°C.

Temper To moisten and mix clay, plaster or mortar to a proper consistency.

Temperature A system in which internal vibrational modes have equilibrated with one another can be said to have a particular temperature. Two systems A and B are said to be at different temperatures if, when brought into contact, heat flows from (say) A to B, increasing the thermal energy of B at the expense of the thermal energy of A.

Temperature coefficient The average percent change in output voltage per degree Centigrade change in ambient temperature over a specified temperature range.

Temperature cut off A temperature sensing method which detects heat rise in a cell at overcharge and switches the charger off or to a lower rate of charge.

Temperature error The maximum change in output, at any measurand value within the specified range, when the transducer temperature is changed from room temperature to specified temperature extremes.

Temperature range, compensated The range of ambient temperatures within which all tolerances specified for Thermal Zero Shift and Thermal Sensitivity Shift are applicable (temperature error).

Temperature range, operable The range of ambient temperatures, given by their extremes, within which the transducer may be operated. Exceeding compensated range may require recalibration.

Temperature range, operating The range of ambient or case temperatures within which a power supply may be safely operated and meet its specifications.

Temperature range, storage The range of ambient temperatures within which a power supply may be safely stored, non-operating, with no degradation in its subsequent operation.

Temperature sensor An electronic device which provides a voltage analogue of the temperature of the surface on which it is mounted. A thermistor is an example.

Temporal analysis Refers to the time taken to complete a series of movements compared to the position of the foot, or part of the foot

Temporary refuge A safe area where workers can take temporary refuge during a potentially dangerous plant upset.

Tensegrity An array of tension cables and compression rods that supports a structure; invented by Buckminster Fuller student Kenneth Snellson.

Tensile strength (TS) The maximum engineering stress, in tension, sustainable without fracture; also called"ultimate (tensile) strength".

Tension A stretching force that pulls on a material.

Tension ring A support ring that resists the outward force pushing against the lower sides of a dome.

Tented via A via with dry film solder mask completely covering both its pad and its plated-thru hole. This completely insulates the via from foreign objects, thus protecting against accidental shorts, but it also renders the via unusable as a test point. Sometimes vias are tented on the top side of the board and left uncovered on the bottom side to permit probing from that side only with a test fixture.

Terminal A point of connection for two or more conductors in an electrical circuit; one of the conductors is usually an electrical contact, lead or electrode of a component.

Terminal block A type of header to which wires are attached directly instead of by means of a connector plug. Each wire is inserted in a hole in the terminal block, and then anchored by means of a screw.

Termination voltage The maximum voltage which can be tolerated by a cell during charging without damaging the cell. The cell voltage at which the charging process should be terminated.

Terrestrial radiation The portion of the natural background radiation that is emitted by naturally occurring radioactive materials, such as uranium, thorium, and radon in the earth.

Tertiary winding A third winding on a transformer.

Tesla (T) Unit of magnetic induction: 1T = 1 weber/m^2.

Test coupon An area of patterns and holes located on the same fabrication panel as the actual PCB, but separate from the electrical circuits and outside the board outline(s). It is designed to reflect the technology used on the PCB, such as smallest plated-through hole size, any blind or buried vias, etc. It is cut away from the panel and can be embedded in a clear plastic to prepare it for destructive testing .

Test data Recorded information, regardless of the form or method of the recording, of a scientific or technical nature. The term includes computer software or data incidental to contractual administration, and usually does not include financial and/or management information.

Test requirement The stimulus, measurement, power, loads and any special test equipment or procedure essential to validate proper operation of a device or some predetermined design control or product specification definition.

Testable A requirement or set of requirements is considered to be testable if an objective and feasible test can be designed to determine whether each requirement has been met.

Theory of inventive problem solving Theory of Inventive Problem Solving (Russian acronym is TRIZ) is a structured methodology developed by Genrich Altshuller for problem solving and innovation based on analysis and codification of technology solutions from millions of patents.

Thermal breeder reactor A breeder reactor in which the fission chain reaction is sustained by thermal neutrons.

Thermal capacity The amount of energy required to raise the temperature of an object by one degree Celsius. Expressed in Joules/ Kg.

Thermal coefficient of resistance The change in resistance of a semiconductor per unit change in temperature over a specific range of temperature.

Thermal conductivity (ê) For steady-state heat flow, the proportionality constant between the heat flux and the temperature gradient. Also, a parameter characterizing the ability of a material to conducting heat.

Thermal energy The internal energy present in a system as a result of the energy of thermally equilibrated vibrational modes and other motions (including both kinetic energy and molecular potential energy). The mean thermal energy of a classical harmonic oscillator is kT.

Thermal expansion An increase in size due to an increase in temperature expressed in units of an increase in length or increase in size per degree, i.e. inches/inch/degree C.

Thermal fluctuations The thermal energy of a system (or of a particular part or mode of motion in a system) has a mean value determined by the temperature and by the structure of the system. Statistical deviations about that mean are termed thermal fluctuations; these are of great importance in determining both rates of chemical reactions and error rates in nanomechanical systems.

Thermal imaging A photographic technique which displays the range of temperatures of a warm body in the form of a colour spectrum. Used as a design verification tool for detecting hot spots in battery and other equipment designs.

Thermal management The means by which a battery is maintained within its operating temperature limits during charging and discharging.

Thermal power The total core heat transfer rate to the reactor coolant.

Thermal protection A protective feature that shuts down a power supply if its internal temperature exceeds a predetermined limit.

Thermal reactor A reactor in which the fission chain reaction is sustained primarily by thermal neutrons. Most current reactors are thermal reactors.

Thermal runaway A condition in which an electrochemical cell will overheat and destroy itself through internal heat generation. This may be caused by overcharge or high current discharge and other abusive conditions.

Thermal sensitivity shift The sensitivity shift due to changes of the ambient temperature from room temperature to the specified limits of the compensated temperature range.

Thermal shield A layer, or layers, of high-density material located within a reactor pressure vessel or between the vessel and the biological shield to reduce radiation heating in the vessel and the biological shield.

Thermal zero shift An error due to changes in ambient temperature in which the zero pressure output shifts. Thus, the entire calibration curve moves in a parallel displacement.

Thermistor An electronic device that makes use of the change of resistivity of semiconductor with a change in temperature. In power supplies, negative temperature coefficient thermistors frequently are used as inrush current limiting devices.

Thermocouple A temperature-measuring device, which contains a pair of end-joined dissimilar conductors in which an electromotive force is developed by thermoelectric effects when the joined ends and the free ends of the conductors are a different temperature.

Thermodynamics A field of study embracing energy conversion among various forms, including heat, work, and potential and kinetic energy.

Thermoelastic Both stress and temperature changes alter the dimensions of an object having a finite stiffness and a nonzero thermal expansion coefficient. Applying a stress then produces a temperature change; this can result in a heat flow which then changes the stress: these are thermoelastic effects, and result in losses of free energy.

Thermoluminescent dosimeter A small device used to measure radiation by measuring the amount of visible light emitted from a crystal in the detector when exposed to ionizing radiation.

Thermonuclear An adjective referring to the process in which very high temperatures are used to bring about the fusion of light nuclei, such as those of the hydrogen isotopes deuterium and tritium, with the accompanying liberation of energy.

Thermopile An arrangement of thermocouples in series such that alternate junctions are at the measuring temperature and the reference temperature. This arrangement amplifies the thermoelectric voltage. Thermopiles are usually used as infrared detectors in radiation pyrometry.

Thermoplastic polymer A substance that when molded to a certain shape under appropriate conditions can later be remelted.

Thermoset polymer A substance that when molded to a certain shape under pressure and high temperatures cannot be softened again or dissolved.

Thermowell A closed-end tube designed to protect temperature sensors from harsh environments, high pressure, and flows. They can be installed into a system by pipe thread or welded flange and are usually made of corrosion-resistant metal or ceramic material depending upon the application.

Thin film A film of conductive or insulating material, usually deposited by sputtering or evaporation, that may be made in a pattern to form electronic components and conductors on a substrate or used as insulation between successive layers of components. (Graf).

Thomson effect When current flows through a conductor within a thermal gradient, a reversible absorption or evolution of heat will occur in the conductor at the gradient boundaries.

Three terminal regulator A power integrated circuit in a 3-terminal standard transistor package. It can be either a series or shunt regulator IC.

Throttle body Throttle-Body Fuel Injection is a type of Electronic Fuel Injection which positions the injector(s) centrally in a throttle-body housing. This housing contains a valve to regulate the airflow through the intake manifold.

Through hole (Of a component, also spelled "thru-hole"). Having pins designed to be inserted into holes and soldered to pads on a printed board. Contrast with surface mount.

Thrust A force applied to a spacecraft by a rocket engine. Thrust is used to move or rotate a spacecraft while in space. Launch vehicles, such as the Space Shuttle or rockets, use thrust to propel the vehicle into orbit.

Thyristor Also called a Silicon-Controlled Rectifier or SCR, it is a solid-state high current semiconductor switching device similar to a diode, with an extra terminal which is used to turn it on. Once turned on, the thyristor will remain on (conducting) as long as there is a significant current flowing through it. If the current falls to zero, the device switches off.

Tibial valgum A frontal plane deformity where the distal one third of the leg is angualted away from the midsagittal plane more than the proximal end.

TIE Any unit of material which connects masonry to masonry or other materials.

Tight receptor structures A receptor structure in which a bound ligand of a particular kind is confined on all sides by repulsive interactions (note that favorable binding energies are compatible with repulsive forces). A tight-receptor structure discriminates strongly against all molecules larger than the target.

Time compression technologies Time Compression Technologies - technologies to support the product development process, that when effectively integrated into the process, offers opportunity for significant reductions in cycle time. These include CAD, CAE, CAM, PDM and rapid prototyping.

Time constant Time period required for the voltage of a capacitor in an RC circuit to increase to 63.2 percent of maximum value or decrease to 36.7 percent of maximum value.

Time to market 1. Time-to-Market is the cycle time of product development from conception of a new product to initial sale of the new product.

2. Time-to-Market is the dimension of strategy focused on getting products to market quickly as the basis of competition.

Timing Timing refers to the crankshaft angles at which the valves open and close and at which time the ignition system fires the spark plugs.

Tire ratings Tires are rated by load capacity, size and speed capacity. For example, a

P225/50VR16 printed on the side of the tire means:

P = P-Metric (Passenger Type Tire)

255 = Section Width (255mm)

50 = Aspect Ratio (tire height/section width)

V = Speed Rating

R = Type of Ply (Radial)

16 = Wheel Diameter (16 inches)

Tire and wheel dimensions are the first point of information in any discussion of size and capacities. Among the other terms used to describe tires are: tread, shoulder, carcass, sidewall, bead seal, bead seat, tire diameter, aspect ratio, speed rating and section width.

Toe in The amount by which the front of a front wheel points inward or outward. A slight amount of toe in is usually specified to keep the front wheels running parallel on the road by offsetting other forces that tend to spread the wheels apart.

Toe off This occurs at about 60% of the gait cycle. It is the point at which the stance phase ends and the swing phase begins. It also marks the end of the second double support phase of the cycle.

Tolerance design Tolerance Design is a step in the design process (following parameter design) where the determination is made of how much variation is acceptable with a design parameter that will still allow the satisfactory functioning of the product to meet the cuistomer's needs. Often tolerance design is not adequately considered, and the designer merely specifies standard tolerances which may be inadequate or overstated.

Tolerance Tolerance is the upper and lower limits of some dimension or parameter relating to a component part, material or assembly which an actual item must comply with in order for it to be acceptable in procurement or manufacturing. The difference between the upper and lower tolerance is the tolerance spread.

Toothing Constructing the temporary end of a wall with the end stretcher of every alternate course projecting. Projecting units are toothers.

Top down design Top-Down Design is a design methodology whereby an entire design is decomposed into its major components, and then these components are further decomposed into their major components, etc. The constraints are established early in the design flow, and then are passed on and adhered to by the back-end processes.

TOR Terms of Reference.

Toroid A round magnetic core with a hole in the middle. A doughnut shaped core.

Torque A force that produces a twisting or rotating motion.

Torque converter clutch An electronically controlled lockup clutch that is automatically engaged at certain speeds to eliminate the slip between the torque converter's input and output, thereby improving fuel efficiency and performance.

Torque rating A measure of the engine's power capability, whereby the amount of twisting or rotating effort being exerted on the crankshaft is expressed in lb.-ft. of force. Torque is the force that gets the weight of the vehicle moving, making it an important consideration in trailering.

Torque, engine Engine torque is the amount of twisting effort exerted at the crankshaft by an engine expressed in foot-pounds of force. A foot-pound represents the force of one pound acting at the right angle to the rotating crankshaft at distance of one foot in length.

Torsion bar A long straight bar fastened to the frame at one end and to a suspension part at the other. In effect, a torsion bar is merely an uncoiled spring, and spring action is produced by twisting the bar. The main advantage of the torsion bar over the coil spring in the front suspension is the ease of adjusting the front suspension height.

Total effect The change in a stabilized output produced by concurrent worst case changes in all influence quantities within their rated range.

Total regulation band The range of combined regulation tolerances such as the effects of input voltage variation, output load variation, temperature variation, drift and other specified variables. It is expressed as a plus/minus percent from nominal. Also called accuracy limits.

Toughness A measure of the amount of energy absorbed by a material as it fractures, indicated by the total area under the material's tensile stress-strain curve.

Tower The vertical structure in a suspension bridge or cable-stayed bridge from which cables are hung; also used loosely as a synonym for the term skyscraper.

Toyota production system (TPS) A system developed by Toyota Motor Corp. to provide best quality, lowest cost, and shortest lead time through the elimination of waste. It is often illustrated as a "house" comprised of two pillars: Just-in-time and Jidoka. TPS is maintained and improved thorugh iterations of standardized work and kaizen.

Traceability The evidence of an association between items, such as between process outputs, between an output and its originating process, or between a requirement and its implementation.

Traction battery A high power deep cycle secondary battery designed to power electric vehicles or heavy mobile equipment.

Traction control Traction control helps provide smoother, more controlled acceleration by reducing the amount of wheel spin during reduced traction conditions. Traction control utilizes the vehicle's anti-lock braking system and is usually activated only at low vehicle speeds.

Trade off analysis Trade-off Analysis is the process of making decisions when each choice has both advantages and disadvantages. In a simple tradeoff, it may be enough to list each alternative and the pros and cons. For more complicated decisions, list the decision criteria and weight them. Determine how each option rates on each of the decision score and compute a weighted total score for each option. The option with the best score is the preferred option. Decision trees may be used when options have uncertain outcomes.

Trade off study An objective evaluation of alternative requirements, architectures, design approaches, or solutions using identical ground rules and criteria.

Traditional masonry Masonry in which design is based on empirical rules which control minimum thickness, lateral support requirements and height without a structural analysis.

Trailing arm A rear suspension element consisting of a lengthwise member that pivots from the body at its forward end and has a wheel hub rigidly attached to its trailing end.

Transaxle A transmission and differential combined in one integrated assembly, eliminating the need for a separate connecting drive shaft. This configuration is typical in front-wheel-drive vehicles.

Transducer A device (or medium) that converts energy from one form to another. The term is generally applied to devices that take physical phenomenon (pressure, temperature, humidity, flow, etc.) and convert it to an electrical signal.

Transducer vibration Generally, any device which converts movement, either shock or steady state vibration, into an electrical signal proportional to the movement; a sensor.

Transduction mode (direct or indirect) How the sensor acquires the desired information from the material. In general, this parameter is an indication of the ability of the sensor signal to provide information regarding a material property or state of interest.

Transformer Device which transfers energy from one circuit to another by electro-magnetic induction.

Transgranular Through or across crystals or grains; also called intracrystalline or transcrystalline.

Transient A change in the reactor coolant system temperature and/or pressure due to a change in power output of the reactor. Transients can be caused 1. by adding or removing neutron poisons,

2. by that is increasing or decreasing electrical load on the turbine generator, or

3. by accident conditions.

Transient recovery time The time required for the output voltage of a power supply to settle within specified output accuracy limits following a transient.

Transient response Ability of a power supply to recover a constant voltage following a step change in output current.

Transient response time The interval between the time a transient is introduced and the time it returns and remains within a specified amplitude range.

Transistor Semiconductor device used for amplification and switching.

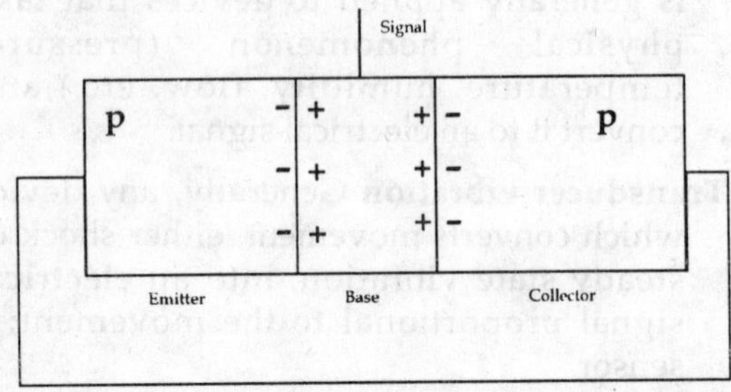

Fig. Transistor

Transition metal A metal having available electron energy levels occupied so that the d-band contains less than its maximum number of ten electrons per atom; the incompletely filled d-levels cause the unique properties of the transition metals; included in this class are iron, cobalt, nickel, and tungsten.

Transition state At the saddle point of a col linking two potential wells, the direction of maximum negative curvature defines the reaction coordinate; the transition state is a hypothetical system of reduced dimensionality, free to move only on a hypersurface perpendicular to the reaction coordinate at its point of maximum energy.

Transition state theory Any of several theories that give approximate descriptions of chemical reaction rates based on the PES of the system, and in particular, on the properties of two potential wells and a transition state between them.

Transition temperature An arbitrarily defined temperature that lies within the temperature range where metal fracture characteristics change rapidly; an example is the ductile-to-brittle transition temperature (DBTT).

Transmission electron microscope (TEM) A microscope that produces an image by using electron beams to transmit (pass through) the specimen, making examination of internal features at high magnifications possible.

Transmitter (Two-Wire) 1. A device which is used to transmit data from a sensor via a two-wire current loop. The loop has an external power supply and the transmitter acts as a variable resistor with respect to its input signal.

2. A device which translates the low level output of a sensor or transducer to a higher level signal suitable for transmission to a site where it can be further processed.

Transpiration The loss of water from plants through leaves and other parts. This loss can be a significant amount of water during very dry periods.

Transuranic element An artificially made, radioactive element that has an atomic number higher than uranium in the periodic table of elements such as neptunium, plutonium, americium, and others.

Transuranic waste Material contaminated with transuranic elements that is produced primarily from reprocessing spent fuel and from use of plutonium in fabrication of nuclear weapons.

Triboelectric noise The generation of electrical charges caused by layers of cable insulation. This is especially troublesome in high impedance accelerometers.

Tribology The science and technology of interacting surfaces in relative motion and of the practices related thereto; the science concerned with the design, friction, lubrication, and wear of contacting surfaces that move relative to one another (e.g. bearings, cams, gears).

Trickle charge A continuous charge at low rate, balancing losses through local action and/ or periodic discharge, to maintain a cell or battery in a fully charged condition. Normally at a C/20 to C/30 rate.

Trickling filter An attached growth biological process in which the microbial film is attached to non-moving rock or plastic media.

Trim Name given to the internals i.e. plug and seat of a valve.

Rather than replace a complete valve the trim is often replaced utilising the existing body and actuator.

Triple bond A double bond is formed when a pi bond is superimposed on a single bond; adding a second pi bond results in a triple bond. The two pi bonds have perpendicular nodal planes, and their sum has roughly cylindrical symmetry, permitting rotation in much the same manner as a single bond.

Triple point (Water) The thermodynamic state where all three phases, solid, liquid, and gas may all be present in equilibrium. The triple point of water is .01°C.

Triplet An electronic state of a molecule in which two spins are aligned. This term is derived from spectroscopy: a system of two aligned spins has three possible orientations with respect to a magnetic field; each has a different energy, resulting in sets of three field-dependent spectral lines.

Tritium A radioactive isotope of hydrogen (one proton, two neutrons). Because it is chemically identical to natural hydrogen, tritium can easily be taken into the body by any ingestion path. It decays by beta emission. It has a radioactive half-life of about 12.5 years.

Trophic level A level in the food chain. The first trophic level consists of the primary producers, autotrophs. The second trophic level is vegetarians which consume autotrophic organisms.

Troposphere The lower atmosphere, from the earth's surface to approximately 12 km. This portion of the earth's atmosphere contains about 95 percent of the atmospheric gases. The temperature gradually declines through this region.

True RMS The true root-mean-square value of an AC or AC-plus-DC signal, often used to determine power of a signal. For a perfect sine wave, the RMS value is 1.11072 times the rectified average value, which is utilized for low-cost metering. For significantly non-sinusoidal signals, a true RMS converter is required.

True strain The ratio of the change in dimension, resulting from a given load, to the magnitude of the dimension immediately prior to applying the load; the natural logarithm of the ratio of the gage length at the moment of observation to the original gage length.

TTL (Transistor-Transistor Logic) Also called multiple-emitter transistor logic. A widely used form of semiconductor logic. Its basic logic element is a multiple-emitter transistor. TTL is characterized by fairly high speed and medium power dissipation.

TTL-compatible For digital input circuits, a logic 1 is obtained for inputs of 2.0 to 5.5 V which can source 40 μA, and a logic 0 is obtained for inputs of 0 to 0.8 V which can sink 1.6 mA. For digital output signals, a logic 1 is represented by 2.4 to 5.5 V with a current source capability of at least 400 μA; and a logic 0 is represented by 0 to 0.6 V with a current sink capability of at least 16 mA.

Tuned intake and exhaust systems Intake and exhaust systems that increase the flow of intake charge into and out of the combustion chambers by varying the length, shape, or diameter of the component.

Tuned mass damper A mechanical counterweight designed to reduce the effects of motion, such as the swaying of a skyscraper in the wind or in an earthquake.

Tuned port fuel injection Tuned-Port Fuel Injection is almost identical to Multi-Port Fuel Injection, except that tuned runners are used to channel air to the cylinder heads. This results in increased airflow to the cylinders.

Tunnel boring machine (TBM) A mechanical device that tunnels through the ground.

Tunnel shield A cylinder pushed ahead of tunneling equipment to provide advance support for the tunnel roof; used when tunneling in soft or unstable ground.

Tunneling A classical particle or system could not penetrate regions in which its energy would be negative, that is, barrier regions in which the potential energy is greater than the system energy. In the real world, however, a wave function of significant amplitude may extend into and beyond such a region. If the wave function extends into another region of positive energy, the barrier is crossed with some probability; this process is termed tunneling (since the barrier is penetrated rather than climbed).

Turbine A rotary engine made with a series of curved vanes on a rotating shaft, usually turned by water or steam. Turbines are considered the most economical means to turn large electrical generators.

Turbine generator (TG) A steam (or water) turbine directly coupled to an electrical generator. The two devices are often referred to as one unit.

Turbocharger Rotary compressor or pump that pressurizes engine intake air. It is driven by the flow of exhaust gases. The increased pressure forces more air into the cylinder than it could normally draw, allowing the engine to burn more fuel and in turn produce more power.

Turbulent flow When forces due to inertia are more significant than forces due to viscosity. This typically occurs with a Reynolds number in excess of 4000.

Turnkey A type of contract where contractor carries out and completes his assignment for a fixed fee; opposite of a reimbursable contract.

Turns ratio Ratio of the number of turns on the primary winding of a transformer to the number of turns on the secondary winding.

TUTU (Topside Umbilical Termination Unit) Provides interface between topside equipment e.g. hydraulics and umbilical.

TÜV TÜV Rheinland Group (TUV - Technical Inspection Association) is an international service company which documents the safety and quality of new and existing products, systems and services.

Twin Two portions of a crystal with a definite orientation relationship; the orientation of one portion is a mirror image of the orientation of the other portion across a twinning plane or an orientation derived by rotating one portion about a twinning axis.

Twin bands Bands across a crystal grain where crystallographic orientations have a mirror image relationship to the orientation of the matrix grain across a composition plane usually parallel to the sides of the band.

Typical Error is within plus or minus one standard deviation (±1%) of the nominal specified value, as computed from the total population.

UCL Upper Control Limit is the lower limit used within statistical process control that define the constraints of common cause variations. When a parameter value falls above the upper control limit, it flags the occurrence of special causes contributing to variation. There is a similar Lower Control Limit and the mean to show the central tendency.

UFL Upper Flammable Limit. The concentration of flammable gas in air at atmospheric pressure above which combustion will not occur.

The figure is expressed as a percentage by volume.

ULSI Ultra large scale integration; a chip with over 1,000,000 components.

Ultimate biochemical oxygen demand (BODu) The total amount of oxygen required to oxidize any organic matter present in a water, i.e. after an extended period, such as 20 or 30 days.

Ultimate disposal The process of returning residuals back to the environment in a form which will have the minimal or reduced negative environmental impacts.

Ultrasonic welding Ultrasonic welding involves the use of high frequency sound energy to soften or melt the thermoplastic at the joint. Parts to be joined are held together under pressure and are then subjected to ultrasonic vibrations usually at a frequency of 20, 30 or 40kHz.

Ultraviolet Electromagnetic radiation of a wavelength between the shortest visible violet and low energy x-rays.

Umbilical Flexible cables carrying electrical and instrument wiring; hydraulic tubing and chemical tubing.

Unbalance That condition which exists in a rotor when vibratory force or motion is imparted to its bearings as a result of centrifugal forces.

Unbalance tolerance The unbalance tolerance with respect to a radial plane (measuring plane or correction plane) is that amount of unbalance which is specified as the maximum below which the state of unbalance is considered acceptable.

Uncertainty range Defines an interval within which a numerical result is expected to lie within a specified level of confidence. The interval often used is the 5-95 percentile of the distribution reporting the uncertainty.

Unclassified Not attracting a security classification but control may be required for other reasons. Therefore official information of a unclassified nature should not be disclosed without formal departmental/ program approval.

Under cooling Cooling a material below the temperature of an equilibrium phase change fast enough to not allow the occurrence of the transformation.

Undershoot The difference in temperature between the temperature a process goes to, below the set point, after the cooling cycle is turned off and the set point temperature.

Ungrounded junction A form of construction of a thermocouple probe where the hot or measuring junction is fully enclosed by and insulated from the sheath material.

Unibody construction A type of body construction that doesn't require a separate frame to provide structural strength or support for the vehicle's mechanical components. Also called "unitized."

Unimolecular Occurring to or within a single molecule; like intramolecular, but can refer to fragmentation reactions.

Union A form of pipe fitting where two extension pipes are joined at a separable coupling.

Unit cell The basic structural unit of a crystal structure, defined in terms of atom (or ion) positions within a parallelepiped volume.

Universal joint A joint that transmits rotary motion between two shafts that aren't in a straight line.

Unnecessary regulatory burden Regulatory criteria that go beyond the levels that would be reasonably expected to be imposed on licensees given that regulations apply to conditions that incorporate normal operation and design-basis conditions.

Unrestricted area The area outside the owner-controlled portion of a nuclear facility (usually the site boundary). An area in which a person could not be exposed to radiation levels in excess of 2 millirems in any one hour from external sources.

Unsaturated logic A form of logic containing transistors operated outside the region of saturation, which makes for very fast switching. An example is emitter-coupled logic (ECL).

Unsaturated Possessing double or triple bonds.

Unstable Characteristic of a structure that collapses or deforms under a realistic load.

Unstable isotope A radioactive isotope.

UPS Uninterruptible Power Supply. A UPS sits between a power source and a device (e.g. a computer) to prevent undesired features of the power source (outages; surges etc.) adversely affecting the performance of the device.

Upstream Generic term that includes oil and gas production facilities; pipelines; gas processing and liquefied natural gas production facilities and receiving terminals.

Can be on or offshore based.

Upstream face The side of a dam that is against the water.

Uranium A radioactive element with the atomic number 92 and, as found in natural ores, an atomic weight of approximately 238. The two principal natural isotopes are uranium-235 (0.7 percent of natural uranium), which is fissile, and uranium-238 (99.3 percent of natural uranium), which is fissionable by fast neutrons and is fertile. Natural uranium also includes a minute amount of uranium-234.

Uranium fuel fabrication facility A facility that

1. manufactures reactor fuel containing uranium for any of the following: (i) preparation of fuel materials; (ii) formation of fuel materials into desired shapes; (iii) application of protective cladding; (iv) recovery of scrap material; and (v) storage associated with such operations; or

2. Conducts research and development activities.

Uranium hexafluoride production facility A facility that receives natural uranium in the form of ore concentrate, processes the concentrate, and converts it into uranium hexafluoride (UF6).

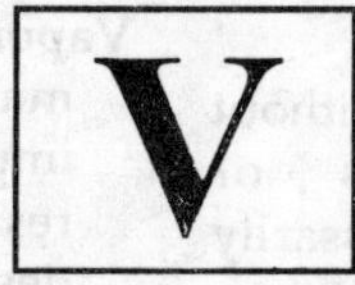

Vacancy An unfilled lattice site in a crystal structure.

Vacuum Any pressure less than atmospheric pressure.

Valence band The electron energy band that contains the valence electrons in solid materials.

Valence electrons Electrons that can participate in bonds and in chemical reactions; lone-pair electrons are valence electrons, although not participating in a bond.

Valence The combining capacity of an atom expressed as the number of single bonds the atom can form or the number of electrons an element gives up or accepts when reacting to form a compound.

Valgus Valgus of the foot or part of the foot means a fixation of the part in the position it would assume if everted. It is a frontal plane fixation in which the plantar surface is directed away from the midline.

Validation The process of determining that the requirements are the correct requirements and that they are complete. The system life cycle process may use requirements that are derived requirements in system validation. Validation is the process of ensuring that the product conforms to defined user needs, requirements, and/or specifications under defined operating conditions. Design validation is performed on the final product design with parts that meet design intent. Production validation is performed on the final product design with parts that meet design intent produced with production processes intended for normal production.

Valuable final artwork A term used in "Streamlined_PCB_Design :" Artwork for electronic circuits which have been laid out and documented in forms perfectly suited to the photo-imaging and numeric-controlled tooling processes of printed circuit manufacture. It is termed "final" because it has been thoroughly checked for errors and any corrected as needed and is now ready for manufacture without further work by the PCB designer . It is valuable because it could be exchanged with a customer for money or other support. Abbr. VFA. (Based on " Valuable Final Product ," or "VFP," a term coined by L. Ron Hubbard).

Valve A device that can be opened or closed to allow or prevent the flow of a liquid or gas from one place to another. Most internal combustion engines use intake and exhaust valves to allow fuel/air mixture into the cylinders and to exhaust burnt gases. Some engines have four valves per cylinder, which increases total valve area for increased efficiency and performance.

Value analysis Value Analysis - an effort to analyze systems and designs to satisfy needed user requirements at sufficient quality (functions) at an optimum cost (maximize value).

Value engineering 1. Value Engineering is a structured methodology for applying value analysis or function analysis to increase customer or user value.

2. A formal technique to eliminate, without impairing essential functions or characteristics, anything that unnecessarily increases the cost of a product. It is a disciplined system for accomplishing the functions that the customer needs and wants at the lowest cost.

Value stream The specific activities required to design, order and provide a product, from concept to launch, order to delivery, and raw materials into the hands of the customer. Whenever there is a product for a customer, there is a value stream. Up to 90 percent of the actions and 99.99 percent of the time along a typical value stream can consume resources, but create no value for customers.

Value stream mapping The process of directly observing the flows of information and materials as they now occur, summarizing them visually, and then envisioning a future state with much better performance. It raises consciousness of the enormous waste of time, effort and movement that occurs. The relevant actions to be mapped consist of two flows: orders traveling upstream from the customer and products coming downstream from raw materials to the customer.

Valve lifter The cylindrical component that presses against the lobe of a camshaft and moves up and down as the cam lobe rotates, opening and closing an intake or exhaust valve. Virtually all modern valve lifters are of an hydraulic design that uses a cushion foil to promote quiet operation.

Valve train The collection of parts that make the valves operate, allowing fuel intake, compression and exhaust. Includes the camshaft(s) and all related drive components, and the various parts that convert the camshaft's rotary motion into reciprocating motion at the valves.

Van der Waals bond A secondary, permanent or induced, interatomic bond between adjacent molecular dipoles.

Vapor barrier A material or combination of materials, which form either an impenetrable zone or at least having a high resistance to vapor flow. For many years, designers were taught that the vapor barrier was a major requirement for insulated walls in order to control the diffusion of water vapor into the colder reaches of these walls, where it could condense and stain the finished surfaces or, worse, initiate the deterioration of the affected materials. When it became obvious to researchers in the 1960s that air leakage into the walls and roofs was a more important source of water migration, authorities began calling for a "continuous vapor barrier." This new terminology was supposed to somehow get rid of all the hidden holes causing so many moisture-related problems. The expression "air/vapor barrier" then began to appear in the literature, raising the hope that this component would finally put a stop to all moisture flow through the walls and roofs of a building. This, unfortunately, did not happen, so it has become necessary to take apart this concept of the "continuous air/vapor barrier " and go back to basic principles.

A vapor barrier is a material that offers a high resistance to the diffusion of water vapor. It is used to separate an environment which is at a high vapor pressure from an adjacent one at a lower vapor pressure. For best results, it is important that the vapor barrier be continuous, but it does not have to be perfectly continuous. Unsealed laps or minor cuts do not affect the overall resistance to diffusion significantly. The vapor barrier must also be located on the warm side of the insulation or at least in a location in the wall near enough to the warm side to remain above the dew point temperature of the indoor air during cold weather.

Vapor barriers (a 4 mil polyethylene sheet would make a good quality vapor barrier) are intended to retard the passage of moisture as it diffuses or migrates through a wall. Air barriers, on the other hand, are intended to stop both outside air from entering the building and inside air from exfiltrating through the building envelope. Air leakage is recognized as a greater mode of moisture transfer than is diffusion of vapor. The Difference Between a Vapor Barrier and an Air Barrier identifies the materials and the method of assembly required to build effective air barriers. Vapor barriers (a 4 mil polyethylene sheet would make a good quality vapour barrier) are intended to retard the passage of moisture as it diffuses or migrates through a wall. Air barriers, on the other hand, are intended to stop both outside air from entering the building and inside air from exfiltrating through the building envelope. Air leakage is recognized as a greater mode of moisture transfer than is diffusion of vapor.

Vapor retarder Will function to limit moisture diffusion through the veneer system. The vapor retarder is commonly placed on the interior or "warm in winter" side of the steel stud wall, but its location may vary depending on local practice and the configuration of wall insulation used. A vapor retarder commonly contains points where minor leakage can occur, such as electrical outlets or joints at structural members, but in general these systems still function adequately. The vapor retarder can be provided as a part of the insulation backing, on the surface of the gypsum wallboard or as a separate polyethylene sheet. Air pressure from wind infiltration, fan pressurization, and stack effect may be sufficient to dislodge inadequately supported vapor retarders allowing for significant vapor leakage.

Variability reduction A multi-part strategy to reduce product variation and make a product more robust or fit to use through design of experiments, design within process capabilities, and process improvement.

Varus Varus of the foot or part of the foot means a fixation of the part in the position it would assume if inverted. It is a frontal plane fixation in which the plantar surface is directed toward the midline.

Vaulting A stance phase modification where the subject goes up on their toes requiring increased ground clearance of the swinging leg. This causes an exaggerated vertical movement of the trunk with an ungainly appearance. Vaulting requires considerable energy.

Vector This has a direction and a magnitude e.g. a force is a vector.

Vector photoplotter It plots images from a CAD database on photographic film in a darkroom by drawing each line with a continuous lamp shined through an annular-ring aperture, and creating each shape (or pad) by flashing the lamp through a specially sized and shaped aperture. The "apertures" are thin trapezoidal pieces of plastic which are mostly opaque, but with a transparent portion that controls the size and shape of the light pattern passing through it. The apertures are mounted on an " aperture wheel" which can hold up to 24 apertures (or 70 on certain models). The lamp and aperture wheel are fixed, and the table holding the film is moved in x and y dimensions (on small photoplotters), or vice versa (on very large photoplotters). A numeric datum sent to the control circuit of the photoplotter is either a D code or an X and/or Y dimension in inches, to the nearest thousandth. If it is a D code equal to D10 or above, the message tells the wheel to rotate the corresponding aperture location into position in front of the lamp. . Gerber photoplotters, if set up by an experienced craftsman, are well-suited for printed circuit artwork generation. Compare with laser photoplotter, which is faster, more accurate and has largely replaced the vector photoplotter.

There are still vector photoplotters in use. Some manufacturers take advantage of the large bed size of the largest Gerber photoplotters, roughly the size of a full-sized billiards table. This enables the production of very large photoplots. An example is Buckbee-Mears, which makes large antenna boards, and the USGS (United States Geological Survery) which has used them in map-making.

Vee or VEE A name for a power net meaning "voltage emitter," usually -5V for ECL circuits.

Velocity This is the rate at which a moving object changes position. This usually refers to the distance it covers in a given time. Linear velocity measures the distance over a straight line, angular velocity measures rotating movements. The former is measured in meters per second i.e. m/s or $m.s^{-1}$. Changes in angular velocity is refereed to as angular acceleration.

The relationship between velocity, acceleration and distance traveled are given as four equations:

$$v = u + at$$

$$s = 1/2\,(ut + vt)$$

$$s = ut + 1/2\,at^2$$

$$v^2 - u^2 = 2as$$

where u is the initial velocity (m/s), v is the final velocity, a is the acceleration (m/s^2), t is time (in seconds) and s is the distance traveled (in meters)

Veneer 1. A : thin sheet of a material, such as a layer of wood of superior value or excellent grain to be glued to an inferior looking wood. b: any of the thin layers bonded together to form plywood.

2. Protective or ornamental facing, such as of brick or stone; in masonry: a single wythe of masonry for facing purposes, not structurally bonded.

3. A superficial or deceptively attractive appearance, display, or effect: gloss, facade.

4. BIA definition: a. "A wall having a facing of masonry units, or other weather-resisting, non-combustible materials, securely attached to the backing, but not so bonded as to intentionally exert common action under load." b. A brick veneer wall consists of an exterior wythe of brick isolated from the backup by a minimum prescribed air space and attached to the backup with corrosion-resistant metal ties.

5. The International Masonry Institute definition: "A single facing wythe of masonry units or similar materials securely attached to a wall for the purpose of providing ornamentation, protection, insulation, etc. but not so bonded or attached as to be considered as exerting common reaction under load."

6. The 1994 Uniform Building Code definition: a. "Veneer is nonstructural facing of brick, concrete, stone, tile, metal, plastic or other similar approved material attached to a backing for the purpose of ornamentation, protection or insulation." b."Anchored Veneer is veneer secured to and supported by approved connectors attached to an approved backing."

Venting The release of excessive internal pressure from a cell in a manner intended by design to preclude explosion.

Verification The process of determining whether or not the products of a given phase of the system/software life cycle fulfils the requirements established during the preceding phase.

Verification function Tasks, actions, and activities to be performed to evaluate progress and effectiveness of the evolving system (people, product, and process) solutions and to measure compliance with requirements. Analysis, (including simulation) demonstration, test, and inspection are verification approaches used to evaluate; risk; people, product, and process capabilities, compliance with requirements; and proof of concept.

Version An identified and documented body of software. Modifications to a version of software (resulting in a new version) require configuration management actions by the supplier, acquirer or his agent, or both.

Very high radiation area An area accessible to individuals, in which radiation levels exceed 500 rad (5 gray) in one hour at 1 meter from the source or from any surface that the radiation penetrates.

VESDA Very Early Smoke Detection and Alarm. Trade name which has become synonymous for all types of aspirating smoke detection systems.

Via Feed-through. A plated-through hole in a PWB used to route a trace vertically in the board, that is, from one layer to another.

Viability assessment A Department of Energy decisionmaking process to judge the prospects for geologic disposal of high-level radioactive wastes at Yucca Mountain based on 1. Specific design work on the critical elements of the repository and waste package,

2. A total system performance assessment that will describe the probable behavior of the repository,

3. A plan and cost estimate for the work required to complete a license application, and

4. An estimate of the costs to construct and operate the repository.

The viability assessment was required by the Energy and Water Development Appropriations Act, 1997 (Public Law 104-206). After the viability assessment was completed, site-specific environmental standards at 40 CFR Part 197 and implementing regulations .

Vibration error The maximum change in output of a transducer when a specific amplitude and range of frequencies are applied to a specific axis at room temperature.

Vicon system A commercial gait analysis system which combines both kinetic and kinematic analysis. The system employs a force plate, goniometer system as well as four cameras. In addition electromyographic equipment may be used.

Video tape digitisers Video tape may be used as a means of kinematic analysis. To improve objectivity skin markers are used and when the image is automatically digitised tracings are more clearly seen.

Virion A virus particle. Viral DNA or RNA enclosed in an organic capsule.

Virtual eccentricity The eccentricity of a resultant axial load required to produce axial and bending stresses equivalent to those produced by applied axial loads and moments. It is normally found by dividing the moment at a section by the summation of axial loads occurring at that section.

Virtual reality Technology that enables users to "enter" and navigate through a computer-generated 3D environment. It allows users to change their viewpoint and interact with objects created in the environment in a way that mimics the real world.

Virus A submicroscopic genetic constituent which can alternate between two distinct phases. As a virus particle, or virion, it is DNA or RNA enveloped in an organic capsule. As an intracellular virus, it is viral DNA or RNA inserted into the host organisms DNA or RNA.

Visbreaker A thermal cracking process unit in a refinery used to break up large molecules into smaller ones.

It is applied to the residue of vacuum distillation as part of the overall conversion process.

Viscoelasticity A type of deformation exhibiting the mechanical characteristics of viscous flow and elastic deformation.

Viscosity (symbol is ç) The ratio of the magnitude of an applied shear stress to the velocity gradient that it produces; in other words: a measure of a noncrystalline material's resistance to permanent deformation.

Visual controls Any devices, such as andon boards, that help operators quickly and accurately gauge production status at a glance. Progress indicators and problem indicators help assemblers see when production is ahead, behind or on schedule. They allow everyone to instantly see the group's performance and increase the sense of ownership in the area.

Vitrification The condition resulting when kiln temperatures are sufficient to fuse grains and close pores of a clay product, making the mass impervious.

Void In a nuclear power reactor, an area of lower density in a moderating system (such as steam bubbles in water) that allows more neutron leakage than does the more dense material around it.

Void coefficient of reactivity A rate of change in the reactivity of a water reactor system resulting from a formation of steam bubbles as the power level and temperature increase.

Volatile A material which will vaporize easily.

Volatile solids (VS) is the amount of matter which volatilizes (or burns) when a water sample is heated to 550EC.

Volatile suspended solids (VSS) is the non-filterable residue remaining after firing the total suspended solids at 550EC.

Volt The (electrical) potential difference between two points in a circuit. The fundamental unit is derived as work per unit charge-(V = W/Q). One volt is the potential difference required to move one coulomb of charge between two points in a circuit while using one joule of energy.

Voltage regulator A circuit which provides a fixed or controlled voltage output from a variable voltage input. Used in power supplies and chargers. Switching regulators , Linear regulators and Shunt regulators are the most common types.

Voltaic efficiency The ratio (expressed as a percentage) between the voltage necessary to charge a secondary cell and the corresponding discharge voltage.

Voltmeter An instrument used to measure voltage.

Volume flow rate Calculated using the area of the full closed conduit and the average fluid velocity in the form, Q = V x A, to arrive at the total volume quantity of flow. Q = volumetric flowrate, V = average fluid velocity, and A = cross sectional area of the pipe.

Volumetric energy density (Wh/L) The energy output per unit volume of a battery.

VPN Virtual Private Network. The use of encryption to provide a secure connection through an otherwise insecure network e.g. typically the Internet.

The encryption may be performed by firewall software or possibly by routers.

VRLA battery Valve Regulated Lead Acid Battery.

VSD Variable Speed Drive. An electrical system that allows a motor to be controlled to run at different speeds.

Waddling gait This describes a form of walking where the subject sways from side to side.

Waiver A written authorization to accept a configuration item, which during production of after having been submitted for inspection, is found to depart from specified requirements, but nevertheless is considered suitable for use "as is" or after re-work by an approved method.

Wall A vertical member of a structure whose horizontal dimension measured at right angles to the thickness exceeds three times its thickness.

Apron wall That part of a panel wall between window sill and wall support.

Area wall 1. The masonry surrounding or partly surrounding an area.
2. The retaining wall around basement windows below grade.

Bearing wall One which supports a vertical load in addition to its own weight.

Cavity wall A wall built of masonry units so arranged as to provide a continuous air space within the wall (with or without insulating material), and in which the inner and outer wythes of the wall are tied together with metal ties.

Composite wall A multiple-wythe wall in which at least one of the wythes is dissimilar to the other wythe or wythes with respect to type or grade of masonry unit or mortar

Curtain wall An exterior non-loadbearing wall not wholly supported at each story. Such walls may be anchored to columns, spandrel beams, floors or bearing walls, but not necessarily built between structural elements.

Dwarf wall A wall or partition which does not extend to the ceiling.

Enclosure wall An exterior non-bearing wall in skeleton frame construction. It is anchored to columns, piers or floors, but not necessarily built between columns or piers nor wholly supported at each story.

Exterior wall Any outside wall or vertical enclosure of a building other than a party wall.

Faced wall A composite wall in which the masonry facing and backings are so bonded as to exert a common reaction under load.

Fire division wall Any wall which subdivides a building so as to resist the spread of fire. It is not necessarily continuous through all stories to and above the roof.

Fire wall Any wall which subdivides a building to resist the spread of fire and which extends continuously from the foundation through the roof.

Foundation wall That portion of a loadbearing wall below the level of the adjacent grade, or below first floor beams or joists.

Hollow wall A wall built of masonry units arranged to provide an air space within the wall. The separated facing and backing are bonded together with masonry units.

Loadbearing wall A wall which supports any vertical load in addition to its own weight.

Non-Loadbearing wall A wall which supports no vertical load other than its own weight.

Panel wall An exterior, non-loadbearing wall wholly supported at each story.

Parapet wall That part of any wall entirely above the roof line.

Party wall A wall used for joint service by adjoining buildings.

Perforated wall One which contains a considerable number of relatively small openings. Often called pierced wall or screen wall.

Shear wall A wall which resists horizontal forces applied in the plane of the wall.

Single Wythe Wall: A wall containing only one masonry unit in wall thickness.

Solid masonry wall A wall built of solid masonry units, laid contiguously, with joints between units completely filled with mortar or grout.

Spandrel wall That part of a curtain wall above the top of a window in one story and below the sill of the window in the story above.

Veneered wall A wall having a facing of masonry units or other weather-resisting non-combustible materials securely attached to the backing, but not so bonded as to intentionally exert common action under load.

Wall plate A horizontal member anchored to a masonry wall to which other structural elements may be attached. Also called head plate.

Wall shear stress In biofilm studies we are normally concerned with wall shear stress. This is the shear stress at the wall caused by the liquid moving past it.

Wall tie A bonder or metal piece which connects wythes of masonry to each other or in other materials.

Wall tie, cavity A rigid, corrosion-resistant metal tie which bonds two wythes of a cavity wall. It is usually steel, 3/16 in. in diameter and formed in a "Z" shape or a rectangle.

Wall tie, veneer A strip or piece of metal used to tie a facing veneer to the backing.

WAN Wide Area Network. A network usually constructed with serial lines extending over distances greater than one kilometre.

Ward leonard controller A motor-generator system which uses a DC motor driving an AC generator to provide a variable power transmission. Used for high power load testing.

Waste Anything that does not contribute to transforming a part to the customer's needs. There are seven types of manufacturing waste: production over immediate demand; excess work in progress and finished goods inventories; scrap, repairs and rejects; unnecessary motion; excessive processing; wait time; and unnecessary transportation.

Waste minimization The elimination or reduction of a waste prior to its generation. This is accomplished by process changes rather than waste treatment methods.

Waste, radioactive Radioactive materials at the end of a useful life cycle or in a product that is no longer useful and should be properly disposed of.

Wastewater Consumed or used water from a municipality or industry that contains dissolved and/or suspended matter.

Water activity A term used to classify aggressive nature of waters based on their chemical and physical parameters and their consequences on the corrosion of metals. Similar to corrosivity index of waters.

Water retentivity That property of a mortar which prevents the rapid loss of water to masonry units of high suction. It prevents bleeding or water gain when mortar is in contact with relatively impervious units.

Water table A projection of lower masonry on the outside of the wall slightly above the ground. Often a damp course is placed at the level of the water table to prevent upward penetration of ground water

Waterproofing Barrier which prevents the flow or passage of moisture through porous materials (masonry, concrete, etc.) due to water pressure. It can be succesfully designed for either below or above grade construction.

Watt An electrical unit of power. 1 watt = 1 Joule/second. It is equal to the power in a circuit in which a current of one ampere flows across a potential difference of one volt.

Watt density The watts emanating from each square inch of heated surface area of a heater. Expressed in units of watts per square inch.

Watt hour An electrical energy unit of measure equal to 1 watt of power supplied to, or taken from, an electrical circuit steadily for 1 hour.

Wave A disturbance which is propagated in a medium in such a manner that at any point in the medium the displacement is a function of the time, while at any instant the displacement at a point is a function of the position of the point. Any physical quantity which has the same relationship to some independent variable (usually time) that a propagated disturbance has, at a particular instant, with respect to space, may be called a wave. In this definition, displacement is used as a general term, indicating not only mechanical displacement, but also electric displacement, etc. In short, a wave is a timevarying quantity which is also a function of position; for example, any time-varying voltage or current in a network is often called a wave. Other examples of waves are: 1. wave on the surface of a liquid, in which the disturbance is the displacement of any particle in the surface from its equilibrium position;

2. acoustic wave, in which the disturbance is the change in pressure from its equilibrium value at any point in a material medium (fluid or solid);

3. electromagnetic wave, in which the disturbance is the change in the electric and magnitude field intensities from their equilibrium values in space. The first two types are known as mechanical waves, since the propagated disturbance involves motion of a medium.

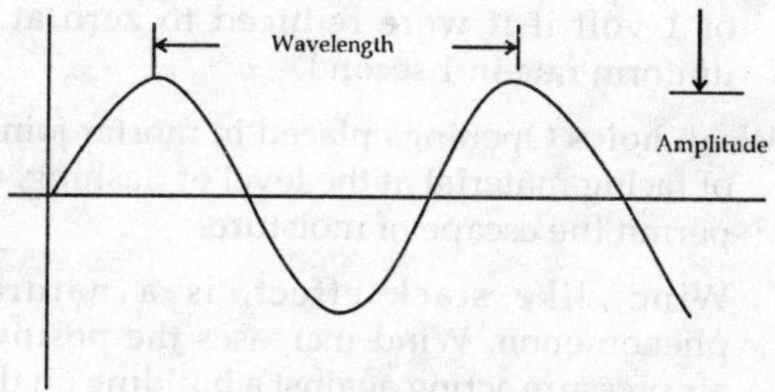

Fig. simple wave

Wave function In quantum mechanics, a complex function extending over the configuration space of a system; its complex conjugate yields the probability density function, and other mathematical operations yield other physical quantities.

Weak acid An acid that does not ionize completely under the conditions of interest. Examples include acetic acid, carbonic acid, and hypochlorous acid.

Weathered/weathering The action of the elements on a rock in altering its colour, texture, or composition, or in rounding off its edges. 1. Exposure to the open air and / or subject to the action of the elements, with the result of alteration of color, texture, composition, or form of exposed objects; specifically : the physical disintegration and chemical decomposition of materials; aging

2. Alternatively used as a noun to describe a slight construction slope designed to throw off rainwater

3. Changes on the surface of glass caused by chemical reaction with the environment. Weathering usually involves the leaching of alkali from the glass by water, leaving behind siliceous weathering products that are often laminar.

4. Response of materials (which were once in equilibrium within earth's crust) to new conditions at or near contact with air, water, or living matter.

Weber Unit of magnetic flux. One weber is a magnetic flux that, linking a circuit of 1 turn, would produce in it an electromotive force of 1 volt if it were reduced to zero at a uniform rate in 1 second.

Weep holes Openings placed in mortar joints of facing material at the level of flashing, to permit the escape of moisture.

Wind, like stack effect, is a natural phenomenon. Wind increases the positive air pressure acting against a building on the windward side, and produces a negative pressure on the leeward side and on the walls parallel to the wind direction. The wind also exerts a suction on flat or low-sloped roofs and a positive pressure on the windward side of steeper-sloped roofs

Weight distribution That portion of the total weight of a vehicle, including equipment and payload, that will be supported by each axle and tire. Proper distribution of total vehicle weight is critical to the service life of components such as the frame, axles, springs, bearings, and tires and therefore one of the most important requirements in selecting the right truck for your customer's particular job.

Weight percent (wt%) Concentration specification on the basis of weight (or mass) of a particular element relative to the total alloy weight (or mass).

Weighting factor (WT) Multipliers of the equivalent dose to an organ or tissue used for radiation protection purposes to account for different sensitivities of different organs and tissues to the induction of stochastic effects of radiation.

Well logging All operations involving the lowering and raising of measuring devices or tools that contain licensed material or are used to detect licensed materials in wells for the purpose of obtaining information about the well or adjacent formations that may be used in oil, gas, mineral, groundwater, or geological exploration.

Well to wheel efficiency The ratio between the mechanical energy ultimately delivered to the road wheels of a vehicle and the chemical energy content of the oil consumed in providing it. It is used to compare the fuel efficiencies of different methods of powering road vehicles and takes into account the refining process, the energy loss in the distribution process (in the case of hydrogen, the energy used to compress it) and the conversion efficiency of the vehicle's power unit.

Wet cell A cell with free flowing liquid electrolyte.

Wet solder mask Applied by means of distributing wet epoxy ink through a silk screen, a wet solder mask has a resolution suitable for single-track design, but is not accruate enough for fine-line design.

Wetland Semi-aquatic land, that is land that is either inundated or saturated by water for varying periods of time during each year, and that supports aquatic vegetation which is specifically adapted for saturated soil conditions.

Wheatstone bridge A network of four resistances, an ,emf source, and a galvanometer connected such that when the four resistances are matched, the galvanometer will show a zero deflection or "null" reading.

Wheelbase Distance, center to center, from front axle to rear axle. Wheelbase is important because it indicates available body length and weight distribution between front and rear axles.

Wheeling service The movement of electricity from one system to another over transmission facilities of intervening systems. Wheeling service contracts can be established between two or more systems.

Whisker A very thin, single crystal of high perfection which has an extremely large length-to-diameter ratio. Whiskers are used as the reinforcing phase in some composites.

Whole body counter A device used to identify and measure the radioactive material in the body of human beings and animals. It uses heavy shielding to keep out naturally existing background radiation and ultrasensitive radiation detectors and electronic counting equipment.

Whole body exposure Whole body exposure includes at least the external exposure, head, trunk, arms above the elbow, or legs above the knee. Where a radioisotope is uniformly distributed throughout the body tissues, rather than being concentrated in certain parts, the irradiation can be considered as whole-body exposure.

Wholesale value The price, based on auction results or retailer trade reports, that a retailer expects to pay for a vehicle.

Wildcat A well drilled in unproven territory.

Window In computer graphics, a defined area in a system not bounded by any limits; unlimited "space" in graphics.

Wipe sample A sample made for the purpose of determining the presence of removable radioactive contamination on a surface. It is done by wiping, with slight pressure, a piece of soft filter paper over a representative type of surface area. It is also known as a "swipe" or " smear" sample.

Wire Besides its usual definition of a strand of conductor, wire on a printed board also means a route or track .

Wire bonding The method used to attach very fine wire to semiconductor components (dice) to interconnect these components with each other or with package leads. The wires might be 1 to 2 mils in diameter and made of aluminum containing 1% silicon.

Wire wrap area A portion of a board riddled with plated-through holes on a 100-mil grid. Its purpose is for accepting circuits which may be found necessary after a PWB has been manufactured, stuffed, tested and debugged.

Wireframe A geometric model that describes 3D geometry by outlining its edges, similar to a "stick figure".

With inspection Masonry designed with the higher stresses allowed under EBM. Requires the establishing of procedures on the job to control mortar mix, workmanship and protection of masonry materials.

Wobbe index A comparative measure of thermal energy flow through a given size of orifice. A measure of the interchangeability of gases used for combustion.

Gases which have the same Wobbe index can replace each other without a change in the relative air-fuel ratio at the same fuel metering settings.

Wood A common natural material strong in both compression and tension

Word Number of bits treated as a single unit by the CPU. In an 8-bit machine, the word length is 8 bits; in a sixteen bit machine, it is 16 bits.

Work Energy transferred by applying a force over a distance; lifting a mass does work against gravity, and stores gravitational potential energy.

Work angle In arc welding, the angle between the electrode and one of the joints.

Work breakdown structure (WBS) A product-oriented listing, in family tree order, of the hardware, software, services and other work tasks which completely defines a product or program. The listing results from project engineering during the development and production of a materiel item. A WBS relates the elements of work to be accomplished to each other and to the end product.

Work hardening (strain hardening) Increase dislocation density in metals through straining a material with an applied stress; degree of hardening may be manipulated through recovery and recrystallization.

Work in process (WIP) Items between machines or equipment waiting to be processed.

Workflow systems/workflow management systems Workflow Systems are systems to support the coordination, communication and control of business processes by means of information technology for the purpose of improving and better managing these processes. Workflow Systems automate a business process, in whole or part, during which documents, information or tasks are passed from one participant to another for action, according to a set of procedural rules.

Working standard A standard of unit measurement calibrated from either a primary or secondary standard which is used to calibrate other devices or make comparison measurements.

Workover To carry out remedial operations on a producing well with the intention of restoring or increasing production.

World Wide Web (WWW) Graphical hypertext system linking many Internet computers.

Worst case tolerance analysis Worst Case Tolerance Analysis - The assembly tolerance is determined by summing the component tolerances linearly. Each component dimension is assumed to be at its maximum or minimum limit, resulting in the worst possible assembly limits. It is a very consrvative approach to tolerance analysis and is not the best approach to tolerancing since that it caters to combinations that are extremely unlikely, rather than focusing on a more probabilistic approach.

Write To record data in a storage device or on a data medium.

Wrought alloy An alloy that is suitable for mechanical forming below melting-point temperatures.

Wrought iron Commercial iron that contains less than 0.3% carbon and 1.0 or 2.0% slag, giving it ductility and toughness.

Wythe 1. Each continuous vertical section of masonry one unit in thickness.

2. The thickness of masonry separating flues in a chimney. Also called withe or tier.

X-Bar Chart X-Bar Chart A quality control chart that monitors the mean of the process. A sample of n parts is collected from the process every so many parts or time periods. The mean of the sample is plotted on the control chart and a determination is made if the process is "under control" or not.

Xmas tree Collective name given to the valves pipes and associated fittings assembled at the top of a completed well used to control the flow of oil and/or gas.

X-radiation Electromagnetic radiation of the same nature as visible light, but having a wavelength approximately one thousandth that of visible light.

X-rays Penetrating electromagnetic radiation (photon) having a wavelength that is much shorter than that of visible light. These rays are usually produced by excitation of the electron field around certain nuclei. In nuclear reactions, it is customary to refer to photons originating in the nucleus as x-rays.

X-ray crystallography The use of the property of X-ray diffraction by crystals to determine their physical structure.

X-ray tube A device that produces X-rays by the impact of high-speed electrons on a metal target.

Yellowcake Yellowcake is the product of the uranium extraction (milling) process; early production methods resulted in a bright yellow compound, hence the name yellowcake. The material is a mixture of uranium oxides that can vary in proportion and in color from yellow to orange to dark green (blackish) depending at which temperature the material was dried (level of hydration and impurities). Higher drying temperatures produce a darker, less soluble material. Yellowcake is commonly referred to as U3O8 and is assayed as pounds U3O8 equivalent. This fine powder is packaged in drums and sent to a conversion plant that produces uranium hexafluoride (UF6) as the next step in the manufacture of nuclear fuel.

Yield The permanent deformation which a material undergoes when it is stressed beyond its elastic limit.

Yield strength The stress required to produce a very slight yet specified amount of plastic strain; a strain offset of 0.002 is commonly used.

Young's modulus A modulus relating tensile (or compressive) stress to strain in a rod that is free to contract or expand transversely in accord with its Poisson's ratio. The relevant measure of strain is the elongation divided by the initial length.

Z

Zapping A desperation measure to revive a shorted cell suffering from dendrites. A very high current, low voltage pulse from a large capacitor used in an attempt to vaporise the dendrites.

Zebra battery A high temperature Sodium Nickel Chloride battery delivering high power.

Zener diode Semiconductor diode that has a well-defined turn-on voltage for conduction in the reverse direction.

Zero adjustment The ability to adjust the display of a process or strain meter so that zero on the display corresponds to a non-zero signal, such as 4 mA, 10 mA, or 1 V dc. The adjustment range is normally expressed in counts.

Zero offset The difference expressed in degrees between true zero and an indication given by a measuring instrument.

Zero point The electrical zero point where zero millivolts would be displayed. Used in conjunction with the slope control to provide a narrower range calibration.

Zero power resistance The resistance of a thermistor or RTD element with no power being dissipated.

Zero suppression The span of an indicator or chart recorder may be offset from zero (zero suppressed) such that neither limit of the span will be zero. For example, a temperature recorder which records a 100° span from 400° to 500° is said to have 400° zero suppression.

Zero voltage switching The making or breaking of circuit timed such that the transition occurs when the voltage wave form crosses zero voltage; typically only found in solid state switching devices.

Zone Any group of crystal planes that are all parallel to one line, called the zone axis.

Zone refining A metallurgical process for obtaining a highly pure metal that depends on continuously melting the impure material and recrystallizing the pure metal.

Zooming In computer graphics, causing an object to appear smaller or larger by moving the window and specifying various window sizes.